THE CROSS AND THE CANTINA

THE CROSS AND THE CANTINA

PART 1

STEPHEN C. SOLTESZ

KJV

Scripture quotations marked KJV are from the Holy Bible, King James Version (Authorized Version). First published in 1611. Quoted from the KJV Classic Reference Bible, Copyright © 1983 by The Zondervan Corporation.

ESV

Unless otherwise indicated, all scripture quotations are from The Holy Bible, English Standard Version® (ESV®). Copyright ©2001 by Crossway Bibles, a division of Good News Publishers. Used by permission. All rights reserved.

NIV

Scripture quotations marked NIV are taken from the Holy Bible, New International Version®. NIV®. Copyright © 1973, 1978, 1984 by International Bible Society. Used by permission of Zondervan. All rights reserved. [Biblica]

NKJV

Scripture quotations marked NKJV are taken from the New King James Version. Copyright © 1982 by Thomas Nelson, Inc. Used by permission. All rights reserved.

Print information available on the last page.

Rev. date: 03/06/2017

To order additional copies of this book, contact:
Xlibris
1-888-795-4274
www.Xlibris.com
Orders@Xlibris.com
744771

Birds with invisible strings
High on a telephone wire
Making a tight knit of the town
To protect us from those who conspire
What if these were God's
Greatest creations
Our American mountains?
More precious to him than Tivoli fountains
The hawks fly above
Our enemies to forewarn
Our very church pews
Where true environmentalists are born.
If you had worked that hard through the years
At quite a steady pace
With the blue skies, the winds, the foliage…
Are we really in a race?
Or is His sovereign voice most high
Whispering to the angels
"Take the log from their eyes."
Take your time reading
This book
I'll tell you the truth
I'm a good cook.

ACKNOWLEDGEMENTS

They could be brief, they could be long. Many to whom they go are mentioned in the ensuing pages of literature. I mark them on the side of the devout: Those who did not set me aside since this project began, an intellectual and comedic discus throw.

I'm missing an important Family Reunion to cross the finish line and get this on the market, though I'd like to thank one and all the extended descendants of the Felipe and Dolores P. Aldrete family, for their painstaking connection to my immediate family in a recession often difficult (especially Herbert, Flavia, George, Dee, Frank, Maria, Rita, Gina, Ruben, Patsy, Cindy, Jamie, and John). They float a clean and healthy version of female and male wisdom both, for which to pursue higher goals.

My immediate family and I share an unconditional love: Without them, financing the project would be impossible. Steve (my father), Lydia (my mother), Diana (my sister), her husband Robert, my nephew Ryan and niece Alexis have all been more than generous. We manage our own as a unit nucleus and extended with the passage of time; in our lives, in America, and in the bottle The Cross and the Cantina represents for our home, The Golden State.

My computer help started with my father, breaking me in on a lap-top. From there it progressed to assistance from Jack Walker (a fellow Bruin Alumnus), Office Depot and staff, and last but not least Wayne Kuhn, who came by while the family was down and helped finalize the project.

I have two Parishes (Holy Name of Jesus Catholic Church and Trinity Episcopal Church, both of Redlands, CA) who continue to offer their prayers. Knights of Columbus #3109 was there for me at key moments of frustration and immobility, and there is no way but to sense a loss in participating at Mass in one location or another.

Other than U.C.L.A., U.C. Riverside and World College West, of Riverside and formerly San Rafael, CA, respectively, were instrumental to instill a drive in me and complete academic training.

My banks, in town, neither the locations where I shop, have ever put me in danger. They're great.

I have friends, too, that whether or not danger is the subject, have backbone, confidence, and loyalty, though know how to steer the tests on the subject to far perimeters for my artwork and soul. Gary, Jackson, Greg, and Frank, thank you.

While I'm at it, going along through the culmination of this project in day-to-day life, my physician, Dr. Mohammed Osmani, and his staff deserve a huge thank you. Mary, my nurse, Alyssa, the receptionist, Theresa at her computer in the back, Sarah (readying for her own trip abroad) and others in and out of their practice's doors have kept me strong enough to work, exercise, drive, etc., and bring no foibles to the routine of monthly medication. Beeman's Pharmacy has been tremendous support, as well as my current psychiatrist, Dr. John Benson, who really understands transitions, faith and freedom.

Fed Ex-Kinko's did outstanding work facilitating the possibility of including my pen drawings along with the text, in order to bring out more of my vision(s) into good light.

Last but not least, thanks to my editor, Mike Foley, for a voice from the university and beyond that I could rely on professionally. It has been a long haul of many events. He carries to hopes of aspiring writers well.

FORWARD

Beginning comprisal in December 2006, from my first peaceful residence since before 9\11, this saga is a personal response to God of my critical thinking skills that I hope you never forget. Lest those reached by this pebble in the pond, thrown by the unknown intellectual I am are not far off, I keep this moment rowing in my metaphorical rowboat astounded. What an idea, that after unforeseen circumstances forever rocked our American waters, I was left alive by God to minister the Gospels in this way and others, not from the vantage point of the Cloth, but of the artist left to succumb to the tide of the drift the Western world would be caught within. Allow me to attempt to bring some clarity to what often may be a cloudy picture for you. I'll tell a story or two, and talk to you as if it were lunchtime after class on the Continent, and if you will bear with me, I will take you back in recent historical time there, briefly, as well throughout what I hope you find an intriguing look at current California.

I tie together the themes of conversion, mental health, patriotism, art, academics, and deep thought by submerging philosophically to grab the pennies at the bottom of the pool. So, the reader must hold breath while I do the diving, coming up to the surface with laughs at the years on these pennies. Don't attempt to learn diving until it is obvious Christ's word rests as protection between you and society: You learn if you are still able after the summer heat of the Surge, and here you must decompress yourself from robotic action to grasp, as I hand it to you. We often use the word 'sticky': that's just California, learning her way as nature paints us a new one. Is there a concise, conservative framework to the book? You mean cussing, sex, curses, and lectures? Not a bad question from those far from the Catholic altar at Communion time, her pews when more troops are sent by the President. So, I'll tell you what I'm thinking about. When the country can't cover its war wound and compete in the working world with you at the same time, those are the terms for starving artist, not USC's or Columbia University's.

We will have a definition of diagnoses' common problems steering in the end not away from religious roots but honing in on where God is not only not against, but meeting science. I leave a personal stamp on this example.

So, in the end, this is not just a discussion of Christ, to an individual disciple, in conversion from Episcopal to Catholic, this is a discussion of psychiatry and patriotism. This, during a day and age when the successor to the British Crown marries the Duchess of Cambridge, a Roman Catholic. Though I'm speaking of psychiatry and patriotism to the American flag, troops come and go, as Popes change, and patriotism to our allies who fight with us in the NATO Alliance is welcome. Mighty White of me. I understand how you feel and what you say—I don't wear a sign on my head that says, "I've spoken Italian for thirty years, why, since Reagan and Pope John Paul II were talking." You just see tan arms, white face, four eyes, and smirk. And you smell tobacco at times. What do I do for a living, while corporations that could use this 'gift' are making friends around the globe for the red, white, and blue? Why, to so many, our national debt is a mathematical miracle in growth (know what I mean)? I'll spell out what's on my mind, in these ensuing pages of *The Cross and the Cantina*, a cross, people, held high, crossing a generation and a frontier symbolically, and a cantina, in symbol of global citizens who have witnessed these modern events in the States and abroad. All the while, America discusses not just change, but how to get along with each other, while enjoying our religious freedom once again after a war Holy to some, to others a mystery, and yet to others the products of patterns, a topic in all fields of study, not just military science and economics. If we inch any closer to solutions, the Crow can save a trip and enjoy the California Spring, right here on the back-porch of Los Angeles, the Inland Empire. We talk about separation of Church and State in the U.S., though yes, I'm reminded by how I feel about both day to day occurences and that a Knight of Columbus, such as myself, does not have separation between the two as he worships, and not all can be Knights. Enjoy after pulling some wine from your own supply. Bet you can't have just one. This all started eight years ago when my new psychiatrist recommended I keep a journal of all events in my life, yes, a concept conceived of between his office walls (that's $100 an hour to the artist in residence!).

PART I

Oh, geez, was the subject childhood? You mean with the New American Miracle in front of us, from three to thirty? And you want my story, from the 60's, 70's, and 80's? With Ebola being treated in Texas? The New Yorkers sitting down, the San Franciscans standing up, the Los Angelinos spinning the roulette wheel, it landing on the Deep South, all froth with urban germs, and European countries agreeing to fight the Muslim funk with America?! Politicians at the beck and call of minorities in poverty who command the ground in the market? My childhood you're talking about?

It happened fast, people. I crawled out of one neighborhood (an Air Force child) into another, searching off-grounds to find a 15-foot snake of God-knows-what breed in the gullies, while Dad smashed a tarantula in the yard with a shovel when I had been frightened. The guy across the street that my sister had a crush on captured Black garden snakes and put them in milk cartons, hiding the cartons in the hedges. I caught fish bare-handed in the crick back of the domicile, following the bed upstream where catfish slipped through my four-year-old hands, a born wrestler with nature's heat. Yes, I can stand the sun, much skinny as I get. Don't ask what you pull out the other end if you fatten me up, but one and all, take your turn at the crapper in the morn.

Yes, I met the Southern lasses in their childhood. I was companion when men were assassinated: I kept them warm and American. Louisiana, Arkansas, Nebraska, before Ohio girls-- I'm an advanced student who spoke Episcopal English on occasion to them, so that now, in the future, their checks don't bounce, whether or not mine do. New York is Canada—Reds. Arkansas is America. No, Mom, I don't want more apple pie—I want more shrimp, enchiladas, an extra Paul Bunyan burger. The beans are good, though.

Oh, are you an M.D.? I'm a doctor, too, if you catch my drift, and clean as a whistle. It's our neighborhood camper, not yours! And we're better teamwork on the athletic field, if you're talkin' your Preppy manure. Learn to relax. Stirrin' up

Katrina and gettin' back Sandra in exchange. You don't move a muscle without a compliment, and I've always had my stats—don't need your opinion.

Yes, well-studied. Taking a nap in kindergarten while the other classmates played catch-up, then a little Four Square. Before fightin' over a toy soldier o' mine with another military child, chasin' him in his house 'til he cried, searchin' for his mother and findin' her buck-naked in the shower, gettin' slapped and walkin' outta there as the little blond prick finally opened his eyes enough to learn the alphabet! A happy, well-fed childhood in a Second Lieutenant's household, espoused to a Tex-Mex secretary that was a true not flower, buster, but garden, in bloom, with an older sister that you don't mess around with on the "gift" exchanges to boot. I mean she'll shut'cha up!

So those are nerves of steel, when it comes to reading Classics Illustrated' <u>The Last of The Mohicans</u>, <u>The Deerslayer</u>, <u>The Pathfinder</u>, <u>The Three Musketeers</u>, <u>The Time Machine</u>, <u>Twenty Years After</u>, <u>The Biography of Lew Alcindor</u>, <u>Charlotte's Web</u>, and <u>The Story of Bart Starr</u>, if you want to go back to when it snowed in America, least in the parts I used to reside in. I came out to California after being hailed popular with my class for winning spelling bees, having escaped the incubator as a jaundice baby and the oxygen tent on midnight emergency runs as an asthmatic nearby Nebraska corn fields (as well as a patentductus arteriosus: the closing up of a valve of the heart that is supposed to close at birth). They liked me.

On to California, a land of literacy problems, jealousy, and a version of hard knocks plus minority issues. I'm part-minority and knew how to handle it from family reunions in my childhood and teens. It's the land of Spanish missionaries. They don't have to hydrate as often and aren't being sneaky about that extra beer: they just work hard and move quickly.

Let's get this straight—my father and mother know how to show a child compassion at holiday time, always worked for it, played favorites because she's a sister, though teach me manhood is not fist-fighting, but work ethic. Gotta lotta hot-shots in California on that note, that did not understand ladies like my sister growin' up and got wrapped around an axle about it. My Dad and I have outlived a few of them, Jew and Gentile.

Tellin' you, Connecticut on vacation in the snow is a sight for a part-Hungarian youth, eating goulash and bell-peppers while learning cards from the cousins, before getting my first football as a gift. Toys?! Jesus! Second Lieutenant not in the mood to divorce from a Tex-Mex cuisine, you kiddin'?!, Football, train set, G.I. Joes up the yang, Play-Do set, Basketball hoop, plus an issue plastic jeep with Rat Patrol soldiers, the covert company of G.I. Joe! The whole nine yards, man, what Military paid back in the day, when Dad was 350' underground in a missile silo during the Viet Nam War, with buttons at his fingertips aimed at "European" cities. I mean tough shit if you don't have his Naval Academy brains and have to squeeze tea-lemons in 'Nam proper—Tough shit! We're the brains, we're the beauty, you want to talk balls, earn it! Screw you, McCluskey!

All right, all right, I'll take off the mask and tell you there was a tincy-wincy bit of corporal punishment involved...!!! (he thinks he's a bare-handed Babe Ruth). She's worse: She thinks she's Queen Isabella of Spain, Evita of Argentina, a native Mayan woman swinging that cross-the-face slap from ancient Latin American days when aliens made the first airport landing strip (I'd rather get in the ring with Sugar Ray Leonard)!

My sister had been sneaky at thirteen, still dressed like a Tex-Mex schoolgirl, the ancestor of those with make-up and tattoos, I mean a real live you-know-what! She kicked my ass twice when we got home from school and got me in fights with the neighborhood bully (Gone were the day of her being my protector in the neighborhood while walking me in the stroller—She yelled at a neighbor for accidentally spraying me with their garden hose on one of our excursions)! She loaded up on me a third (and final) time at the age of four to pronounce her turf one day in the foyer. I AM. Ducked her right and twisted her entire reality with a dead-center gut punch. My mother rushed in, this being the first altercation she had heard: "Stephen, did you do that on purpose?!" I actually did not have the expression in my vocabulary at that age, unfortunately, and then it came after I acknowledged anyway, without further explanation, a healthy conscience-clearer of an open-handed whip across Prince Stephen's face. Never again. I later heard word that when she was young, my mother dragged a girl by the hair for blocks and got her good on her parents' front porch for spreading rumors when the family had, poor as it was, tried to be of service (my grandpa letting people have free groceries and what-not when it was that day and age for Mexican-Americans in Texas). You've heard the song, "Ojos Verdes?" That's the true story of a Tex-Mex mom with green eyes. Her first husband was part-French(need I extrapolate?). She took my sister back to my grandmother's and dealt with hometown divorcee stigma (much what I went through at my age now after several romances in my twenties and thirties, but no, not divorced). She worked around her older sister's house in San Antonio later while the sister paid for her Secretarial School, paying all of it back when she got a job as the Base Commander's secretary at Laughlin AFB. That's where she saw my father with his well-built East Coast legs in shorts one day, and no, he didn't even get to see the foreign section of the field but once(Morocco), and then that ring went on that won't now come off unless you know how to cut metal.

They were transferred to Louisiana and I was born. I had a Black nanny for one week, before other arrangements were made because I kept crying when she would hold me (I'm a man now, unlike other Americans, in other words I didn't cry when I saw President Obama inaugurated)!

During about that time frame, my Aunt Dee (mother's sister, youngest of eleven—none of them twins) and her husband, the late Colonel Richard Spring of Choctaw, Oklahoma (Uncle Dick was a trainer of pilots for Boeing's jet fighters) came to pay their respects to my Dad and Mom, the Major and Mrs., upon my birth. They set me down to crawl around on the rug and disappeared to look at family pictures in the living room. When my uncle came back to the kitchen area, I was sound asleep on the rug. He went to get his mixed drink and his glass remained with only ice. I had crawled up on the bar stool and drank it down.I went through stuff when I was young after the incubator, wearing braces on my legs to correct them. I grew up to excel at many sports requiring strength and agility. I like long distance walks now. Oh, can you, chico? Walk two miles downtown and back at night for another bottle when you've already had a couple?

To continue, you can pick your friends, not your family. That's why our grandma (grandpa died at 49 years of age), aunts, and uncles trained their children how to be a good, 'affectionate' cousin at an early age in our first reunion town, Del Rio, Texas—that's why what? So that hopefully when marriage came along for the new generation, they wouldn't pick a total dick for an In-Law!...But shit happens from time-to-time! We used to probably make their motel staff chlorinate the pool three times a week, we splashed so much in the humidity, coming from California, Louisiana, Oklahoma and Arizona to lower the water level with our own Cannonball Run, while the elders sipped Margueritas under the umbrella tables. They kept us on soda pop the whole reunion.

At Grandma's, my Aunt Chello (Lola's sister) had made a pin~ata on the Fourth of July with red, white, and blue crepe paper. I was the first one, what age I can't recall, up to bat, but no, not quite through puberty. I whacked it open first crack of the bat while blindfolded, the candy exploded into all directions unto the lawn, at night, with just a porch light. My cousins did the honorable thing, protecting my teeth as usual, and grabbed all the candy off the ground in a mad rush, before the blindfold even came off! (For those that have served, like their uncles and ancestors, from the American Revolution to the Civil War, the Spanish-American to World War II and beyond—Somalia, Iraq, Viet Nam, Korea, and Afghanistan, Dishonorable Discharges have never been a problem in the family. Probably not a good idea to talk candy with the children though!).

To make a long story short, yes, I've traded gut-punches with a lefty from outside of Cajun country. Only lesson in the Civil War you need—they don't wind easy because they have a little chub to pad the landing of the blow. On a brighter note, nothing like watching a 19-year-old Louisiana blond wrestle a Tex-Mex debutante from town the same age on the couch all afternoon. Oh, the warmth of first rights (You needed an explanation for the humidity in Texas parts)!

Staying out of that Kissin' Cousin rap isn't easy: we now have four cheerleaders from 16 to 60 years old. No, not Dallas, but they won't complain—they beat the Super Bowl Champ Seattle Seahawks this past weekend as I recall these things on Columbus Day 2014. Which brings us to the part of the story where I must confess it's been fun, but I've been wearing glasses for near-sightedness since seven years old, shortly before moving out you-know-where, y'all, the land of True Guts and Glory, liberal California. Where after a blissful fellowship with Santa Claus, I began what seems like an astute observation by the aristocracy in the Golden State that, why, added up academically, this is a life-long study of racism! Chore, got it, my ball in the university system. The Jewish doctor has castled on the board as a 'child within' issue, and the sen~ora at the Church feels it is an esteem issue. Don't show them the Rorschak results of mine unless you want to see their faces turn red!

So how does one overcome a succinct institutionalized racism? Despite racism in high school sports in America (now, it's a different generation), I was able to appeal to people who were involved in the American Field Service, a societal structure involving several cultures, as well as credit where credit is due academically, career-wise, and socially(my parents' military service was put into proper perspective as well, and I was honored for excepting my taming). I was selected to study with a family of a Law Enforcement Officer (don't call him an old paper-shoveler—he earned his job and took the bus with a .45 around his shoulders on the ground every day, waking at 5:30 A.M.), in a city known for its organized crime, La Camorra. That city is Naples, three hours south of Rome by train (approx. 8 hours north of Reggio Calabria—home of Undranghetta—and 14 hours north of Sicily, home of the Black Hand), where although not allowed to study by a Fascist Party principal, I did bring a men's team basketball tournament trophy home to the school snack bar, as their starting guard, playing Center on an European Semi-Pro team after. This was even though I only stood 6'2": My explanation? The coach himself was growing, and had a bit of a panz' (stomach). I'm retired from hoops now, knowing both American and European systems, which may simply reap what they sow from Achille's Heel to trade to Cleveland. I was quick, and don't say for Pope John Paul II's timeframe—there is no translation. You play in this America, I played in the Reagan days. That's the train station in Naples, where the court is we presided on. Is not where, cap on backward, you grab your crotch at the pick-up game to show your name.

Changing the subject, true, not all cousins speak it, but I do, and not in the home: I learned it later in life to appreciate my heritage as part-Hispanic. I've taught it in California to young and old through brief stints in the classroom. Speaking of young, yes, actual incidences of racism I remember going on, to the point it got hard to tell the truth. My peer group would often not pick me on the football field, make me play in right outfield, cut me from the basketball team even though I had a proven track record outshooting their starters in two-on-two games, and hey, is this the walk home, the bus stop, or the neighborhood proper? Let's bring the older group too and have fun with him. So yes, I know the problems in the city now, but I've seen it on my own plate the other way around. Plus, now we've seen Black youth get it. Get home, stay in, study, sure, when practice for soccer comes, the coach won't stop them from throwing rocks at you. Yes, I made it to Italy—the year they won the World Cup of soccer in '82 vs. now-champ Germany. America has <u>never</u> won amongst the men: Women only. I think that's a special statement on behalf of peaceful people as myself.

You had a question: How does a Mexican-American family from Texas serve in the American Revolution? Civil War, maybe, but the Revolution of all, the origin of the dropping of the ball in a free New York City? Well, let's fill it in first for the era we crossed the border into the Lone Star State, and see if we can fill in the history proceeding after.

Villa times: Point of take-off as Americans, or had we been here the entire time protecting the frontier?

In 1878-1923, "Pancho" Francisco Villa inhabited the parts of Southwest Texas—on the other side of the border. His full name was Jose Doroteo Arango Arambula—not a relation to me and mine save for perhaps three-four generations down the line on the swing-back. Yes, I have a cousin with the surname Villa, and I don't ask: She is a former accountant for the L.A. County Sherriffs' Department. The books are kept well, so are the men, and the Mexican Mob of today's scenario are shut down by domestic authority.

What makes 1900 Texas different? Why was Villa glorified? Villa wasn't: It depends if you're a South-O'-The- Bodie Mexican sympathizer, common to California, Arizona, and New Mexico, or a shelter-building Texan who says she/he's got an international child prostitution situation to monitor at the border. I could say while President Obama is drinking Chai with Gwyneth Paltrow in L.A., but I like him—he's got a sense of humor and understands the short end of the stick in America (I'm so short, I have atrophied an inch since being a starting guard for an Italian semi-pro team!).

Villa, a joker or plain killer, showed up on occasion at the riverbed nearby Piedras Negras (come on, Eagle Pass, Ciudad Acun~a—where the original version of the movie El Mariache was filmed for $9,000.00?) Not one senora, but several used to show up frequently to do their families' laundry there from Piedras Negras. He would then move to the Cantinas with his men. One of those women was my great-grandmother Aurelia Guerrero Martinez. Aurelia was a schoolteacher, whose mother's name was Juana Martinez, along with her father Vicente Guerrero (born 1845, died 1900), but not related to the President. Aurelia had a daughter, Dolores "Lola" (Poole maiden) Aldrete. She was English by her father and Mexican by her mother. The Aldrete name comes from Spain ('from the ranch or small farm' is what it means). She was born in 1894: 1910 was the year of the Mexican Revolution. 1910 was also the year that her father, husband of Aurelia, Edward "Fish" Chubbuck, died. Word has it he was a womanizer and an alcoholic, who had moved south of the border for ranch land from Pennsylvania parts with his brother, Truman Hill Chubbuck. They didn't move directly, but via both Chicago and South Dakota. They married Aurelia and her twin sister Dominga, who were both said to be elementary school teachers.

He, Edward, was under much temptation in this territory, as it was a time-frame of rough patches between borders, meaning an historical movement of cattle-ranching, cattle-calls, and cattle-rustling, intertwined. He was found, word has it, with his throat slit, face-up in San Felipe Creek, the Mexican neighbors' side of the creek in Texas parts. Other word has it he simply drowned. 1854-1910, Edward "Fish" Poole, father of Dolores "Lola" Aldrete, who had been situated in a convent school until married, while he carried on his marriage to Aurelia. From 1854 to 1870, at the age of 16 (his original name Chubbuck with a name change we've guessed in Mexico), he views the Civil War time-frame from Pennsylvania, though as there is no official record from 1870 to 1910, he disappeared about five years after the Civil War, when still in Pennsylvania, an age at which he was too young to serve: He did not defect.

Family stories tell us he was married to Aurelia on March 10, 1893. Between the ages of 16-38, during Civil War reconstruction, he was trying to make a living, having survived the war. He could have had multiple properties, but by Ciudad Acun~a. There was never any indication of him being with other Chubbuck family members. The actual location of the family was Piedras Negras, Coahuila de Zaragoza, the name of the state in Mexico. He most likely ranched cattle, sheep, and goat. Cattle-rustling, gambling, womanizing and just drinking. Drowned, or throat slit. Lola's English side in Texas parts, but before how the English came to Texas, let's say her husband, Felipe Aldrete, who sired eleven children (none of them twins), died young, too, of natural causes at 57. The women were destined to survive the men. Felipe came over on a ferry. Grandma (mine) Lola and Great-Grandmother Aurelia came over in 1904-1905, Lola between ages eight and eleven. We don't know exactly how they arrived. Did they swim the very river bed where Villa watched the women wash their clothes, with younger siblings clinging to their skirts? Was it a ferry? Horseback? I'll cross when Christ beckons my elbow, when the Lone Star State is in the sky with its jewels. Felipe crossed at 15 years of age, in 1903 (or 1907 by ferry), and married Dolores in 1911, less than a year after her father's death (He was sixth of seven from a Civil War Pennsylvania's family of British descent). Felipe himself has a surname, as I mentioned, from Spain, some feel Aldrete de Leo`n, who helped conquer the Moors. He was a poor, though generous chef, family man, and grocery store owner, a firm disciplinarian of his well-reared eleven and happy children, whose husbands, descendants, siblings and cun~ados (brothers-in-law) served in World War II and American wars ensuing: CONSCRIPTED AND SURVIVED, ALL LIMBS ENTACT.

Felipe (looking at his family tree) cannot be traced back farther from his father, Reyes. Baptized Jose de los Reyes, he was born in approximately 1860, son Felipe in 1888. So from Reyes to Aldrete de Leo`n, this Mexican line of Aldrete is a current mystery. Reyes' wife was Ramona Tijerina. We know the name Aldrete, but no record dating back to Spain other than huckster genealogy at large inc. In 1696, Bernardo Aldrete published the first history of the Spanish language. We do not have direct lineage to him, but the name is the same. Apparently as well, during the time-frame that the Court under Cortez was torturing Moctezuma for sacrificing virgins to adulterous gods, an Aldrete was in the mix as an interpreter or treasurer. Lineage to ourselves, not traceable after the conquest of Mexico, though we resurfaced after the siting of the Tilma (1531) and again during Pancho Villa's Revolutionary years (1910-1920).

Our English ancestors came over as Puritans. Edward Fish Chubbuck "Poole" was mid-point between Puritan and Presbyterian, before switching over to Roman Catholicism.

His brother Truman Hill was in the Civil War, from September 14, 1864 to June 13, 1865, my great grand-uncle. He served between his eighteenth and nineteenth years of age, recruited in Delmar, Tioga County of Pennsylvania and assigned to Company M of The New York 15th Engineer Regiment.

This engineer regiment handled ditch-digging, bridge-building, and military comradery to support the troops. The platoon did not participate in battle after building the bridges. They are honored at Gettysburg, having participated in all the campaigns of the Army of the Potomac ending at Appomattox. Now the story evolves that at the outbreak of the war in 1861, Edward was only seven, and remains in his Pennsylvania home until he reaches the age of sixteen in 1870, well after the conclusion of the war.

At the end of the war, Truman was mustered out, and attended university in western New York from 1865 to 1866. In 1869, he is listed in the Chicago City Directory as a co-owner of Chladek, Chubbuck, and Co. perhaps selling insurance with A.B. Chladek.

Truman's father, my great-great grandfather, was Levi Chubbuck, husband to his second wife, Cornelia Emily Hill and the mother of these seven (Cornelia's father's surname of Hill, her surname of Hull). Their first son was Charles Hull, born in New York, later a salesman and foreman for various in the Wisconsin home of his sister Susan where he died of consumption (tuberculosis) in 1878. The second child was named Cornelia, or "Nick", who married into the Van Horn family. The third is Truman Hill, named after his maternal grandfather, though who later assumes the name of his eldest brother some time after his brother's death, but changes his surname to Pool (or Poole), thus he becomes Charles H. Poole.

The family moves to Wellsboro, Tioga, Pennsylvania where the fourth child is born, Susan Marie. She marries two sons of Colonel Lemuel Bostwick Platt II, first Zephaniah, with whom she raises her family in the St. Croix River valley of Hudson, Wisconsin. Upon the death of Zephaniah, just two years following that of her brother, Charles, she marries Lemuel Bostwick Platt III and removes to Burlington, Vermont.

After the death of Lemuel III, Susan Marie occupies herself as the Director of the Home for Destitute Children. Now known as the Baird Center, it was founded in 1865 to help children orphaned by the Civil War (Vermont had the highest per capita losses resulting from the war). She resided on College Street, in the shadow of the University of Vermont, for the first 30 years of the Twentieth century.

The fifth born, Mary Francis, died in infancy in 1852. Then came Edward Fish, sixth, my afore-mentioned maternal great-grandfather. The seventh, and last but not least, is Arthur Burtis, who at the age of 24 is living with his sister, Susan, just after the death of Zephaniah in1880. At that time his occupation is printer, learning the newspaper business. Within five years, he has moved to the Dakota Territory and becomes the editor of a newspaper, witnessing the birth of the railroads cutting through the prairie. In 1910, he is the Postmaster of Ipswich, Edmunds County, South Dakota.

Born in 1575 in Hingham, England, and died there as well, was Thomas Chubbuck I. His son, Thomas Chubbuck II emigrated in 1634. Thomas was part of a group of 14 who came from old Hingham that sailed out of Yarmouth, Norfolk, England, the first week in May 1633 on the Elizabeth Bonaventure, with Captain John Graves, Master. He came with a bride of three years, Ann Hobart, sister of Edmund Hobart, and his famed twin brother, the Rev. Peter Hobart, the first minister of the Hingham Puritan congregation who built the Old Ship Church, the oldest church in continuous ecclesiastical use in the United States, and the only remaining 17th century Puritan meetinghouse in America.

Last of nine children, is John Chubbuck I who lives out his life in New Hingham. His middle-born son, John Chubbuck II does the same, bringing us to the year 1750. Son of John II is Jonathan, who midway through his life, marries a second time and moves to New Hampshire. Son of Jonathan, born in Massachusetts and moved to New Hampshire with his wife is Simeon Chubbuck, middle of seven children. Simeon is in the military in the Revolutionary Army. A patriot in the Revolutionary War and would have been 20 in 1776. He would have been 56 at the onset of the War of 1812. He moved to Madison County, New York, dying at age 86. Simeon had eleven kids, though some didn't survive.

The next ancestor in line is John Chubbuck IV, my third great-grandfather, born in Bedford, New Hampshire. The father of Levi, he ends his days in Madison

County. Levi is the eldest of ten. He is my second great-grandfather, born in New Hampshire, moving to New York, then taking his Chubbuck family to Wellsboro, Tioga County, Pennsylvania in 1848. Wikipedia says, "Tioga County resembled upstate New York more than it did Pennsylvania with its population primarily consisting of settlers from New England. This resulted in Tioga County being culturally very contiguous with early New England culture."

From this seemingly pastoral and genteel setting, a pioneering spirit emerges as the sons and daughters of Levi Chubbuck embark on their own unique adventures!

I was a hard-working youth, helping my parents as instructed around our various quarters, have even up to this day, weeding their rose garden in California and transplanting plants in ceramic pots and the ground, to show you diligence leads to beautification. If it's a "Don't Go There" town, by all means beautify the garden: that much asexual people can handle! (So California's not what you think: 50,000 abortions in my home county last year, or is the priesthood blowin' whacky weed smoke as he reports it on the pulpit?! Strange but true version of a California Catholic sermon: The priesthood didn't grow up here from the age of eight.).

I had friends male and female throughout childhood, with whom I got into and out of things with. California's 'boys will be boys' and 'wife-beating' occurrences have drawn nation-wide attention, why yes, leading to the electric chair for one young man. I remind you at this time in the story that although having genetic and environmental problems leading to schizo-affective disorder as a legal, medical diagnosis, under volunteer observation alone, I hold the rank of Eagle Scout, am a member of The Order of the Arrow, was selected twice to serve in ambassadorial fashion abroad with other selected students by the American Field Service of New York, New York, and possess a Bachelor's Degree in Humanities and Social Sciences, with an emphasis in Hispanic and Italian Studies (a major I created myself), from the University of California at Riverside, #2 across the U.S. this year of 2014. On the engineering side, the Highlanders were researching the proper creation of security cameras (Downtown Redlands, my hometown, is now under 24-hour video surveillance, though I don't know if that's my Alma Mater's intuition. Moot or 'mute,' can't you hear, don't go there downtown!).

On the Social Sciences end, I turned in paper work, totaling to 30 units between two papers, on the Mexican pueblo and the Sicilian Mafia. I passed an F.B.I. background check and Federal exam to earn a spot taking the Census for the 2010 count in Redlands, CA, now under Federal Oath to gossip all I want, but not to tell you if Landon Donovan still lives here or not! Let alone members of the Villa family. Last but not least, I'm a Third Degree (out of four) Knight of Columbus, under Supreme Knight Carl Anderson and Pope Francis of the #3109 Council at Holy Name of Jesus Catholic Church, which has plans in motion to build a larger church in Redlands, combining the parishes of the North and South sides of Redlands. It is packed on Sunday, as Redlands goes about rebuilding, but is not influenced by the Church alone. If you think I'm provocative and tell a good tale, try the Priesthood at Holy Name of Jesus. We have out-grown our own pews between our roots and our converts from around greater San Bernardino County. Am I a convert pissing on paper looking for attention, or did Lola win the Bingo game in Southwest Texas under Father Sepulveda, where I attended my Masses on vacation not far out of the cradle?! Moot or 'Mute?!' It ties like that, Episcopal baptism or not!

My Uncle Cristobal was an interpreter in Japanese in World War II, leading Civil Rights walks in Del Rio before becoming an attorney and going on to the Subcommittee of Agriculture in the U.S. House of Representatives under a prominent Senator. Uncle Phil and Uncle Herbert served in World War II, Herbert killing a Nazi in hand-to-hand (as my father recalls), before, as my mother recalls, returning after the war was over to cry in Lola's arms, having protected his country in the most brutal argument you can offer a soldier. However no, killing a man as a Catholic with a strong conscience is a minor shadow of the costs of war. Herbert, over 90, is the proudest and oldest World War II Vet of East Los Angeles, now an incorporated city.

My Uncle George, the second-to-youngest between my mother and Aunt Dee, earned his way through college by serving in the U.S. Army and now, as an engineer from L.S.U., owns and operates a shoring company with his two sons, which moves houses damaged by Hurricane Katrina from their foundations and back, after rebuilding the foundations. The eldest child of Lola, Flavia or 'Flossie', married Armando Figueroa, who owned and operated the slaughterhouse in Del Rio (where they prepare the beef for the market), keeping the live black bull and cows on the grounds of the business. Their son is U.S.A.F. Colonel Frank Figueroa. He's a grandfather now. Herbert's son, Ruben, served too, as well as Cousin Joe, Aunt Odelia's oldest son, who used to train the R.O.T.C. men at a Texas university. Cousin Psilkey, an in-law, served in Iraq. Lastly, Second Cousin Matthew Dixon, a Gunner Sergeant from Camp Pendleton in San Diego, is back from Afghanistan, Third in Command of forty men, having hunted the wild game at Pendleton after, so that it was safe to train new Marines, given a Navy Cross upon retirement. We saw him in my immediate family in San Diego before Afghanistan and at his retirement ceremony. He graduated from Sacramento State, and is now relocating to Texas with his hazardous waste removal company, after siring Lola's great-great granddaughter with his wife from Latin American parts. A few others are married to Active-Duty U.S. Armed Forces, and the rest, first or second cousins, we file a Kissin' Cousins Status Report on annually!

The Hungarian side of the family has been described in the past, as well as the Tex-Mex side then and now. The Hungarian is succinct about his West Coast ties: All day long California can push mental health issues, after being afforded an extension on racist issues. Some of the Hungarian side has relocated here, operates separately by California daylight while the Tex-Mex heritage pulls strings. But when both California and Texas fail the Intellectual, how can I tell my East Coast family has been praying for me at church all year? It's not a Jewish moment (as an uncle of two of them), it's not a Naples moment (as an Italian speaker), and it's not the prototypical juxtaposition of the California Latino and Tex-Mex incongruities. It's the East Coast Army and wind: I walk downtown Redlands, and it's as if their whole coast has moved in, as those of us who know the East Coast in person can see Redlands through East Coast eyes, the beauty of an environment with trees of no disease known to Democrat politicians. I walk, and as I advance in my stocking socks and Nunn-Bush shoes, the town is quietly reminded the Yanks of Now march with me, Democrat or Republican. That's Hungarian existentiality on the East and West Coast. If you can't hold the axe, people, you don't chop the firewood. Cute models, skinny legs and all, though no, the Holy Winds, the Holy Spirit, Il n'y a pas du difference (A polite reminder to the Californian of summers past not to push it with the intellectuals)! Are you sure? The cards, where the bridge game is at? Is it money, California? May they teach you a lesson on it, hard work, and assimilation, while retaining your heritage? Are you sure U.S.C. or U.C.L.A., let alone Stanford, has the backing their Alumni at East Coast universities do? Do I look like a sitting duck to you? Do you know my name, and can you spell it in the cold, not the heat? I walk downtown without lunch in my stomach for a bottle, come on, West Coast, let's play 'Spirits!' Normally, people of Education, A Connecticut Yankee In King Arthur's Court applies elsewhere... does it apply to the West Coast in modern times as well? Do you want me to read and scrutinize it for you, while maintaining my past at the Roman Catholic Church? Can you hear that bell, Note Dame? That's not your campus: That's the Congregational Church in Redlands, and that's me getting my wings in the West Coast for Civility.

Tex-Mex, Hungarian, Italophile, Californio: What kind of Youth Group or Scout Troop produces that when growing up, and where are they now?

All of us in the Episcopal Christian Education group of the 70's and 80's, stemming from the Parish that dated oldest in town (1887 on its cornerstone) were not unlike private school youth, whether public or otherwise, and the few that weren't excelled at another career eventually, gifted or not, regardless of formal college or not. Our exuberant voices, our grades, our athletics, our enthusiasm for both Acolytes and Scouts, with only a cherished female threshold going forth to the Altar Guild of the New Generation. Is the choir the party? It's the choir's Friday and Saturday, the Parish' Sunday with the Priesthood, and you talk about ass-backward academically and monetarily, just to get the choir to suit up in vestments on Sunday! You don't have to hand all that to a Catholic choir...but the Catholic Choir, a teetotaler nonetheless! You want a camp to talk about religion in the mountains at the site? <u>That's</u> where the Priesthood shows its human side, by turning a blind eye so the older, 'soft' boys have space to flirt with a younger, toned model, who squeaks out of these 'maturation talks', while the girls talk more make-up. A culture under Yanks of heterosexuality, in disintegration, save for the underground pot the female gravitates to, in hopes of improving her grades. All do that in California: That's why we say, the Priesthood and their wives in the Episcopal Church aren't worth a rat's ass, and have their jealousy issues about up-and-coming male achievers, academically and athletically. That was the status before the Anglican/Episcopal divide (The Episcopal Church now has an Interim Minister that is genuine about Christ's love aand goals for his flock). True balls. Broken. I've been fluent in Italian for 32 years, Duchess, I understand. The grass is always greener on the other side. Tivoli, or Ireland? That's how Pippa's hair split, American side. All affectionate females, don't get me wrong (back-rubs, hugs, kisses), but slow enough on commitment to let all of Mexico catch up to Texas! Was that the battle or the war? How's about I be the America interpreter/ Episcopalian/Catholic, to spread good will, and you pop out another bun? No wasn't a tower in London, was a California downtown, Congregational Church ringing the bells, since they saw the acts, but yes, I have my wings (Downtown Redlands is an interesting Queensbury Rules dialogue for the London Irish Rugby team)!

We can only put up with London so much when he's loaded, and trained, with the Prince's Plate (Don't let Redlands see there were plenty of others). So where are they now? They're cultivating their new youth, as prescribed, and that's all the money they get to do so for blowing theatre out their White asses. But sure, why not, a play: The deck is stacked White, some Jew. It makes for un-performable Republican politics. Some of the Whites are worse problems than the Mexicans on the liberal end. See, he, the proverbial male Episcopal side, thinks Prince William's the stronger of the two, so why, Queen, doesn't he do the grunt work? When 'Grow Up!' doesn't work (a common problem in Europe), teaching the angle in which you see the arts as spoiled shitless is an option, though don't forget, they had to help the amateur both with his younger sister's period and getting his first girlfriend. That's why I recommend you remember which acolyte is which, because some can't speak it: English. Let's not forget the peephole between girls' and boys' cabins I have to fight him for after he balks at campfire skit time, because he can't <u>read</u>. But he's straight. A photographer. Real estate. Construction. It doesn't translate, though, until you get abroad. Then on the subject of rats, it's as fun as the movie *Rattotouie* (animated comedy story about underground rats, who float in the river from a French restaurant to a 'central location'). In other words, take it with a grain of salt if you don't have the 'clarity' for the humor in this!

My father drove the four of us (five, including Samuel Schulz Soltesz—the family's farm dachshund) out from Ohio to the West Coast, circa the latter month of 1971. We cleaned the back windows in the station wagon of dog snot, played the License Plate game with numbers and letters as the other cars passed by, passed snacks, napped, and took pit stops. I don't even recall us stopping at one hotel, but maybe there was a rip-off of a Holiday Inn somewhere along the way! I say rip-off because I never got my love life back until my 30's! So, West Coast girls—Ya' THINK?! Fallacy. Fallacy in motion. How, do you ask, does an American male sacrifice all that, just for a higher order of Fileo, or family love? Oh no, don't get me wrong, there's been sex, but I suppose I should take my mother's side of the family's advice and stay single.

You might know that if you're an American who speaks Italian, the European women visiting the West Coast pick on you! A world of fun. Seriously, we support the Constitution, American Capitalism, Religious Freedom, and Equal Opportunity Employment: To each according to his own means, and that's a Holy War, people. My family isn't far from each other: We have new, adult members, and despite a problem with drugs in the California environment, I came square with the medical field over fifteen years ago, and square with science NOW, during war, having donated seven pints of blood to both military and law enforcement. Toxins now are out of my system after the Forrest Gump routine two miles downtown and back for a bottle of sasparilla on foot. See? He told me to pull my head out of my ass, and I facilitated the procedure. Fileo. In the U.S. Military, retired or not. Now how can I be a blood donor, after that? I never stuck a needle in my arm. I never had gay sex. No, I'm your boy! I love my country. I'm here. How many fingers am I holding up? 2? 3? Add the hand up a second time—TEN!

That's if you begin to take a proper headcount of the number of Italians I know by name in this world. Yes, Rudy Guiliani, of South Brooklyn High. I'm 'Stefano Sottese,' of Liceo Scientifico Ottavo and Istituto Classico Potano, Fuori Grotta and Vomero (respectively), Napoli, Italia! See the difference?! Two, Rudy, Three? Ne dieci! Ten!

But yes, when it comes to jeans, Hungarian and Tex-Mex, California is currently in a lengthy discussion on the subject (genes) with God the Father and La Virge`n, la Santa Maria. My mother's first name. Yes, she's Maria Lydia. He's Stephen Arthur. I'm Stephen Christopher. My sister has a touch of her first father in her, but we don't tell her: Diana Lynn. A little patience? Next time, get us to teach the subject, California has holes in the hull, from the environment to their school system, and back to their ability to handle a sexual rapport in adult language. But, hey, don't knock it, K.C., look, there's your baseball back. Down the pipes. Yes, out-of-staters, try a little more military-style Fileo on how to raise the two of them.

Do you understand Hungarian in Episcopal and British influence now? Let it exude, or let it work, but let it serve the Lord Christ. Like father, like son. I'll buy a dozen. He had his hey-day too, in Hungarian plays as a youth on the East

Coast. No one is Master but God, Honor Thy Father and Mother. That's a tough lot, sure, but it's better than Hollywood Boulevard vermin. Nothing like a small town theatre's safety to air out my own artistic enterprises, whether I act, volunteer, or not. "The choir is shy of the Public Eye." I am not. I greet it with a smile. That's how a mixed ethnicity and religion family works on children. God is proud of his birth experiments-- he'll let you show off a bit too. Assimilation? Which culture? Yes, I'm tri-cultural. Thank you. The Marines are not, but I have my pride they took it to them. No, I'm not Mr. French of Family Affair, I'm Mr. Soltesz: If you leave it on the counter, I'll cook it and feed them, all ambiguities intended, no what I mean, Anglican? No, thank you.

The Anglican Church, before its division, was always strong support on accomplishments. A Tuskegee Airman was on my Eagle Scout Board of Review. His son had gotten me involved in the high school chapter's American Field Service, where at about the same time as making rank, I was selected to go abroad for a full calendar year (The Airman letting me know that I couldn't make out with my host-sister like camp or schools…after all, I'm not a European exchange student, I represent the flag of the U.S.A. in a foreign home!) It wasn't a difficult problem to trouble-shoot—they stuck me with a wine-drinking, bright, Italian son of a cop as a host-brother, us sharing separate fold-out cots from an older model of European couch, going to sleep watching the detective Colombo (Peter Falk, in Italian language). Thatcher was alive, yes, Pope John Paul II, and Raffaele Celozzi as well, my host-father of the Naples Central Police Department, who hung his .45 in holster on the coat-rack at 5:30 P.M. every night as we would sit down to dinner (prepared by wife Rea Speranza), pouring a glass of what his brother Michele grew on their family vineyard in Puglia. Challenges galore outside of the home, one and all. I got stuck witnessing an Episcopal Gospel to smokers in the park at night, friends of the older son. Neapolitan boys love to tease: "Are you a virgin? Do you masturbate? No? But you must: if not, your gonads will burst! Hey, 'Stefano,' how much would you pay to sleep with this female friend of ours?!" They didn't go too far with it, without getting a little Jesus in their secular bones and reminding myself and themselves that Celozzi was Law Enforcement. Nonetheless, they had their freedom at night to travel the city and buy cheap, underground packs of smokes from connections to the Contrabandieri, the fast boats of the bay that bring cigarettes to the shores of the city illegally (American brands). So I didn't smoke then, but I started later in Mexico, in memory of them and our mutual Neapolitan enthusiasm for humor. That San Gennaro, the Patron Saint of Naples, must have been something. They have a parade each year in which the Church carries a glass relic, similar to a lantern, in his honor, and according to superstition, if blood appears inside the glass relic, the whole city will be blessed with good luck (As if they aren't lucky enough to survive conquest attempts over the centuries, preserving their city's history and extraordinary cuisine)! A given

Sunday, as the world know now, is a Naples victory with tenacity at Stadio San Paolo, listening distance from Celozzi's old balcony at the apartment building. I would run to the store for fresh bread and put it in the basket Rea would lower to the ground floor by a rope. Fasci`tama magna! Tenga 'na ftend' e fam, ia so murrend'e fam! I'm hungry, three times of over in dialect. When the city is not understanding Americans, the son recalls O Cappa Mort', or a dead head. It has other connotations: What page we're on now in dialect remains to be seen. Give me ten minutes at the table to be babbling fluently in it again! It's infectious. So good infection, bad infection, depending on whether or not you can keep clean, showering when you can share the bathroom with others. California has a joke, 'Hotel California…' Naples really can dispel myths as to which cities are which in the Western world, it's dynamic, and the people don't pull punches on affection like Americans. No, not all of Italy is like that, and it's not easy to relocate, even if you pass the Federal Citizen's Exam of Italy, so I reside here, and not many true Neapolitans come my way. Hence, 'See Naples and die.' It stays with you, you develop a different perspective of the night (such as Paris), but you reside Stateside.

I returned home after the year, having travelled the country well from beach to city to mountain, by train, and began my higher education at a Community College (my parents felt if it was good enough for the neighbors, it was good enough for me, despite my scholarship for foreign study the year before and my acceptance to U.C. Santa Barbara). I excelled even more than I dreamed of in high school, where I held a B average except in P.E., now making the Honor Roll with a 3.75 and having my stories read aloud by the professor to other Freshman Composition classes. Socially, I had a male friend, Irish, who use to come over to my house after school, for pasta and wine, but no real romance since Malin, the Swedish lady I had met in Val D'Aosta by the Italian Alps on vacation, though it be brief.

I don't know if it was the Irish musician's curse (he did not do well in school that year), or the mark of Malin (let alone the whole group of exchange students), but brief romances are almost a cult amongst the well-travelled to me: There's so many of them, the price to pay for not being the Village Idiot who married his high school girlfriend. Oh, I'm sorry, you two are Jewish? I apologize. I understand ("We're not Jewish!" "Get DOWN, Dick, get down, these are Neapolitans!"). Does that explain the ethnocentricity I perceived when I got back home to Redlands, California? We now return you all to Sitting Duck!

In any event, many, whether in Naples, we're from America, which has military bases in the city proper, and without getting pick-pocketed, they also yes, we'll be true to flag and hit the ground running if you visit, even if the camera crew is good back-up in a bar-fight! Sorry, Batman, that's a $250 fine for parking here, and that's a dated Italian/English dictionary, I still don't understand what you're trying to say: "It doesn't translate." Tell Robin that's where the expression derives, 'You either speak it or you don't!'" Naples? All the flies you want, no shit, still my void to fill. My home away from home. Cooking Italian brings it all back to my apartment in Redlands. Do I long for other places? All travelers need to stretch, but that's where the story starts.

For the good year at Community College, a school in San Francisco offered me a half-tuition American Field Service scholarship for another trip abroad, this time to Morelia, Michoacan, Mexico, and after to return to the Bay to continue study. We stayed with host families, this time mine an attorney, his society wife, four daughters and three sons. Keeping my mind on the professor's lessons wasn't easy. One of our teachers, our prototype for the Mexican male, got a few of us into shenanigans, and the professor who escorted us from the Bay, happy with his flat shared with his wife, began not so much as to support the group, but pick and choose his arguments with bright students. He saw it differently, and now I understand: He handled the driving on the field trips in the surrounding area, logistics for lodging, local police and families, as well as complaints from the home campus, not to say San Francisco played Nanny. If they did, we all grabbed a cab, spun out to La Commercial, and watched James Bond's Octopussy (or should I say Broccoli's?)c in Spanish. It was better than Sophie Marceau's Il Tempo Delle Mele (The Time of the Apple) that I saw in Naples, one of the Braveheart star's first flicks at sixteen, about a sweet sixteen birthday party. Several times, Octopussy. We even had a girl travelling with us that looked like 'Midge' from the movie, so in sum, we formed our own liberal cult, before going off to village stays (some with Indians, some with Mexicans). Mine in particular landed me at a ranch owner's townhome, as well as the cabin on the side of the hill at his 150-acre avocado ranch, on the outskirts of the pueblo or village. By the grapevine in Southern California, it is in the High Sierras of Mexico, a location that now has problems with cartels. At the time, only ranch conflicts amongst the male workers occurred, before the owner hosted the whole group of 10-15 people from the Bay at his unique one-story abode with a central plaza. I dated one lady at a pueblo party one night, but it's not what you think: the town borracho (drunkard) poked jokes at America the whole night before she walked me outside, and I went home after a kiss, which shot through the pueblo grapevine, I not to date her again. We celebrated both a single Mass and The Day of the Dead together as a Bay group, also travelling to the top of the Pyramid of the Sun (Pyramide del Sol) outside of Mexico City (Teotihuaca`n, precisely), where, while within her confines, as a capitol, we witnessed the original Diego Rivera murals of Spanish and Indian during that time period of Mexican history. Yes, he painted, if I recall correctly, at the top of Palacio Municipio's (Municipal Building in Mexico, D.F., the formal name of Mexico City) second floor, or the top of the steps, on the wall: a ceiling-reacher. Not an easy balancing act during that period.

That is where my concept, recreated in pen-drawings now by me, of the abstract began, with a merge into Mexican village mysticism. To say I'm a mystic to this day, well, as a Catholic would put it, such is the Holy Mystery. Yes, subtle though connecting Latino moments with the old Michoacan`, in the environment of Redlands, CA, now still much timeless in its pragmatism and not as rewarding to the artist as other areas of California, though popular for its beauty and safety. If city life in Redlands was ever as inexpensive as Naples or Morelia, people would have a blast, but no, a dollop of the expensive side of the Continent is how Redlands sees itself, with its Real Estate prices that set the business strategy in town. As if the Bay Area can talk back on that note: all of California tends to price its citizens out of a good time while infecting them with germs in the meantime, so more Mexicans come here to teach productivity from South of the border, to a California stuck in a bear-trap of an unfair recession due to P's and Q's that are an embarrassment of a protocol to more American states. No, Redlands is not a sure-fire communication to its own, but tough times have taught a few of us to be just like the professor: pick and choose our battles if we want progress, urban-wise and in our lives.

This was the 1983-1984 timeframe, and yes, there were banners at the university in Morelia that translated, "Yankee Assassins Out of Grenada and Central America!" There was also, besides movies, the Portals of the commerce area with cafés, restaurants where we would exchange cigarettes if we spoke Spanish, and outdoor theatre houses at night. Some places in the surrounding area (Uruapuan`, the home of the famous park) are rather tropical and there is a lakeside, popular village or two in Patzquaro and Erongaricuaro, where they paint tropical settings on homemade bed-boards, working well with wood. Copper works, guitars, Michoaca`n has Indian villages that specialize in artisanship, as well as the weaving of <u>robosos,</u> or blankets of blue and black stripes that babies are kept warm in.

The thesis statement of my description of village life, perhaps pertinent to that timeframe alone, is that despite U.S. corporate involvement, such as Pepsi, the people seem very content in their agrarian rhythms. No, they have bluer skies than San Bernardino, the latter which reminds many Hispanic travelers of Northern Mexico, NOT the Inland area of Michoaca`n, in between Guadalajara and Mexico City. Universities in our parts of S.C. are connecting all the time. We read Nobel Prize winner Octavio Paz, Carlos Fuentes (what we can handle of his cussing in <u>The Death of Artemio Cruz</u>), Jose` Carpentier of Cuba's <u>The Lost Steps,</u> as well as on the Huichol Indians, in <u>Peyote Hunt</u>. Yes, politely put, if you learn your Spanish in Mexico, other windows may close to the greater Latino world, though you have an understanding of the arts that prospers. I anglocize this topic as I am writing this non-fiction work covering other topics as well.

My Alma Mater in the Bay (I have several Alma Maters) did not continue its Environmental Studies program upon my return, neither International Economics (but instead, an International Development subject that was begging the issue of student grunt work, amidst tough times, for students in the pocket book). So, I returned to the Inland Empire, where I met conflicts with my father, for cultivating some recognition of Hispanic culture in me, and both he and I began a series of mental issues, for which I saw treatment for eventually from a psychiatrist. He was in the midst of his own issues in life that, though timely in my rise as a savant of sorts, is now in the past. My disease is from ethnocentrism, ostracizing, difficult family nucleus, the strains of abstract creativity calling me artistically (and intellectually, to develop critical analysis skills which are keen to this day), as well as religious issues on an organizational, collective and theological pathway: So, I contracted schizo-affective disorder. Though genetic, it may be avoided if environmental pitfalls are side-stepped, as opposed to a shit-in-the-fan Sabbath. I receive medication now for anxiety issues, a slow-release tranquilizer administered once a month that does not prevent me from driving. Yes, other cultures might have (and have, actually) handled issues of feminine mystique with more respect for the mind and body I am, a fairly handsome, well-chested man. It's the community's burden on their own day with God in the future, I love them, but efforts of less starkly American proceedings must be made the next time I have visits, for information reasons, from possible U.S. Foreign Service recruiters. The exam itself wouldn't have been an issue. My middle initial, in Stephen C. Soltesz, stands for "Crunch-Time." Whether or not, judging by the difficulty of a professional exam or battery of psychological tests, I feel like I'm on something, due to stress and atmosphere, after travel to test site, I'm only getting better with age. There's much self-centeredness I've always dealt with (relatives say I'm the same way, but the point is others have warmed up to my humor quicker, and Christianity pays the toll, not the rent). My brother-in-law, who suggested treatment along with his law partner, is the same way. America does not rank number one in freedom between the three countries I've resided in, and I get tired of placating both parties with votes when the ignorance is used as defense to prevent the arts and society from living their lives. We would say oppression, but how severe? Repression is more like it in times of little employment, as those who are working to keep things going are such luxurious pressures that the rest they are serving get bottled up. America, in the meantime, has failed its course in philosophical improvement, and her relationships often fail. 'Pobre Steve, tan lejos de Mexico, tan cerca los Estados Unidos.' 'Poor Steve, so close to the U.S., so far from Mexico,' a portrait of a persecuted artist who adopts Catholic ways when the Episcopal Church splits, predominantly because he was rejected. Again, he may leave the Catholic path to seek reconciliation with society economically, when he's by now used to narcissistic replies about his age, as the fact is, they can't handle inclusion of their intellectuals in cultural society: They don't have the know-how and they've waited to dump it on my age. They've balked since the

Iraq War began, they now have a President who is giving them foreign problems with Allies (what comes around goes around for not paying heed or paying period, your intellectuals). They also give birth without sexual freedom. I had mine for a while, though follow a Catholic Christ who has me along the trail of the Church's tribulations in these parts, and it's not all me (I may air it out in a document though have a Vet's son tune I sing to the public in person, not a leftist-style diatribe. Don't call it sneaky: Relationships fail and talk of the challenge to the American institution of marriage is not mine. Philosophically, I say, we never 'take the bus' together anymore.). Look, it hasn't been a banner day for American Catholics, as if Italy shot the moon against us in a game of Hearts. Some said the Church bit off more than it can chew with its plan to build anew, another that they lost the big election statewide, as if to say, "We tire of your company." Something they say to us. So the theory is, you cut off free-thinking to stem a revolution, without properly arguing the thought (and without economic etiquette), you risk a revolution, but perhaps from the top –down this time, to teach you a lesson after Iraq about what God thinks of your command of the ground when your citizens have too many restrictions from loving each other, and you're a clown on the topic of starting a family young, when fresh. You're a clown, not clever, and we have the wisdom to guess what sex in the upper class, that you protect, is about.

Yes, people, history repeats itself. Just as there was racism in youth, in high school, and abroad, there was some in college. I'm very White, though never deny I speak my ancestors' tongue of Spanish, so go figure. I'll tell you about the guises I've seen in the collegiate world, the masks of the Baby Boomer generation. The professor is absolved: For the most part, he grades knowledge, man-hours, and behavior, not race, though his Student Council throws the fight...I've actually always tested higher than I was graded, save for the Italian language, which is the same equivalent over the bell curve. So let's talk Mexicans, not Brownies, but well-behaved patriots that don't fit the Tom-boy mold because they're English is higher.

When at a high school social, from a certain conservative mouth, the words came, "Spanish, Mexican, what's the difference?!" Do you know it is incidences like this (requiring quite of bit of air-time at a party to list the facts of the two countries' different identities and histories, no? So it's left unsaid), over desert, while forcing us to compete perhaps for females' attention circumstantially and cordially while others are freed up to other tables, that warrants me exchange student material? Faut pas' off the charts against the Mexican mix/intellectual, people. I've said in recent past, Europe is almost worth the risk even today, despite the rise of radical Islam. A healthier, less static arena for social twists.

Next, San Francisco-raised people, in the Eighties, were infamous for throwing a dart at Southern Californian roots, and the 'lower conscience.' It's my belief that it's a guise with the very same roots in anti-Latino sentiment, but this time we can all play, the whole region borders Mexico. I wonder if I would've wound up being a psychiatrist myself if the social arena in the academic world, regardless of where hosted, can be labeled The Fubar Festival—people would rather take out the whole side with a screw up then be forced into any circumstance whatsoever where they would have to utter the word 'Congratulations.' On the flip-side is, "Whoa! Right on! Hey! Get out!" All synonyms to congratulations that would be taken as requests to leave the party, but the smile communicates all: that's it, you've been honored.

So, where does this pick up? High school, racism, then an exchange student to Naples, experiencing the negative attitude against Americans in Naples, Italy, but enjoying my time, learning Italian at 19, drinking wine, and meeting other American exchange students (as well as foreign!) who can do. Returning home and being told by my military mother, "Yes, Stephen, you've been accepted to U.C. Santa Barbara, but if community college is good for the neighborhood kids, it's good for you." So, I stayed home, went to Crafton Hills College, was awarded a scholarship (after making the Honor Roll) to World College West, an its exchange program to Mexico. It was an existential school of spoiled Americans and foreigners, many pretending they had a work ethic like mine (who was working, not dating, every week-end in high school), but they were fun-loving despite frivolities: It's just economic disparity started there. Some are not qualified to comment, but the staff will support rich, spoiled children. And foreigners. Mexico was not about Spanish, as a language to incorporate in our careers: It was about their feelings as ethnocentric Americans crossing culture with the Mexican male, and I was crossing my second culture at 21 years old. The other exchange was Nepal. The "money" exchange, Nepal, who were catered to academically when we all got back by a rather dry staff from Northern Cal: Professors, not vineyard cultivators. Dry? There's a drought in Northern California to the day. Not an ounce of European savoir faire in the lot, plus a Chicano Studies "peace-pipe" smoker. True blue it was not. It went under financially, although supported by the American Field Service, and reopened as a business school in Downtown San Francisco by the fire station, under the name of El Presidio College. Professors who know more people than people skills suck.

So, was that sex-ed in Northern Cal? I'll take my athletic career next time instead. Believe me, other than wildlife and a brief description of her 'nature,' she can't add military dollars worth a damn (though, oh shit, can she pour them!).

Yes, to go further into detail with slight redundancy for story lines, after a one-month orientation in which we were instructed to read the 350-page book Aztec, we got on the plane to Mexico City. The night we landed, I had sex with an Italian from Rome whose family had relocated to Hobart, Tasmania. Her California Mexican hotel roommate walked out of the room muttering, "Mamma!" The others commented they heard us from their rooms. We then ventured into Morelia, Michoacan from the hotel in Mexico D.F., to join our host families for a more conservative dialogue, or so we thought: The male peacock on the university staff was a pot smoker. The host families were bullshit-run with jealousy if it wasn't about their Mexican male son's pride on all fronts—nothing compared to the manners of the Italian sons in the old country of Italy: nothing. No wonder the country has problems. The women were sometimes polite, sometimes sanctimonious, but no, I lived with three sons and four daughters, two sixteen year old gorgeous twins, and one gorgeous one my age who did not know what her lips were for yet. Complete perplexity. I would say pseudo-Catholic, but the American Catholic Church doesn't have much of a sex life either: They're not south of Rome—which will write a better script. So that's why the High Church and the Low Church exist, and if only America could learn its place in this without under-pouring and pigeon-holing. Mexico and America, asshole neighbors who reflect each other perfectly in asexuality and affluence: Two upper classes who need their asses kicked.

Next: A visit to the historic villages of Patzquaro and Erongaricuaro, where the copper goods and artistic bedroom furniture are hand-made. A Los Angeles Rams front-lineman had a daughter in our group, who did a polite job of seducing me in a Patzquaro hotel room, having eyed me the whole study season. We had an intruder in the shadows: My male roommate who 'needed to potty.' The potty is in your all's room. Buy anything for a look at a naked woman but intelligent communication.

We studied our Spanish grammar, Mexican history, Indians, politics, and read the signs at the university that said, "Yanqui assasinos afuera de Grenada y Centro-America (Yankee assassins out of Grenada and Central America)!" We climbed the Pyramid of the Sun, dined in the underground restaurant in Teotihuacan outside the Pyramid, and did our research on our village stays, stoned or not. My paper dealt with contradicting the book Global Reach, which talks about corporate oppression. The pueblo, or village, doesn't mind good Pepsi, and despite economic disparity issues similar to America, the haves and have-nots, it is a happy agricultural place with "smiling" livestock.

I vacationed in Puerto Escondido, a small surfer shanty south of Acapulco, as well as taking a fourteen-hour bus ride from Morelia to a seedy hotel in Acapulco, having acquired really bad runs on the way from having bought some Indian candy in the bus station in Morelia. I spent three days rehydrating in the hospital ward in the seaside city, before being released again to join my female friends at the restaurant overlooking the Acapulco cliff divers, who were performing that night. I enjoyed an exceedingly classy and well-priced chicken dinner post-hydration, which was all they said my stomach should take on. We were all pretty fluent by year's end, some venturing to Cancun. We got back from the villages before departure and watched 'El Maleficio' on T.V. with our host mothers and sisters, the older one who tried to steal my sister's Vanity Case while I was in the pueblo. I caught her. She had the easy opportunity to laugh off my getting her goat, as she pulled her tampons out of my sister's case before returning it, while her brother asked me, "?Son tuyos, Esti`f?"("Are they yours, Steve?"). That's how it rounded up with the host-family, ever Mexican-proud. The grandmother used to serve me an occasional hot meal, a true saint to the lawyer's family. One of the twins ventured in my room one night and danced with me, but held back from kissing me. The middle-aged brother explained it to me after I rowed a boat with he and friend on the lake at their cabin there: "Look, I study <u>Mexican</u> law: By our law, since your grandmother was born in Mexico, you're entitled to citizenship here." But America doesn't allow Dual Citizenship.

Yes, Jose Cacho-Vega of Ario de Rosales, the pueblo, and family Vega de Marin of Morelia, I know people in the Latino world now, even though we don't keep in touch. I don't keep in touch with my European friends either. Iraq (9/11) meant starting over.

I believe my other most memorable moments of the Mexican sojourn, as a young intellectual growing in his language skills at early cultivation where (1) the owls at the bird sanctuary outside of Mexico City: live owls, in cages, and not affectionate with Americans. There's a metaphor there: Don't mess with Mexico either, they have their wisdom; and (2) The Diego Rivera murals, cited at Palacio Municipal in Mexico D.F., amidst the poverty-stricken children with trumpet-blowing brothers collecting money, whilst I collect inspiration from the tall murals for my own artistic aspirations. Yes, I have a portfolio of 29 drawings as an artist and writer in Redlands, CA, my new start. So, I tell you, the sex of a Mendocino County girl upon my return from my Mexican education in the arts and sciences, or my athletic career: She <u>is</u> my athletic career, she spoke fluent Yugoslav and had a tight figure. No, my turf in the college scene, not fraternity turf or acting turf: You speak one or you don't.

Yes, after Mexico I returned to Crafton Hills College, receiving criticism from my professors, students, and parents. Now an open mind, they would not place me on the Honor Roll again. But the Inland Empire was just building. The future of then: The City of San Bernardino goes bankrupt. Escaping poverty in the I.E. is a mystery to one and all, but yes, I got tired of the dry side of Redlands, including their university. We have academic/economic issues as such that <u>now</u> the business world is being educated in, the critical thinking that I had mastered at 21. They got a "D" in how to cultivate an otherwise qualified interpreter. They gave him mental problems. A full thirty years after receiving my degree, the university commenced with a dialogue on certification for interpreters, in Spanish alone: Italian they are scared to comment on. So it's an adult world after all—if the university has answers for your career that's great, but they don't help everyone with their needs—they're not a cure-all, especially socially. The State system is becoming more and more popular for less curves, more instruction and inclusion. Inclusion is not the U.C. System's forte`.

So, what happened next, upon the actual timing of an imposed fall from grace, a coup of sorts involving the imaginations of countless, jealous, academic minds? Now that the Sophist debate on I.Q. was over every time I opened my mouth in Spanish to my mother, my father (an engineer) would begin to bitch and moan about my (at the age of twenty-two) late nights out with the men on the other side of town. No, seriously folks, does it have to be Crafton Hills College on a virgin's terms if it's not the Fire Dept.? Yes, The Great White Whale of a Republican the Major was in my youth consistently stood up as a lady's man himself, giving me his version of a cross to carry, exhausting the lead on the field we had. A Naval Academy graduate. True. A closed system all its own. A goat for a mascot. Not the normal Republican infighting have I been subjected to as a writer, let alone before you get to the Democrats' brand of narcissism in motion. Money doesn't solve everything (They haven't solved a damn thing for moons. And now, the lower and middle classes have caught on and gone President Obama).

"I will not scrutinize my fellow Americans' social behavior like a pain in the ass, I will not scrutinize…," 100 times on the chalkboard, and before you know it, a new election will roll around. Is this dad hiding from the times in his house on the Hill with the rest, or a mental issue? We've had to played at both tables, people, the military has to learn, one table at a time for Junior, he can't bat that high if you sit on him. Don't accuse your son of being an insufficient traitor when you dominate the conversation at a 'family' dinner table. Stop losing your children opportunities with your own words, and learn the nature of the state you're transferring to next time.

But the sex that comes with marriage? C'mon, it's DINNER TIME. I would love it if I had my economic freedom. What else can I <u>not be</u>, now that Air Force stuck us with this script, written by some pussy who pays for it, personally, when south of the border? Protect that? That flag? I guess that's where the Mormon job coach thought I might want to apply to be a janitor at the high school…they don't publically eschew dating, whether or not they have a concept of what they're doing on one when there. See? High school, not college. The town suffered from arrested development and my parents cordially followed suite for the undercutting version of the status quo, just to keep going. No, it's Redlands, not corporate necessarily, its idea of adult life being a bar-fight downtown. There are other college towns on the West Coast handling business period, while the University of Redlands expects $45,000 for a Master's Degree on the subject to get your start, interpreter or not. Now don't blame it on the Rabbi and his folk this time, gentlemen, you look the Major's mouth inside like a gift horse, and <u>did not serve either!</u> The Army asked me to, when I returned from Italy. "How would you like to get back to Naples?!" The home town had failed to put anything socially viable on the plate, even though I was in good shape, and the Europeans had taught me their aesthetic perspective. In sum, I no longer bowed to the Queen, and just got on with college and my family's roots in Mexico. After all that work, yes. California was trying to recruit while wiping its ass. That's how much you need 1) a more Republican state; and 2) an over-haul on your idea of young leadership in the Party at that. What a prune.

Where was the story of the time? I suckered Dad into buying a $2500 VW Camper and escaped in it. He stayed home, ate, worried, bitched, attended church, and saw a psychologist. YES, YES, Mexicano, the psychologist was another large Gringo like Dad. Madonna Mia! Meanwhile, reading the recipe for the pie in the sky over Redlands, the horoscope says an environmentalist software company will rise--that outsources when the subject is a translation. What a team of Old-Timers—no wonder I was interested in Judaism later in life!

Yes, Knoblock of the 53rd Polish Regiment couldn't have done it better than a Hungarian this time Like Abraham's Isaac, I am alive to continue psychiatric treatment to this day, the Commies the psychiatrists are in getting their patients into full-time jobs and off SSI. Yes, the other Republican youth were under the impression that was all the free will I could muster. "What are you getting at, Hoss?" "The Isaac Complex prevalent in the Republican Olde Guarde, that must be solved, as other than lining up their own behind Mormon youth, the Elders cannot pull the votes to win California. They have served into the net at election time, loyal as we are."

Yes, one night, in the upstairs of the home, I broke it to him: "I'm a Chicano now." The response, with the upstairs window open to the neighborhood's theatrical air, was overwhelming, as if this was his big debut, and the upstairs pulpit was his stage: He bellowed to the open wind for the Santa Anas to carry like crows, "A CHICANO?! A CHICANO?! YOU"RE NOT A CHICANO!" So I'm not. I'm not a Chicano. 'B' in Chicano Sociology in college, but still Hungarian, regardless of fluency in Spanish. See, professor? A chess game with the Naval Academy on the subject of Chicano power. Zoot Suit? I would've had to have been there that night to understand, though that remains the trickle of understanding, that night, that I do understand.

Shortly after that I got tired of him backing me up when I attempted to make progress: I took the camper and joy-rode first to Big Bear, where the mechanic ripped me off, me sleeping in the camper at night while he attempted to repair it by daytime. The transmission was a 1963 Special, the year was 1986. $2500 was the cost for this artwork of a German time machine. Felt like I was sleeping in the couchette on the way to the Italian Alps. "Woodie" the mechanic in Big Bear with long hair, who can forget. The Italian bakery hired me to wash their dishes in the meantime. What status after Naples, the dishwasher for Belotti's Bakery, a couple Italian-American bro's who won't hesitate to take a chain-saw to a pine tree for firewood in their free time! Still a successful Big Bear bakery to this day, why, training for Big Bear Big Time while Tanya Harding was in diapers. How many Hardings now? The mountains don't understand the language, but they're as shifty as Napuleta`n. All before you get to Arrowhead, where the college Psychology 101 professor owns her home while practicing only God knows what faith, hiding behind her statue of the Virgin Mary in her office down the Hill. The future of then: The fire of the mountains, post-2000 A.D., where the whole lot of them got themselves packed together at the Orange Show overnight, in evacuation. If it's not crime, slander and racism, people, the payback is still yours to collect.

I left Big Bear after being mistreated and over-worked to head down the I10 for Santa Monica, CA. No sooner did I get there that the transmission stripped again. I pushed the van to a parking lot and paid to park it there for two weeks, sleeping in it and walking Santa Monica by day. It <u>will</u> disorient you. I pushed the camper all by myself down the ocean's boulevard to Venice Beach, and found a spot to land it by the side of an apartment building. While collecting unemployment, I paid the manger her $40 a month to camp there, 20 feet from the Venice Strand, where I would jog and do pull-ups on the lifeguard tower, often swimming a quarter mile to the buoys and back to shore. I collected my food from Ocean Park Community Center. My sister was pregnant by her attorney husband with her first, and I was swimming upstream from job-to-job in a high rent district. Finally, after the boy was born, I touched base with her. We met at a Santa Monica Italian café and caught up with each other. I felt sick. I took a chance and got in her car, somewhat uninvited, and she drove me to her home in the Valley. I had a sit-down with her husband and his law partner, who thumbed through my high school yearbook, keeping conversation as such, the partner with a thing for an Alum from Redlands High. They asked me if I'd be willing to see if the doctor at UCLA NPI had any ideas. I stayed a month at NPI on a volunteer basis, free to leave anytime, but I had nowhere to go: All bridges burnt. My father and mother picked up the $10,000 hospital tab, I was placed on S.S.I., and relocated to a board and care in Lawndale, CA, in-shore from El Segundo, by Torrance). Is my relationship with my dad a torture, all facts included? Past your bedtime, in the present day and age of Rick Stevens: We have bumps, but it's an unconditional love coaster now through threats to our country. Unconditional. He bought the camper. Later the truck. Later a Camry. And we would love it if I had my economic freedom from SSI, and a happy marriage, but yes, insurance as well!

I became a part of their lives in Lawndale, rooming with them, smoking with them, going on excursions with them and walking four miles back and forth with them to day-time occupational therapy with them. The guys and I dated the girls there, watched T.V. together (the Miami Vice era), slipped out at night to shoot pool in the bar, and were up early to get to Occupational Therapy. Despite shenanigans, I worked hard. I got an 'A' in Algebra at El Camino Community College so I could transfer eventually to U.C.L.A. I rented an apartment beforehand with a friend from Palos Verdes, continued to party, and worked at the Community Center as a volunteer, before landing a busboy job at The Spectrum Club, where I once saw Byron Scott dining at lunchtime after training there. Kurt Rambis had his summer training league there, though I was now, as medicated by the doctor, far removed from my semi-pro days. I was 26. I returned to UCLA and lived in Westwood with an Austrian, a Brazilian, and a Jew in a two-bedroom apartment, all attending UCLA and going to café's and clubs in between classes at UCLA Extension and the main campus. I did well after the Italian Dept decided I should simply start over. I received A's in Italian 1-6 before moving on to more advanced classes in grammar and theatre. I played Betto di Signa in Puccini's <u>Gianni Schicchi,</u> my one stage appearance in Los Angeles at UCLA, for the high schools in L..A. Betto di Signa has the opening lines of the play.

I pledged a fraternity but was shut out from full membership. Race was a sticky issue, as the year before, M.E.C.H.A. had squatted on Fraternity Row in protest of a Tequila Sunrise theme party. I'm not narrow, but others can be. I was a little close to a fellow pledge that had some affiliation with Hispanic roots as well, a true hot-head. However, they offered him a bid too--it's not as if, as a twenty-six year old student, I had all the choices in the world with whom to socialize in pledge season. His parents were actually Trojans! The Alumni Advisor for the fraternity had certain childhood jealousy issues with me of jealousy though had been a friend in high school back home: Give me two scoops of spoiler flavor for this apple pie! They stuck us with dues, chores, and attitude. As an older professional now, I feel the Bruin dialogue had a variety of holes in it: that's why the ship doesn't sail all the way to NCAA football finals like Ohio State. Their career counseling is amateur. They should offer more than a minor in Business to complement the extensive work the Languages Dept. performs. Over again? A more studious life in their Film School writing.

Can you compare from here to there? Oh, San Francisco is much more the social opportunist, despite their latitude on other regions at first, than L.A., who succinctly pretends the same status, though no, not one to bust it wide-open California-style, and L.A. Windows for adequate, equally-yoked female company are a rarity. It's a locals only dialogue with a low vocabulary. Some aristocracy that can't share the opportunity, but can collect all year and then donate. Social cache'? They don't speak it. It's a dichotomy from here to there on how a global corporation should be managed, as well as who should get paid what. My last offer: To move to L.A. on a $1,000 a month while working full-time to sell translation services to the company's clients. Bullshit. Forty hours a week is more than $250 in L.A., at your rent prices. Grow up, L.A., start paying your grads what they're worth instead of being a baby/film industry kiss-ass. <u>LOTS</u> think they have talent that don't.

In sum, to continue, after an otherwise social first quarter on the right track with foreigners of <u>maturite</u>`, a fraternity fiasco with younger male students the second, I got a job or two on campus in food service, to have some cash, as offers of a career nature? Later in the year, I became an interpreter for a businessman from Milan (the late Nino Vincenzi) at his son Gary's office in Marina Del Rey. At the end of six months, well-after food service work, he offered me full-time work at a reasonable salary if I would leave college behind and work alone for him. I turned him down, determined that if not at UCLA, I was going to complete my degree somewhere. I took a business class at Santa Monica City College and met a Japanese student not quite fluent in English, helping him in class groups. He offered me a room to rent for cheap in Hermosa Beach, so back to The Strand I went, while looking for work. Nothing turned up, rent went up, and I moved from L.A. back to the Inland Empire in 1989. From there I continued my food service career with Marriot Catering at the University of Redlands, a food service career while renting yet another apartment, throwing parties for old friends. I attended San Bernardino Valley College, in Speech and English classes.

Shortly before the end of the semester, I took my monthly SSI/Social Security payments, my latest paycheck, my savings and my room deposit, to buy a round-trip airplane ticket, flying out of San Diego. I revisited my host-<u>family</u>. At the train station arriving from Rome (where the plane actually lands, met by the Italian military), I took the 'Metropolitano,' or city subway, to Fuori <u>Grotta</u> and began to carry my internal frame backpack through the streets from <u>Campi Flegrei</u>, the stop of the <u>Metropolitano</u> <u>and</u> train there, if I'm not mistaken, that winds up by the stadium in <u>Fuori Grotta</u>, if you don't continue the train ride south to Reggio Calabria, and, then, 14 hours away, Sicliy. Sicilia. My host father Rafaelle bumped into me in the street while on a stroll, inviting me back to the old apartment. We walked together.

Rea ("Mamma") was not pleased to have a surprise guest, but most likely, things had changed with the years. After Sunday Brunch, we shared a Roman apple, glass of wine, and cigarette on the old balcony while belting out <u>O Sole Mio</u> one more time to the neighborhood! I was able to see Fabio, whom I hadn't seen since his visit to UCLA to give me his Italian synopsis on fraternities and the L.A. nightlife, socially, from a perspective of a more worldly man, my host brother now having the serious responsibility of escorting Italian exchange students from Italy to New York City. I stayed for three days, before taking the <u>Circumvesuviano</u> (the train which will get you to other parts of <u>La Campagna</u>, the name of the region of Italy holding Naples as its capitol) to Castellammare di Stabia, past Pompeii. Stabia, or "Lo Stab," is on the shore, not far from the island of Capri. I had an old friend there, now married to the lady he was just dating when we met on exchange through Intercultura, or Italy's A.F.S. organization. The perfect Italian woman? Not easy to find in all of La Campagna. The other side of "No one knows what it's like to be the man behind blue eyes" is that my friend had a younger sister he never told me about, who turned out to be non-stop entertainment, complete with her guard-dog (a Boxer) that roamed their four-room flat. This is the Cuomo family, Nicola and wife Olga, Carla, Dirk the Boxer (a guard-dog by breed), Guiseppe the younger brother and more, after the rest of the family, immediate and extended. Perhaps a humble Neapolitan establishment in the suburb, but, uh, are you a Californian testing them? Not really, though please, the dog will escort you, you will not rest as guest long: No Duchess, no Duke, Napuletan—I'm the Neapolitan here. I was able to see other friends, donated a pint of blood at the hospital for a brioche, didn't eat them out of house and home, and got directions to my Grandma's home country, Hungary by train, after laying a few kisses on his sister. She was not one for the Church. Youth.

I took the train from Lo Stab back to Naples and caught the transfer to Venice, by the Yugoslav border. I took the train through Yugoslavia at night, listening to Aerosmith while viewing the rustic countryside in the shadows. On the ride, I had the privilege of meeting a ballet dancer (Hungarian) who had been performing in Italy. Her and I shared a few laughs and cigarettes. She was travelling with her brother and another relative, but she was the only one who spoke Italian on the train. We, about eight-nine hours out of Venice, arrived on the ground in Budapest, where she reunited with her Hungarian male. She asked me to get in a taxi with them, but I wasn't completely comfortable with the discussion. I went, instead, to the snack bar in the train station and bought a pizza, when three Turkish young men, speaking Hungarian, connoitered me, none of them speaking my Italian, however. I had tried previously to walk to a motel in Downtown Budapest outside the station, but they had no vacancy. The Turks introduced me to a very tall, young, attractive Hungarian woman (more my Anglo hue than they), who did speak Italian. I rolled out my down sleeping bag, and the two of us kept warm under there in plutonic fashion. So, her and I chatted in Italian that night, watching the happenings in the station. I never saw Budapest by day, low on money and needing to just get all the way back to Rome's airport to fly back stateside. She grabbed the train with me, rambling on about her boyfriend in Florence. I gave her a back-rub and she purchased Espresso and sandwiches. However, at the Venetian border, the Italian train official requested she exit the train before crossing into Italy, apparently too low-funded to get back to Florence via Venice. The moral: Teasing the nice guy doesn't pay!

I got back to California and petitioned the University of California at Riverside, after Valley College in San Bernardino. Before I go on with the story of my Highlander education, an anecdote has been jogged from my memory, during the time-frame my sister, state-side in 81`-82`, was dating her new man, a New Jersey Jew. I had been offered the volunteer opportunity by a neighbor of my host-police officer to translate casual correspondence for him, from Italian to

English, destination of the correspondence Scotland. Upon complete translation of communications from Naples to Scotland, I got an invite to dine with the letter writer and his wife, approximately six months into my Italian stay. His wife served a great meal of fresh beef, mouth-sized cubes in a red sauce, while we poured a glass of wine and let the conversation culminate. He was German. I had asked, as I clearly saw he was not Italian. "What race are you?" (In Italian) "Ariano." "Ariano? Where is this?" "You don't know the Arians? We're the Superior Race." "The only race with a claim to superiority, according to the old Testament, would be the Jews." It was after that he nonchalantly drew a Swastika on his paper napkin to educate me. He had a peg-leg: he had lost his leg in World War II. This was one of two, possibly three real-life encounters with Nazis. No, they don't think like I do. I don't recall staying for dessert, eventually sat down later with the cop, who changed the subject at dinner. Aberdeen, Scotland has issues, I've read, with the Neapolitan mob. It's not so much be careful who you translate for, as provide the salt for the wound with your diplomacy, now that you have learned where in Europe you have been stationed. They did not drug nor poison me; a different day and age under Pope John Paul II, isn't it? "No, I say, love your enemies." No, I was polite, not tantrum-minded, and no, I was not translating a code. He needed peace after waking up to a fallen Germany to accept the reason God gave him for the war-loss, and so I walked it for him. I was encouraged to study German by Austrian musicians once while vacationing in San Francisco. I chose to be a writer in the English language instead. Maybe someday I will: Be, as a Pope Benedictine Catholic, commenting in the beer garden with a dictionary and grammar book. Time flies. German troops side-by-side Americans in the Middle East now, worried about Isis, a national threat.

In any event I finally heard from U.C. Riverside: I was accepted, and moved to Riverside in November 1991. I wrapped up my degree in December 1992, and substitute-taught Special Education and Bilingual Education for a while, as well as working as a delivery driver for Domino's Pizza, socializing and beginning my first writing project at a local café, See's. Shortly after a rather brilliant one-night stand in Riverside with a red-head went nowhere, I packed my bags for Santa Barbara, bound and determined to get my California stride back.

Now with a Toyota truck holding a tinted-window camper shell, I parked behind yet another café and stole into the youth hostel by day, sometimes laying down money to sleep, shower, and park there. Soon a man at the hostel hired me to be his co-janitor at a night club on Upper State Street. I then moved into the hostel full-time and began to date the foreigners, while the guys living inside worked on our pecking order. However, my boss was The Man. Nobody went over his head, and not either the owner of the hostel, as he was Le Bef: The Beef. There were that many zeros in the rape department. I worked for him, and if someone had a problem, he taught them the difference between adults and children, getting one young guy on the ropes in a hand-to-hand before, naked, his shower towel slipped off him. The young "gent" stood in the middle of the foyer, threatening with a baseball bat while his Wanker swung from side-to-side. The crowd coached him down before he got hurt and got the bat. My boss had won a college football scholarship in Louisiana, which he had to pass up for getting his girlfriend pregnant. He was working in Santa Barbara and sending Child Support payments back to Louisiana (Needless to say the Europeans, men and women both, LOVE California Theatre!). My response: After work, pour me another. It's a meat and potatoes janitorial job in the club at the best beach in the world. I have a concept of Virgin Mary's Court in the sky: Santa Maria, Santa Ana, and Santa Barbara. It is a matriarchy in the sky, my people, and don't go over the Lord's head on the subject of his mother and female flock, especially the Communion of the Female Saints. He's all that Jesus and more, don't pick it, are you on good terms, or do you want to think about awhile, because it's a life of service to your Savior, not a Campus Crusaders For Christ bar-b-que with the Baptists—unless you want to bar-b-que with the European girls!

Yes, I fell in love of sorts with one of the ladies, but it was short-lived, and she flew back home to finish her Master's in Psychology in Trier, Germany. We didn't have enough time together but kept in touch for several years, before she married a German native in the construction field. That's beer, that's brains, she wasn't satisfied with her first German commitment, I cut in California-side and offered her heart off of our platter, and last I heard, she is an executive in the Department of Personnel of the The German Railway, making sure no Nazis slip into the train system—trained personally in Gentile sex by a bona-fide uncle of two Jews. Six foot tall plus, the both of us, with tans to make Catherine Zeta-Jones and Don Johnson jealous. I'm a military officer's son in America: I held court with the Allies from Europe, while the Europeans managed the money.

The hostel moved to a new location while I was back in the I.E. celebrating Christmas with my parents. When I returned to the new location the next tourist season, I was hired as their janitor, and within a season's time, promoted to Activities Coordinator. I organized after-hour keggers and excursions to the lake in the mountains of Santa Barbara, driving ten some-odd foreigners in my camper to the mountain lake of Red Rock. They would swim and picnic all day. We drove home at dusk. We bought a keg. We all had an over-drive and academics wasn't the only subject: Italian, Belgian, German and Irish women vied for the spotlight by offering their interpretations of what an international relationship is about. I loved the elective not offered in the U.C. system. You simply can't what? Keep a good man down. Additional electives offered in out-of-state ladies and The Chicana Female again. Not a string in sight that does not belong to a foreign bikini, and it's yours to pull. The Boys of Summer, in order, Larry from Louisiana, Lyes from France, Henry from Oregon, James and Richard from Britain, Dale and

Brent (from Britain and Australia, our faithful owners before selling the place), a couple more Aussies with no names, a slew of New-Schoolers from foreign parts, plus the S.B. Community College Locals.

Henry and I got into it one night, him dragging a Boom-Box with Rap music playing into the middle of the after-hours bar-b-que. I turned it off, telling him it wasn't proper, and he got a gruff tone. I shot back with "You start something with me, you'll be lookin' at the bars from the inside, my brother-in-law's an attorney, a Jewish one, my niece and nephew can't be slowed down by Hen-Dog of Oregon. You're about as far from L.A. as you are from jail, you punk-ass, new-school out-of-stater bitch-boy!" The sermon was actually in polished English at the Prince's testosterone during wartime. He crawled to his room. I later apologized myself but told him we had a business to run. He wisened up and started acting with class, just being a stand-up comedian at the parties, telling jokes about bacon-strips in the men's underwear! We got on with it. There's assholes who won't take counseling on the subject of the law, and fighting goes on, so I'm succinct, but approach me on the topic of violence, if you want to see what it's like to be made into a country turkey verbally. It's Betto di Signa, in the Italian language, all day long. Plus a few Jewish law books (Is that Hagadah or Agadah? Halakah) for you to pick up before the chess game. Which is not chess in the Inland Empire with the hedon: The husband on meth with twelve teeth tells his wife when she's moving into checkmate. Have you ever played five meth freaks at once in an old home? Chess, the adventure. My advice: Let me retire from that too, and just talk football where we started!

I went back and forth from the I.E. to Santa Barbara three times: and on the fourth, when I had quit smoking, the hostel owner announced he had married. I made the staff again, though there was an argument at the tables on the patio that night with a Northern Californian hussie and her boyfriend about their marijuana legalization beliefs. An Aussie traveler put a large knife on the table point-down. I ran down to the liquor store, got some whiskey, parked my car in the lot and turned up the Rock N'Roll. When I reentered the patio the owner's wife said he was on the cell, and the culprits had left their table. He told me to just go to bed, that he didn't mind a liberal establishment, we would always be friends but frankly, his wife shies from the American truth of the matter.

I came back to my studio apartment in Redlands, getting involved with a mental health group and dating a couple bipolar girls, before hooking up with a sweet woman from Santa Cruz who kept in shape hiking in the mountains with her dog. Nature, love. My first book had been completed my first year in Santa Barbara ('93: at this point in the story 2000 A.D.) and I won an Honorarium with two speaking engagements at U.C.L.A. from the Chair of the Dept. of Psychiatry. He didn't put an extra effort forward, however, to get me published and off the economic chopping block, and the publishing companies (as much as Central

California loved it, foreigners and all) turned me down. The lady from Santa Cruz, an environmentalist, did not want to talk Christianity to such an existential level. The metaphysics of the discussion dealt with feelings of an actual anathema, but unspoken.

So many have similar sins, but do well, that I'm not worried anymore. Politics were bound to happen, in fact, the year after my love from Santa Cruz broke up with me (who had been looking for something in life more viable than the pub and hostel combined), 9/11 occurred. After she broke up with me, I opted for Rehab to deal with the pain, even relocating to the Rehabs in L.A. (Cedar-Sanai, San Gabriel Valley Medical Center, while Loma Linda Behavioral Medical Center acted like the stuck-up, ethnocentric, small-town). I had two young, male Jewish roommates in L.A. apartments stemming from Cedar-Sanai, but neither understood adult responsibility yet, so my efforts to be a part of screenwriting in Los Angeles fell short logistically. Upon returning in 2003 again back to Redlands, I rented at a couple of places while putting together an application for the University of Redlands' Screenplay Bachelors' Degree. I was accepted, but the laws had changed: No longer would Financial Aid pay for a second Baccalaureate degree. I turned forty about that time, volunteering my efforts in grounds-keeping at my parents' and other rentals (a Green Thumb), before landing an apartment that is a mile walk from Downtown Redlands. I love it. I can walk a safe street (and I've seen the walk from the Garth House in Beverly Hills to the Rite-Aid in the neighborhood, as well as the company provided: Enough to wake up the Martial Artist in me—the only mention of it I will make, as I am retiring my feet and hands to take the doctor seriously) late at night without worry, and have a roof over my head, complete with a $ two-dollar box of food from The Blessing Center in town. The food will last a week for that (I wouldn't mind helping them, nor the Church, as I rebuild my career—a homeless shelter is in the plans, though needs donations to construct).

I haven't dated seriously since 2000 A.D., and now it's a different climate for such: more conservative, but yes, even sexier than ever. I've been abstinent long before my conversion to Catholicism in 2008, the year before I was awarded a Federal assignment or two with the U.S. Census Bureau. Conversion was very real, though Catholicism can stall from time to time. Both Democrats and Republicans occupy Mass time. The whole pretext, by the way, of abstinence was to be able to bring more to the plate for a woman such as my friend from Santa Cruz: to be able to take her out on the town, to be a provider.

The politicians from Republican to Democrat cut our arms and legs off on the subject. They're addicted to talk about medicine, but don't know athlete from obesiosis (Is there a difference? My Godfather Catholic-side is an elder, 300-pound former martial artist of native Mexican heritage). The state has been a loser since Bush' first term. War didn't cost: it gave the Democrats excuses to act like idiots on the subject of romance. Schwarzenegger too, an affair (or, uh, a few). Listen, before California loses its ability to court period, turn the cheek and let's keep searching for jobs. The voters in both parties eventually have to get off their high-horses, having sold their souls to the medic when their men needed them. An atrocity. Up and down the Coast. No, S.F., can't afford the road-trip and roots is interesting, except when the bourgeoisie play it for the money alone and don't share a dime. So the Lower Class is learning: It's our fight, cheap wine and smokes or not: Stick together if you can't afford a drink at the Pub.

Well, we did that. We stuck together behind the scenes. And Fall of 2014 came.

I kept my apartment peacefully in Downtown Redlands while my father enjoyed yet another of several consecutive Naval Academy victories, in their tradition of playing West Point. CIF Division II had the East Valley Wildcats in town play-off bound, and he gathered the news for me on the Final vs. Carson on T.V. from Los Angeles for the State Championship.

I had progressed not only as an athlete, often walking two miles back and forth for my beer as opposed to driving, but as a person in appreciation for Golden State annual events, and, as a psychiatric patient in appreciation of the Christian espri`t offered in the area for those caring to practice anywhere in town. East Valley brought home the first State Championship not only in the rugged sport of football, but any sport, to San Bernardino County's doorstep in the first time in county history.

My father passed away at the age of eighty-two (complications of a stroke), after living a full retirement from both the Service and the county. We remember the good times, though will always feel it. After over five years of Catholicism, I returned to worship in the Episcopal Church, where his funeral was held and our memories of him remain. Despite what one may read further on, as I wrote profusely as a Catholic, I currently prefer the reflections of the Episcopal church, all theology aside.

We had not participated past flag football, as I've had eye-ware beginning at age eight, though I often took them off, broke tackles, and played on the same terms. He gave me his Navy P-Coat, with his initials embroidered on the inside, before passing. It's fifty-nine years old.

I'm fifty-two. I've seen an overall progress in the town's attitude since we relocated here in 1972. Congratulations: The rest of the state was speechless, bankruptcy or not. Was that, "Redlands, The Friendly Place," as the sign says, or "You from the Bay?! You from L.A.?! Screw you and the horse you rode in on!" Who know what goes on in the vicinity of The Emerald Necklace?

Clear enough that the Gipper in my life is now looking down from the sky, I take you to Part II, which began in 2006.

Enjoying My American Religious Freedom in
Redlands, California, During Wartime

The world spins, America spins, The Southland spins, and my city spins, though my life is a Christian life. This is where the pressure comes from to do something with one's Christian life, though plain and simple, The Lord works in mysterious ways to provide for us even from Above. Our life is one of prayer, devotion, and sharing his message and Spirit, with whoever comes our way.

Our role selection in the Christian world changes through the course of our lives. It's shuddering some of the hard-to-reach people to whom I have been a conduit to the light of Christ. All levels of schooling, many work capacities and organizations, though of late...admittedly after succinctly showing people how Christ, of whose greater body we are a part of via Holy Communion, reacts to motion, I got confused by my break from it all. At first I thought this was a stiff lesson about failure, a need to change faiths, my non-improvable shortcomings, why, and my early retirement due to not being needed and expendably so. No, Christ gave me a break. For over a year.

He ordered some changes in the meantime. There were fires, storms, deaths in battle, earthquakes, a vast reeducation campaign, millions of new births, thousands of medical community announcements and medications, and election after election. This entire time I wondered if The Father was upset that people insisted the world worked this way: a documentation of event after event. That's how it's done. That's life. And talk about our results.

Christ made sure I had my privacy while this false message of the Media's biological clock was going around, with hundreds of millions gossiping about excuses to avoid Him, His Holy Spirit, God the Father's other thoughts besides Ten Commandment training camp, and last but not least, Mary.

I was definitely somewhat confused by being closer to the Trinity and Mary. Inexplicable, uncontrollable changes happened to my mind. How I thought, what I thought, how often I thought it and even the personality of my voice as I spoke all adapted to this new closeness. I conjecture, with a healthy introduction to the actual modern-day concept, that angels were taking exams. It had nothing to do with a degree and a job! But they had an exam nonetheless. Could angels often have to do things that people do so they can relate better to the human forms they are not? God is running the show away from the show, and apparently the angels thought we should repeat the patterns of similar time frames in history when people argued on certain issues. But when Christ investigated in private about Christians who desire things like more change the better, He found out they were just trying to have respect for the idea of a Second Coming in case He asked. The angels thought we were acting out like we wanted a Second Coming, so we got a lot of repeats.

In the mix-up and in between exams, I actually was kissed by an angel! She's cu-u-u-t-e! It was long and drawn out, I conjectured it would go on all eternity, and then the thought came, "No, don't you understand?! She and a host of others have a mad, passionate relationship going on with people, and she's here to serve."

"Stephen, not only is it Susan, and Susan's influence, but it's Susan Keppler: The family changed their name from Keller to this, prior to Peppler, after Helen died! Explains the eternal problem with hometown romance."

"Gentlemen, if you would not mind in the future performing more proper decisions, less elections. I've heard the Priesthood come up with some doozies to get us all out of it another week. I've gained ten pounds in the meantime, and that is 208: it still makes me one of the skinniest men in the County! I had no idea the war would drag out this long, but if that's the same woman you've been married to the entire time, she hasn't changed, she still talks to me like her first language is German when I buy my food. Just serve me another beer and move on with it."

"Oh an 'evil spirit,' mother, is that what it's called? You've got more of those in this town than the spot in the Holy Land where Cain killed Abel!"

I get the feeling there's an exciting conclusion along Christian guidelines to what I'm going through, personally. That is, even though they kind of like me. Cold feet for commitment here is what I call it. Almost not sure what we mean by 'don't discriminate,' as if the story in the beginning wasn't all true, hence crossed lines and confusion, over-corrections and pressure to monitor in order to validate historical reputation… Claims it's over their head, but when we get down to earth it gets snobby, not Christian, and appearances kept, but no progress for me. All, whether Catholic, Jewish, or Protestant, all culpable of this same neurosis-causing character defect. They seem to portray a similarity between arrested development and mental illness. Ten to one someone did a study on such after a patient got caught in a rut due to the others getting to keep the job when they should've sought psychotherapy first. Crap. Benchmark processing for others takes precedence. The corner. No, it's simple. This is not true. Call it 'smoker's theory.' I hate keeping it on ground zero just to address nicotine cessation. You can get out, but the city's Celestial paradigm is dysfunctional.

--"Casey. Casey Haus. How do you say that in French?!"

--"Ca-si'. Ca-si' Hose."

--"Ca-si' Hoser!"

--"Ha-Ha-Ha-Ha-Ha-Ha!"

--"Okay, group huddle. It's 'Messianic Judaism,' not 'Messianic Prudaism.' C'mon, one and all, back on the field! Hubba-Hubba-Hubba!"

--"Oh, was this about Ireland, Scotland, England, Wales, the Dutch, the Belgian, the Swedish? Stephen, you're a Hungarian who speaks some French. This is the proverbial guillotine in the Olde Days, traveler: You're a good, well-behaved son to the man who wields the axe! <u>Entendez-vous</u> (Understand)?"

-- "<u>It's</u> your father, not you. You get confused. That's why we implore of you to take your pill—we're not teasing you in the sun about your sex drive!"

--"<u>Je suis Hongrois</u>!(I am Hungarian!)"

--"So, Steve, relocating to France for a family of four was easy, if your parents had the right job qualifications!"

--"I would drink more of the psychiatrist's beer. And less of his coffee, if the subject is quitting tobacco."

--"Wow, can you believe the California Democrats? I've seen some people try to put surfing out of business to get more votes, but that's the winner! All that was included plus a free sign-reading class from Senator Wiener!"

--"I'm sorry, I can't accept it. Apparently some of the deceased Mexican ancestors are in Purgatory, and I'm their lucky bingo number to win Jesus' game—when I can take the job, I'll do it!"

Paradiso Canto XXXIII......*The Divine Comedy by Dante Alighieri, Fourteenth century Fiorentine poet*

In essence, it talks about Dante's experience with thought or sight of Mary in mind. It is thorough in my experience, and the difficulty lies in describing the levels on which it takes place, mentioned in a previous Canto. Canto XXXII 15-18 of *Paradiso* writes of Hebrew women who, from a level of the Tenth Heaven, descend from the Empyrean. This is where the Rose, the metaphor for the Virgin Mary, lay. They divide the tresses, or long hair, of the Queen of the Empyrean. Verses 19-27 talk of these Hebrew women as a wall between ecumenism and ALL of its churches, not unlike flights of meaningful stairs each and every one, all leading to the Rose...the Bethlehem mother of Ben-Joseph the Messiah, in whose womb the Holy Spirit planted the Christ Child. Depending on the denomination in ecumenism, this may be a Christ now to come, or the Christ already having had come to Earth. The heritage of Hebrew women who descend from those who were there with Mary, from Immaculate Conception to Crucifixion/Resurrection, remains a constant regardless. Hebrew women, braiding Mary's hair while Jesus either comes or goes amongst churches post-Crucifixion, matter more than the churches themselves to a God more in command of Old and New than New and Old. Mary.

This is a mouthful one could explore at great duration on the subject of Mary's prayer levels alone. Experience on the subject is just forming. Sometimes early morning thought brings these feelings close in, without saying a Rosary, randomly, though in a way where one is comfortable imagining Her counsel, in a way that directs the very morn's environment. It's beautiful to feel acceptable, despite masculinity's legacy and faults, to still carry forth with purpose in mind.

The thought of Angel Gabriel's strength, like a rock of unified conscience for deducting reasons, indicates something. The theatre I attend draws off a colloquial root, and this is why I have difficulty interpreting a crowd on opening night, for conversation. Albeit historical, in its religious sense, The Divine Comedy reaches for contemporary, pragmatic applications. It reminds me to note that at said time, I literally had to 'steal' my way, or path, onto my college campus the last year of attendance to even hear of this literary work written by Dante Alighieri. Strange times—contemporary Inferno on Earth! Courts' issues, politicians corrupt, a riot in the Southland's capitol of sorts...I survived and was awarded my degree. However, the memories of such are hard to nurture hence the purpose of said prayer devotion to Mary, among other reasons: to learn to heal myself, then perhaps others. Discovery of said levels requires cross-referencing with the New Testament, particularly Paul's letters, in order to properly understand and process daily "Purgatory," or society today.

The realm in which this study commences is a garden in the small city of Redlands, which has as many roads as historical figures, from philanthropists to doctors to land owners. A realm that is tranquil, though outside the gates the aforementioned colloquial wheels can trample an unprepared heart. The city calls for growth. I call for maturation, modern maturity, not to be confused with the periodical of pop-psychology! Hence I am steadfastedly interested in dialoguing here. So I seem to be shook on one level to wake up by the population after experiencing anti-Semitism, as a Gentile, not raised a Jew, but having relatives. If said traditions bridge with Alighieri's The Divine Comedy, how many 'pasture days' will set roadblocks to anything but a suffering existence in said quest?

Hence I have my parents' garden, with statue of St. Francis of Assisi presiding over this essay as sanctuary. The road from Florence to Assisi is precarious, psychologically speaking. It is little wonder that, as mental and physical health are a knot Dante alludes to, the history of the local hospital recently came out, complete with photos beginning in the 1890's.

To continue, I note Paradiso Canto XXXIII, verses 82 through 100 talk about a single volume of love. In the great beyond we are placed within, this single volume is surrounded by fragmented or frayed ropes, compared to the solidarity of healing power through love (or the volume), in the ropes opposite a knot to be untied. That is, in order to free the flow of love from said volume. To diffuse and heal. The sin ties a knot while love works to dispel the oxymoron of issues leading to closed circulation. But this is the first passage, together with the thought of praying with Mary in mind, and oh, the confusion one is in. It will be a trip to Los Angeles for me, not a vacation to San Francisco, and coming soon, it is indeed something of a psychological business trip, to see the Jewish adamant-tight hold on said loose talk being unleashed, adamant with wealth and only Jewish spontaneity and wisdom. Lest I presuppose why this is my course,

I'll cross-reference with some relaxed (as opposed to interpretive) reading of The Jerusalem Bible...after dinner in the sanctuary by dim patio light.

Now, after a somewhat Kosher sandwich, I, let's say entrapped, just an early prayer with devotion to Mary, for I am tired after a long, natural week to hold the fort. Nothing holds the fort like a natural constitution, with nothing more than a stiff Pentateuchal divide...my mother is Hispanic, though has a knack for making Gentile meals taste kosher.

In prayer I say:

> "Blessed Mother Mary,
> I thank you with deep
> Gratitude for seeing
> To it that I could weather
> The long week. Please,
> As we in family rest this Sabbath Sunday, pray the
> Healing Service brings constitution, in preparation for my trip
> To the busy city. Be with me." Amen

It is obvious to state after reading this Canto that we think in levels for a simple reason: Depending on your discipline of choice, a particular interpretation of Dante may unfold. Undoubtedly the history and figures along the road from Florence to Assisi, between the time of Dante and Da Vinci, would produce some sort of research miracle to patch the old flask for new wine. But that's counseled against in proceedings. On one level I am dizzy, light-headed and fragile, on another moving king-sized beds so that the parents can sleep more comfortably. There are non-believers in my path that are full of energy and noise. Are all the roots of this 14th century author here to withstand onslaught of conscience in his Florentine inspiration, or as well is his inspiration a universal appeal to the Light from Heaven for all of Earth? I have rested on the latter, though a somewhat flighty conscience it produces (or is it simply Southern California summer at its peak temperature?) After all, is not laughter the best medicine in colloquial American terms, when praying to Mary, and is that not what angels do...fly? Not in planes ethereal, but through the stubborn proceedings of human conscience?

The immediate response is "We'll just pick up a new bottle of wine, and let it go." The well-retired Christian said, "I'll forego the history lesson, bake a turkey, share the wine with you, and do my own flying and driving," and Mary probably wants, "Yes, more of your wine, a healing service and keep the humor—it's better than a recipe for the holiday blues." This seems to be the first time in a while I got a jump on that blues thing despite not having a flu shot yet. I suppose angels know how to leap too and would be amused if given license to teach humans to do tricks!

THE SECULAR WORLD

--"My friend, you are in Redlands. Captain Fitzmaurice country."

--"Can we smoke?"

--"Smokers stay in their bounds. This is a dangerous urban environment. Legacies have ceased to exist after years of owning the field. This not is, but was, Fitzmaurice during Iraq, the early years of the conflict."

--"Duck, smoker, duck!!!"

--"Hey, Boss, Sicilians comin' to town."

--"Chicago??? New York? Florida? Frisco?"

--"No. Olde Country. What the hell you want me to do?"

--"Get an interpreter while we're eatin' Planters at the Malibu Film Festival, for one."

--"I did. I called the agency."

--"Who'd joo get?"

--"Some Old School guy with a baby face by the name of Fast Cash Forty."

--"Fast Cash Forty? That's his name?!"

--"Takes his cut after the job and runs down to the liquor store in Inglewood and buys a forty-ouncer of Olde English 800 every time. You got spades, he's got hearts."

--"You mean, he's takin' no curves in this town from Gay Society?! How the hell does a thorn like that get stuck in my crotch when I run such a beauty of a porn ring?'

--"No, Boss, you should ease up on the good bottles yourself. He's drivin' in from another county and takin' the bad boxes you got with the Sicilians that you could easily break, along with their balls, prior to warfare, with him. Run. And look Ma, no cab!"

--"Tab, please! I'll be polite. As long as he ain't a UNESCO Liberal Studies graduate."

--Psychologist: "Young lady, what was it about the sexual experience with the writer

that led you to break up?"

--"I had a menstrual moment and got worried about our future."

--"Please don't put on an act like the editor of the new book. It's a serious occupation, and the writer has to pay for editing, why, sometimes to self-publish! This is Capitalistic America still—not a Socialist state by a long shot!"

--"Oh. Do you think it's apropos to say he's still good masturbation material, even though no substantial talent?"

"I think you can spit his penis back out now and get closure."

--"It's for intellectual reasons, the masturbation..."

--"And you want a prescription for?

--"Skunk, sinsemilla, immaterial."

--"Well, possession is nine-tenths of the demonic law, if you plan on possessing him, but this isn't a religious discussion, it's a medical one. Try tobacco, without dragging a Red herring, and take your own initiative."

--"Please?!"

--"How much?"

--"A $100 baggie: this is a once-a-month appointment. I do work, after all."

--He's gonna help us out with what this is like if you're a Junior High athlete, not a Junior High musician like Frank Sinatra.

--"Sinatra? Jr. High? Didn't he study business?"

--No. Music. See, you need help. You don't speak it."

--"Didn't know our version of fine cooking and your version of Italy were getting you into fights with the California Politburo."

"Oh. Politburo. And all their songs to the 'petite chateau (petite cat).' Nice touch. True Los Angeles and White. Oh. Terriers and USC vs. UCLA and No Cal?"

"YOU'RE ON, BOY SCOUT!"

"HAVE YOU EVER FASTED IN CALIFORNIA? THE
QUANTITY OF FOOD AROUND YOU IS AN ILLUSION
AND THE EDUCATION, OR LACK THEREOF, IS A FACT!
INTERPRET AT YOUR OWN LEVEL THE TRUE DREAM!"

I had a dream. We had our dog in the backyard, and the neighbors' daughters and their brother were still young again. This is all while I was trying to concentrate on writing and being a good son, with my loving family. The crazy daughter gave some drugs to the dog. He began to bother me, his owner. There was another dog too, the one that I saw on the walk outside. It got excited when it got a quick peek past the fence's gate when opening and closing, about the big backyard. Three Italian girls then came for a visit. They were from Europe's version of the American Field Service. They were on their way to hear Liberated Wailing Wall, a Messianic Jewish music group (I thought at first when I heard the name the genre was punk, but no, seems I recollect the rest of it going with the tambourine). I was having a hard time speaking Italian at first, as it was Fall, but warm, and my father had surprisingly built a pool in the front, the street actually with stairs from the bottom sign leading up to a diving board at the pool. When I smoked a cigarette with one of the girls, the other two got to the point where the one who smoked seemed infatuated by being put on the spot.

As I snapped out of this dream in early A.M. hours, it was as if a female demon left my body and I rushed to the apartment's restroom to free myself of my post-5:00 P.M. coffee intake.

--"Can you be professional?"

--"Professional? I'm already employed being an adult. Because of my occupation I can't always be professional in America. I'm only here because this is where my occupation has me. Personally, I love it. Troops home again, the cooking, the whole nine yards."

--"Okay, getting to the point...define economic submission."

--"What is economic submission about, historically and evangelistically? It is humility to Cortez, since one was not unlike a good, straight, but humble, friar who serviced the wounded and showed the Queen's manners to Cortez' frontsmen."

So, jokes aside, 'tongues,' 'foreign languages,' what's the difference? It is more about 'tongues of fire' to an interpreter, when tongues of fire oppress the soul, which, in conversion, has attained immortality. I say oppress in a similar sentiment to Paul as he informs early in Romans that we are slaves to Christ, not the world, in following our Lord to redemption. It can be a slavery of certain factors while within the context or grasp of the spiritual ritual.

What ritual is this? It is the blind foot-soldiering of the Roman Catholic Gospel during a phase of a convert's mission in which his Church symbolizes a cradle Catholic with autism to protect, adore, and have patience for in its infancy. All the while at work, the tongues of fire show demons of fire the difference in celestial and hellacious temperatures (the Devil and such all around, unleashed on Earth, sometimes piercing believers from within their own version of form). It is the celestial conquering through the convert. In this capacity people of the State do not have to convert American-style, but be won over gradually at times in the Book of Love of Mary's heart, poetically speaking, and maintain a stronger allegiance, through favor and grace, with their new friends, The Roman Catholic Church, American sector, and its constituency. Footprints take time.

I had another dream last night. There seemed to be some sort of festivity involving movements, and then a group of fifteen found its way down adobe or stone steps, I'm not sure, to a terrace. On the terrace, something seemed to be going on. People were moving toward and back from a Hidden Staircase, going further down. I started back up the initial steps, then scarily heard my nephew yell, "Uncle Steve, help! Help Uncle Steve!" and I woke up, at the point where I turned back to head for the terrace anew.

It seems as if the Sioux Indian Dream-catcher hanging from the ceiling fan cord by a leather loop is working, for the simple yet difficult task of getting me to dream while sleeping.

--Hmmm. It's something about Western society's structure from end to beginnings in philosophy and its course, as it appears in the sun-dewed now.

--No, do you even know what 'hermaneutics' means, as a California attorney? What high school reunion was it for you when you learned the word 'polyglot'? 'Dumb' and 'shit' are connected linguistically in one word, 'dumb' and 'nigger' aren't, because he needs to escape your 'now' scene to attain higher goals in the Democracy America is historically, since. "I'll have the chorizo con juevos, please."

--"Jesus, the birthdays in America are so off the mark, they're like Caesareans! My sister?! When she's premeditating communication with me by phone on my birthday? We get a string of events that Holmes and Watson deduce are a tough girl with multitasking hands who can ruin my career and my health, unless I cosign her bullshit! Forty-nine today. Felt like a meth freak at my interview after the phone call!"
--"Button popped off at Christmas time!"
--"Jesus, is that all? The Algonquin Cree won't lose a customer at the Pow-Wow Parlor then!"
--"Cree? They still selling goat cheese at the snack bar?!"
--"No. Some thread for my hoopskirt."

--"Nothin'! The local Madame is disappointed!"
--"Well you knows what that means—the day after your shot from the Muslim physician. Good syringes in both cheeks of your American ass at that! Goes right to your balls! No spermatoids for the time bein'. And you won't need more sun!"
--"Thanks, Frenchie! We Americans are learning more about psychiatry and reproduction from you generous donors of our favorite statue all the time!"
--"Yea, give it a week straight, L'Amerique! That way you know if she wants kids alone, or can appraise the family jewels!"

--"The 'truth', after 9/11? Makes Americans hungry, because they only hear it from their mothers, and they're great cooks!"

It really doesn't give you anywhere to go. This state never paid collegians for its failures under Grey Davis, and has now banked all on the new generation. A huge gap was created psychologically. Psychiatry at best has been a stall game that won't come forth with the truth because of politics and religion, though in the wake of political blunders, it became clear we pushed around the patient, played favorites with other interest groups on or extending from campus, and kept collecting. The patient was character-assassinated in a war on drugs, with a label of alcoholic to prepare the arrival of newcomers to the state (military, investment, foreigners, and the usual apple cart from Southern parts). To say, their option, after college, is the State Department, stipend-wise, while pharmaceutical companies create an extract from conserved cataracts stored in a large freezer before liquefying to steam into a new pill powder: It is the compromise they are forced to take to escape the labyrinth of solitude, imposed by discrimination, which is at once subtle, overt, as well at times media-oriented. Why, on top of this, California also labeled them liberals.

After they were requested to only school in certain locations, where success is not taught. In fact, not until recently, it has always been the group therapy box in the California classroom. Some go crazy, some just acquire such selective listening that the school retests their attention spans at forty-three, like me. Now, now, now, a conservative religious element of America acquires an enemy, and goes to war. They, the patient, in said manner in the know the entire time, are right, and they vote for George W. Bush, Jr. Is it a war…even amongst our own? Out of the classroom one emerges to engage in a society much more lazy though with the same bad attitude. Work is aspirin at noon, lunch is restriction, and wages become a sign you are willing to make a sacrifice so that you have a defense system, while you view 'gossip writing' on television. Art and literature become a retreat to Classics therapy, the art world becomes thieves period (!), and yet another divorce in Los Angeles film happens.

Whose vicissitudes are these? Beyond culture it is and an understatement at that. The state, you say? Interesting. I find it hard to believe, after all this, that the state has any charisma to get me a job, if and when skills are finally taught in safety, without drugs, after said problems sent deals being made both local and abroad to privatize all this as much as possible. The eighth-largest economy in the world? This, perhaps, is only population-wise, and, as well, these are housing costs to keep out riff-raff plus a few major corporations, the latter for which military service is a nice touch on a resume. Hard to believe? It is a post-smoke-blowing session of a gambling phase of California's denial that there is no culture war, though the rest of America? It is dark out there. Enough venting until exhaustion. Digression is a role one plays so others get to talk therapy instead of group therapy. My vacation pen-break from the demon-hunt this is, with blessing by Mary.

--"No, Southern Cal doesn't get in line when the subject is other cities' rivalries with San Francisco. We have a 200' x 75' crate of antique miner's picks leaving from Long Beach to Latin America via New Orleans this A.M. Southern Cal entrenched deep on this one. C'mon, San Francisco, what was it this time, 'she had her things', as they say, translated from the Italian? You have Italians in your city, they'll explain. Keep it in bounds with, 'We're all Americans here, not all that to worry about!' until Monterey Institute opens up the Star-gate again!

--"Jesus, what the hell is he doin' in the discussion?! A voice from pot-smoker past!"

"Oh, get your own cigars! Jesus, next time line up the Scouts nationwide in suits, rank or not, who speak it and take a photo so the Bishop doesn't have to go through this. And line up the ones who put on their chorties right after Mass for their separate photo shoot! You got a Paparazzi with the foot-speed to handle that one? Before we're so backed up we put a star on the brick walk in Redlands, an hour and a half east of Hollywood off the 10, for Andrew McCarthy?!"

So, in the end, it seems with that left to bear the state wants to constantly organize the economy for its key issues such as freeways, pollution, and crime-fighting, all good, legitimate reasons California is popular off the subject of education. In effect, thorough-faring in County and community issues becomes a lot of information, but hey, that's collective conscience. This is as long as C-Span has the final word, ay? If one connected the dots on all four of these levels they could afford to get out… they could afford to go out in their own general areas, that is. Religiously, I see, several counties are involved in one diocese. See, these are two paradigms mirroring each other in comprehension of organization and governing as a concept. The third slant in this pyramid is culture, and on that note several other pyramids appear, all with the same state, religion and culture angles to the Maslovian scale of basic life ingredients. To climb a pyramid is to operate a business, i.e., if dissatisfied with the university one has to improvise legally, and keep moving. One has to be centrally focused yet flexible. All this is worth going over, and one can have rhetoric, but do they get the time?

As opposed to 'political meat,' isn't the Gay marriage issue simply throwing a sandbag in the road as the milk-cart is passing through during wartime?

--"Trying to quit smoking?"
--"Yeah. You suggest Nicotine Anonymous? Twelve Steps?"
--"No. The Case of the Picky Beer Drinkers. Read that."
--"You see that Unnecessary Roughness?!"
--"On Tampa? Penn? Oh, probably Scots-Irish!"
--"They're fightin' on the field!"
--"Oh, shit, Lumpkin, then Owens, got'em stirred!"
--"You see that Falcon?!"
--"Shit, we don't have'em dark chocolate! We better not be recruitin' UNICEF!"
--"I lost you at Lumpkin! Must be his uncle that threw to me in flag football!"

--"All right, Bring 'em over! It's about food! Suit'em up and show'em up and pay'em! Don't ask how I know them from their Uncle Osbin!"

--"New bones, not old bones?!"

--"More new bones than I dumped on his Uncle's plate! Back then?! Communion with Black Colonels!"

--"YES! Look at that! Want to see if he can still throw to you at 49 years old?!"

--"Okay, ready? PUNK-ASS NEW SCHOOL BITCH, HE'S A MIRACLE, BACK FROM THE STORM AND NEEDS HIS 'G'!"

--"Okay, wait up, I'm goin' to the guitar shop to get a new muffler pad!"

--"...WORLD WAR TWO VET'S FAM, NEEDS TO PULL HIS STRING, YOU DIDN'T MEAN TO DRIVE LIKE CHANDLER BING!"

Mary Thinking, From Abstract to Reality, As Applies To The Now

We now have, as a recourse of the Iraq War and policies' effects, a narrow interpretation of me with which to face my Christian own. Within the course of this time we have 'recycled' painful memories of a past of temptation, as well as impossible settings and circumstances. Every party I ever associated with, from family to organization, seemed culpable to a certain extent of leaning me into the winds of persecution. Still not convinced I have been off the pathway of persecution in this state, I continue reaching forward with conscience that is now, hence, 'spiritually fire-treated,' (by fact I remain believing after all that!), yet vulnerable. Who knows what obstacles lay for a mind stretched thin, or where strength will come from? Who can foresee what issue will next immediately challenge me, as sniffles run down my nose on this cool November day?

The artwork I made myself for my own walls stands evidence vision has gone deeper, though but what to retrieve? Yet it is only a short story of a love of the past, forbearing signs of what the present wants. I crawled in this studio room in the midst of it, and took long walks to clean my mind, with always a turn on a corner at the end to reset the clock to 'no growth' for me, just movement. At the very end the table was reset to where the past wanted it. I could walk up to two miles with a reminder of terrible news alone on the headlines. No Future For Young Adults, they read, Though Children Look Great. As if it was I personally responsible for failures of decisions I was not a part of because of the way the past needed me, for if truth showed its face, God would anger again, and something would happen. Hide him: Hide Stephen. He draws a line to reason, and we have none yet cleverly spoken excuses. We did it right, as money is here, and food is there, this moves like that, and he often sits still. 'Twas not a self-imposed solitude, 'twas safety from sin. I suppose I get full credit for not leading them into it, even though maybe find it they did. Sin of conscience, that is.

Yes, it started with thinking, not so much out-loud, as in vocally, though with a nakedness that was rather proud. An angel, for not a demon it is, perhaps was cause, of the compelling upon me from an abstract realm to "Please cooperate! Just cooperate."

So shop politely I did, not knowing more than that I was now informed, hurriedly at that, by whoever monitored this abstract realm. If you will. And if you will not, suffice to be content in the pride of your own naked thinking. For long ago stories began their repeat when all nude logic had commenced, and oh, what a reserve of now-concise theories my awe-nurturing nation perspires by, having been said students of Life apparently. Blatant it is no longer, academic America, yet still bold, and ever-innovating for a peace with weapons to withstand the wicked, the pricked, the priceless, the pair, why the Beast and His Bride. Enemy notwithstanding, an historical meld of world trivia delights it was, aye, candies from the Continents and teas from the East, yet a tobacco so Southern even all of our own new, premier historical intertwinement with Spain manifested into a smorgasbord of medicine. A Moroccan proclaimed, "Worry not! My blood is fine and so is yours!," leaving me to know that naked thinking led not like money to merriment's moments of intimate contact, only to a Christ with ritual so hallowing that the depths of our American souls are so deep were as to indeed, "Please cooperate."

En fin, an age of romance anew it was not, amidst which such so ensnared solemn saint transpires to understand more about Mary, getting a grasp for the pull, the empowering purity, and pomp which God Almighty found so select for divine seed. Unto us a Savior was born that left eyes opened to so many chores incomplete on this planet, that the priesthood lost tithes at the hands of a charity so busy that bees burrowed not for more than an unfolding of Earth's spiders, which I promptly killed from end of the valley to home again twice over. The Bible scholars were embarrassed to admit that She had prestige to see them scurrying to the nooks of the manger, lest lamb mew for shepherd and waited as a perfection hosted by Gabriel commenced. "No, I'm not going into rhetorical detail of the manger, I'm a repented Gentile, I know not all to be done, yet to recount a spider so wicked as to wink at Joseph is beyond apostolic reach to unfold!" So Mark must have thought, nakedly, until a resurrected Lord loosed his tongue through a baptism by Spirit that said fire extinguished any such element in him remaining of doubt to proclaim the Word. The hallowed souls showed themselves, Earth reflected with spiders (cobs) which were no longer held at webbed, and sin was purified to yet another extreme.

Mary does not have a tangent a saint cannot be busy for whilst She envelops the abstract womb with the chemistry of Her personal heart. So I see it now. Alluding to the fact that as you approach Christ's creative capacity, in lies a mother who oversees the lessons of the abstract, whereas most seem to ponder whether the Devil drew it. No, Christ in analogy has a bit to tat with every soul. Only selected seem to be those who fall unless it is their walk, a human maneuver in said womb-like wine of a thought. The Blood of Christ now, The Passover Lamb, all relations we are tied to, Mary after Noah and sons, Jew and Gentile, relations, to recommence family politics and naked thinking at the store. Her humor does not touch the ground, but what a rosy cheek mine became when kissed! I assuredly was bent over for in my dizzy peak into art, whereupon if was not angels, was a resurrected lamb indeed telling me "It's in here, but you're not: Out now!"

Then I became aware that she taught me to seek clouds for signs, yet I could not see past the veil. A stubborn version of the Holy Spirit beset me, as well as begat me incarnate from my stubborn father's own flesh. Tenacious broke a bone in rebuttal, not for God's seal on my forehead readable by all who cared to avoid affection and its history, the seal of my infant baptism as well as that of my conversion to Catholicism. To think I, a Hungarian-American, saw not what touched me yet a trail of guesses in which family politics dug its foothold so deep, I too smoked a Southern tobacco, though root-rot it was not. Furthermore, the spiders indeed resurfaced in fewer numbers, as did the flies. So, that's the hummingbird's season, all told, in St. Francis' garden in California.

One would most likely then inquire, nudity covered to the point thought is spoken, ready for public, "Is it so abstract that it cannot be confined to descriptive thought?", to which the answer is no. It takes an effort to confine at least part of it, upon which something besides a dream experience must be occasioned, if you will.

I'm really not going to comment at length about what I've gone through in the past few years but these are ruins, spiritually and psychologically, compared to where I used to be at in life. You do not smoke in the ruins. You move against the pressure to survive their meaning. Depression? This is a war. To make the most of it you have to keep moving toward a better center, for balance, despite all that will happen next. You have to be careful if and when you let an influence in your life. During this time frame we've experienced a lot of insidiousness.

Maybe I'm in a better spot now, though my appetite for nicotine will take a while to quit. Wishy-washy? It is the product of albeit well-meant insidiousness. A city's recent claim to leadership and popularity in academic circles vies to rule the people over God: in competition with the military, who puts more food on people's plates and defends. All the arguments in the world. It's just too confusing and depressing to quit, even though they don't taste as good. Money vs. money. Why did God put me in this road of faith? To get to the other side. But how? What turn? Forward. The Oval Office has been punctual. America has a power struggle in religion, while some are safer alone, and some only know the public eye. In the midst of this, I close my connection with Mary. Angels, though, dance, and teach me how to, even in my small apartment. It is beautiful. Christianity it is alive, but alone. It is so-called isolation, but who wants to talk of it now?

Yes, spiritual. Landed. One bridge at a time, is what Mary would say. The Blessed Virgin She is. Just like my mother's name, but not my mother. And although my mother was very sick, I got her back. To full strength. Despite absorbing some of it. People constantly collecting facts and theoretizing, it is but not truth manipulated in our very motions. Let the Holy Spirit have His way. Maybe He likes me.

What I protest is a gothic interpretation, in its contemporary meaning, of the Body of Christ. Maybe it's how I rolled into this apartment. Seeing a tenant kicked out for marijuana spirits, another evicted for assault, a philosopher or two from another political belief system and class, and last but not least, subcultures in sports and music. That's a flag (we salute it), and that is light: We tend to gravitate toward it…all this was my Purgatory from former rentals and sin (or an apt. managing team having heat spells at studio application time)! Was it better that way? It's in the past, or the costs can be worse. Progress? In all of this, no, it is not really. Christ is unchaining me, slowly but surely.

Many have the wrong story. Is the exercise futile? Or am I on the bridge now? Yes, the bridge back to employment…not volunteer work. The discrimination was fierce, it was, and I have to move on. I'm safe. A second-story plateau I have.

What's exactly on my plate that I need closure for? That is, assuming psychiatry makes progress in treatment and that we have the power to decide how clarity is reached and maintained, with what we have credit for: Abundant faith.

(1)The Gay issue in my former Episcopal Church and their radical Bishopric; (2) A parish that is argumentative and discriminatory (these two, by the way, leave a hallucinatory air, trailing to one's own home; (3) Smoking; (4) Parents offering advice that confuses the issue of my psychologist, or a connection with my parents that renders me criticism when they get frustrated with these issues; and (5) People interrupting and talking down to me.

I truly wish this election I had stuck it out with the Republican Party, though the cloud, clear as it was amidst disasters below, of bipartisan tension, exuded me. I opted for the Green Party to avoid this showdown's temper, bragging, and misconduct of intellectuality. As an intellectual who struggles in life affront the trafficking of triggers to the collective conscience, I know little wonder why Christ blessed me a stubborn Capricorn. Surely we realize He created the stars and thus organizes this entire dialogue, with acknowledgement to writers of such even that He must reawaken from diversity's stagnate vignettes. One playing after another as we lay exhausted, despite a constitution that walks two miles yet can sedate properly with nicotine.

The Bull in us all foresaw cancer, yet obstructive connections lay us prey despite having quit the habit at one time. Libra gained a lot of pounds, Pisces and the tradition of her sober meetings. Yet Capricorn affords none other than the charity of Truth, and Truth has been rocked to the core as he silently awaits to connect for Christmas with the magic of Scorpios. Scorpio, indeed, a different elevation entirely. "To the stars" is not said. A Sagittarius in the midst of all this walks by attempting not to talk to herself, with animal, and licks her lips from morning's thirst calling. Everyone has been warned about Capricorn's coffee making them jealous, and now will Christ have mercy, or warn about over-connecting when in motion? No, St. Francis apparently was a Capricorn, St. Anthony a Scorpio, and they attempt to find in the early year's birth some serenity and peace. Nice to have St. Francis vacation in California to hold the fort.

Is it true, is He alive, or is it merely symbolism so engraved on the sculpture of the Capricorn mindset that it becomes a truism to foundation's layer of forward dynamics? Either or, Scorpio might say, why, banquets are just that. The Capricorn mistake: The actual duration of Alighieri's banquet of ecumenical persuasion and hosting, a cycle of such festiveness, repentance, constitution and longevity embarking now. And Dante's reminder: Christ's wisdom unfold, do not relocate, out of humility, while vignettes are played, and wait for your part!

Did family tip the apple cart with its humor? Is it a tipper in motion, of Republican-Democrat proportions? Continue the dialogue and your own personal amelioration. Humorous Uncle from the South. An aunt with Alzheimer's who will handle exorcisms provided proper conduct is continued. An Arizona domestic control factor on muscle and potatoes. A different Los Angeles that castles you. You're not contagious, though open air is calling. The vignette must be a success. Trail covered. Symbolism intact. The assignment is anew for short days and nights.

They don't remember. It was years back and it has been busy. The proverbial American 'they.' You wanted faith-healing, you got nostalgia (A whole year of it). A judgment of sorts followed suit. It was apparently to clear the slate, for Peter's sake. Now this judgment from above all is ruing over which synopsis of the whole slew of escapades He likes better. He wants you to snap out of it in a gifted trail of processing. He swears its triggers you experience. Last night He pointed out the demons of your path surround you, leaving the trigger in easy access of any public creature, and then they flip the switch attached to the trigger's mechanism so that on and off go cigarettes and other sales. So it's not that you, or I, I should admit, am a fool who is easily parted from his money, it's that the demons have a job, and I don't. They've got time spilling over, all Eternity, and I'm a year older well-spent, all ambivalence intended.

The charges on the judgment: He said my sins are vulnerable, gullible, easily hustled and pseudo-prophetic with an appearance of false, over-sensitive, kind of weak in constitution stuff(if it's not genital herpes, it's something like it, I urinate all the time) and well, I get breaks but no closure. The plate? Same plate at every T.V. dinner it is, starting with minor demons and pseudo-prophetic. Thus led the trail, and thank that Judgmental Lord, He covered it. So credit where credit is due. To worship.

On the compliments after worship, to worship again, the general impression is my youth was well spent.

But wait! Feedback from the angel in a visible frame to me! When I doubled back here, to this center in gravity's mission to harmonize all, the general opinion was "Good luck on starting over, we've been here for 20 years." And when referenced with my youth predating that time, it was "Good luck, that's only a disappearing, now disappeared, segment that cannot count. That generation moved on."

Lastly on the angel's list of comments, "All California Christians have been going through a tough time." As well, "you must have a mental block." Yes, something fierce skinny, somewhat still juicing cute to the pale-white face that used to glow, and then, "Water!" My face is washed. So healing. Just lazy it is and the former funk of gullible, etc. No, that's the exact nature of the mental block: You make wishes, you don't do it. It is the perennial mark left by a critic who liked my abode's ambience for a drink, but really has had about as much of Christianity as I have of fish heads chopped off right there at the wharf. What is the difference between a wharf and a pier? Not sure. The Conservative in me can't have any wisdom until it's all over and our plate is clean of Judgment. Strange, a note for both Presidential and Gubernatorial winners who did not keep the segues of youth's blessings from continuing: "Come now, after all!" this God must be saying, as well, "We need paradox superimposed on a precarious scale of time for saintly effect!" And some are shopping out of town.

Yes, and I live in a house that is very difficult spiritually for the time being. You have to pray. That's what makes you awkward. Praying blocks out the pain.

I know what it is you cannot get about another degree: the money. So, it is a different trail. The high-rollers are in this state. And, judged, no dice can I roll. This card game seems rather insidious as well, but that's just the influence the proverbial they want credit for, not what the Judgment was really about.

Judged we are, though flexible, and yes, let's get it over with: Love was credit, and His power is not a repetitive, ritualistic diatribe, but, 'Going steady!?' In your dreams! You need a shrink with an advanced degree and religious savoir-how-de-do and don't!

Oh, my shrink?! Are you kidding? If she were a vampire, I'd be the Mayor by now. It's not Christian, not ecology, not money, not motion, not politics, it's the best darn dharma earth can spend on stairs to the sofa chair. It reaches right in there and twists the truth right back to how the architects of psyche surround and impose surrender on the great Nerve. The great nerve waged all on me being a matriarch-oriented eunuch, and the rabbit hopped nowhere. The great nerve has been enforcing smoking, or no car. The Great Nerve said ballet and ballroom are risks. He's not a puppet, he came from the fault-line, that's right, San Andreas! HE plugs in after three cups of coffee and calls the demons back. Tangents are a longer term on the Judgment.

"Pass through to the other side. Then you are a warrior."
Eve Ensler
Reconceiving the dream
Don't dream of people
Holding a bad idea of you
Anymore.

Most people need not be alone. They don't need to be fixed, healed, or worked on, you just need not to be alone.

> *Dear Kobe,*
> *Please entertain my son. We'll pray for Lakers issues at the White Elephant room. HE has problems, though I have other priorities to keep the "ciao line" moving. I love him very much, but he hallucinates, well, let's say he's always been a dreamer. Don't let him wing it on the Black thing—he just does that so you don't set limits on beer intake at the party. Please consider continuing your career in the NBA, and just appear on T.V. to play when it's not too much problem. We get all three: 2, 9, and Fox Sports Net, win or lose, even road trip games greatly carry the 205 lb. load Junior turned out to be.*
> *Sincerely,*
> *Major Soltesz, USAF Retired (Father's note: He never sent such a letter to Mr. Bryant).*

--"You're not drop-dead gorgeous."
--"Well, I prefer faith healing to dropping dead."

--"Something important is not easy."
Donald Rumsfeld
--"I personally find the German girl and the American too White to offer a comment, and no my name is not 'the Hispanic vote'—it's Steve. No, I'm not Latino, Chicano, Hispanic, Mixed Ethnicity, Central American or Other, the Marlboro man nor Descended from Spain. I'm Soltesz."
--"And that was 'You Don't Wanna Sleep With Me So I'm Gonna Hit You.'"
--"I would have to chase your sister to understand."
--"Try Manhattan, Gringo."
--"That's not even the Santa Ana chartin' all that way through the backyard, is it? Can you blame Chief Little Asshole for carving a new totem pole? The gardener, he's similar boots, likes his exercise too, and does transplant things, though no, never a barrel cactus like Stephen!"

--"We were hoping assimilation would be through with high school."

--"I was hoping the gaping hole would close, in the line and track roster both, in junior high."

--"Well, get your suit and shoes on for the dance floor, we're going to move this number to 'Cisco Kid'...the others didn't mean to notice how fast time flies!"

--"Okay, you carry well during tests and pop quizzes. She wants five."

--Well, if you put me in the right mood, anything is possible and thank you for the complement!"

--"That's an alcoholic."

--"'Nicht', as you say, that's roots. Assimilation, Bachelor of Arts standing."

Faith and Politics, Senator and Father John Danforth

Chapter One

Our faithfulness in politics depends less on the content of our ideology than on how we view ourselves and treat each to do with the way faithful people approach politics than with the substance of our positions.

Me, personally, as a writer, I'm aware of the fact that we are again trying to rise economically as a nation, while trying to close Pandora's Box.

--"No, correction, I went to the U of R, and after got my law degree at your Alma Mater."

--"Wow, young man, you move quickly, considering I used to baby-sit you! Better than J.P. on the dance floor! We've cut more rug than Carpet-Pro!"

--"Was that the fight?"

--"You, you're surprised how Mexican you are, how White I am, how Dutch we're not!"

--"Okay, don't go back and forth between the blonde and the Italian lady."

--"I'm not—I'm trying to get a job, an occupation."

--"That is like going back and forth in this day and age."

--"That's where you got the impression."

--"A blonde, an Italian, no job, and California sun. No wonder you have no money: It's not the subject if you put it like that!"

"WELCOME....to the British Culture Festival, hosted at Claremont-McKenna! Meet the Irish, Scottish, Welsh and British from around California!

See all the tables on advanced degree options and grab some grub. Sponsored by Guinness."

--"You'll have to pardon the Duchess, she's been comparing wallets from different regions of the world, and even though there's no money in them, California wallets have a nice style!"

--"Heavy points for the big spenders in'training!"

--"No, I want respect for my age...because we're kinda' stretchin' it out here for <u>The Lion, The Loophole, and the Sociology Dept. Budget!</u>"

--"What was that look for?"

--"Uh, for future reference, The Pink Panther isn't involved, but actually feels the same way—is a bit put off by the inquiry as to whether he is Gay."

_--"I don't like the bar... The girls smoked me in the pub. Yes—turned out to be their Olympic event!"

--"Golds, silvers and bronzes lining the pub walls, no doubt!"?

--"Oh, Jesus get me to any pool for my liver you can!"

--"What was the final score? Did the interpreter at least get it?"

--"Yes, he said America 1, Italy 0, our first victory over Italy in the 82 years the teams have met."

--"What's he drinkin'?"

--"He's not—he's smokin'!"

--"Don't want to spoil this 'troops home from Iraq' business with a poor sport rap, Italian-
Americans they be. Hoping this Romney fellow realizes the plan is not to beef up Afghanistan, but focus on the periphery of the continent we are. Really—New York must be squeaking as tight on security as Betty Boop!"

--"You're gay? And working on your Ph.D.? Here, you define the straight world—you're better looking. Maybe she'll snap out of it and I can date her!"

"—Is she picky?"

--"No, it's Nicky. Nicky is the name. As in 'Every rose has its thorn.' See? 'Nicked ya'!"

--"We're just Jews."

--"Oh, I won't over-explain. Am a Gentile, though like healthy food. Whether or not to the dentist's standard!"

--"Hey, Steve, Boys' Night out with the Rabbi and The Bishop! Got the bucks to go see, "Hot Flash?!"

"I'll get a loan and see you at 4:30 at the ticket box! That's center-left of the Box on the field. Giving you a yellow card in this World Cup Game.!"

"Peter 1."

--"Think I peeled through that New Testament part in Italy under Pope John Paul II!"

"Here, the Italian-Americans may touch base later—what does his Dad do?"

--"Steve's dad? Says if you want a dirty old man during war-time, really, he'll give you the treatment. He's a retired Air Force Major. Everyone else in the wood made Colonel... he's the Major behind that Bullshit."

--"Yes, Mexicans to the left, Mexicans to the right—the Major's wife came up with a solution...her nephew, Colonel Frank Figueroa. Could've been a General. So, in the long run, we don't worry if Colonel Mustard disintegrated on the Hill!"

--"On the warpath?! Isn't Patton this time, Sicily: Is the Center-Left Hungarian ex-pat nation, residing on the West Coast, unless you prefer the phone version of the Japanese! We have met in the past and formed an informal accord. Back to you, Italian..."

Mo"ving California Through Italian Film: A Hernia, an Abortion and a Bowel Movement, a documentary by Rip Von Schneid, Ph.D., Early California History, Cal State Bakersfield."

"I got it: our school has a drink with the roommate, yours does Ad-Libs and Sudoko."

"The Perfect Picky Woman, the Great 'B' of Term Paper Conspiracy, by Coach U.C. Regent."

"Is nothing Sacred?!"

--"The roof. Over the building. Costs more than her jeans, but is too much older for most Real Estate folk."

--"She can't hit a note."

--"Wait 'til we get to the Plate Toss. True Greek."

--"What would I say to Sophia Coppola? 'Is this still my glaring hang-up that I can't see?   I can only go by how people look at me and what they say. We can resuscitate any dialogue in the world, but I wouldn't want you to give birth, if you're used to giving such with artists practicing more hours at the piece.'"

--"I Went through tuition hikes with both U.C.'s as an Air Force Officer's son: It's not just their testosterone in combination with lack of domestic mentality discipline, it's their abstract as opposed to literary theology—but the professor can have a touch of it too. I'm a Golden Retriever—I like them."

--"Polish discussion, Stephen. Polish can't date Black."

"That's not a Pole, jack-ass—that's Jim Belushi!"

--"More Jews than Gentiles in Poland, huh? More Gentiles than Jews in Hungary. So is that it? They understand. You want to play the smart one. Okay, they'll play. A friendly, as they say in G.B.!"

--"The Olmos—Morales Foundation Question, for the Gold, students....are we off the hook?!"

--That's all the Duchess aside! Yellow flag! Now continue play, CA, with the C.S. Lewis Foundation!"

"Yes, America 1, Italy 0, Arsenal 5, the other team 2."

--"Question of God," by Dr. Armand Nicoli. Believe that was the hand, Prince Andrew, and have to get back to my American citizenship now, and I am not going to teach the European Youth facts of medical masturbation—am not the "token" of Europe, was my time off, and track record, not track. As in, "What is his track record on the F.B.I.'s items of scrutiny?" I know, it's the subject of hair. You split like a masochist!: This is victory, American soccer's curse is broken in my lifetime, and off the subject of votes, you kick the goal on Italian soil with the exchange students next time—I'm not your international relations nor economics slut! I'm a Hungarian-American polyglot of parties to come, not an interpreter on a comeback trail, but we'll play another hand! For the one in the womb and the throne of the future!"

--"You think this is a wine and cheese party every time the subject of medication comes up. You over-medicate—that's my excuse!"

--"They labor in vain who build it;
--Unless the Lord guards the city,
The watchman stays awake in vain.
It is vain for you to rise early,
To sit up late,
To eat the bread of sorrows:
For so he gives his beloved sheep.
Behold, children are a heritage from the Lord,
The fruit of the womb is a reward.
Like arrows in the hand of a warrior,
So are the children of one's youth.
Happy is the man who has his quiver full of them:
They shall not be ashamed,
But shall speak with their enemies
In the gate."

Acts 19:4

'Then Paul said, John indeed baptized with a baptism of repentance, saying to the people that they should believe on Him who would come after him, that is, on Christ Jesus."

When they heard this, they were baptized in the name of the Lord Jesus.

And when Paul had laid hands on them, the Holy Spirit came upon them, and they spoke with tongues and prophesied. Now the men were about twelve in all.'

****

--"Oh, look at California now: if one blonde goes through it with a good male, the rest are assholes around all blondes!"

--"And you're not discussing Navy service, are you?'

"Are you kidding?! Notre Dame is for the marriage and the cool million, if you can get there after serving in Southern Viet Numb. nearby L.A. County!"

--"Oh, Robbin, it's a great game. While the Film Festival is showing the award-winning, *'The Roommate's I.Q.'*: There's a shin –kick meant well on the field."

--"Oh, 2-2? Quickly?! Is that Robbin? Didn't realize I had missed the Martin Luther Bible Study in Louisiana! That's our two buckets, Katrina and what not! Was Green, not Irish, and I was supposed to tack on, 'Peace!' Welcome, Hungarian director with a tight budget! Sorry! Cameraman needs time with his electric guitar! But he's been paid! Guess the Brits are payin' you now! Jesus didn't know the Duchess understood competition and men as an objective European matriarch. Is that the cat, or did that third goal get my tongue? Fine. The Queen's English in Lent. My, we're getting' too friendly for Americans now, are we? I hit the paper with my piss and Dad took me for a walk. What?! Boxers?! Thought I was only hosting Rottweilers! Guess I <u>am</u> a Dempsey fan! I need my rest, Clint will see you on the bet, but that's all, my beer is finished!"

--"You've heard of a 'boy-toy,' but have you heard of 'Toy-Boy?' He's got a bike, a telescope, golf-bags, motorboat, deluxe microwave, and a guitar he mi-i-i-g-h-t sell, if he gets low…but isn't sure."

--"Electric or acoustic?"

--"It's a Stradavarius being auctioned to raise funds for KUSC, with home-made hook-up invented by the neighborhood's electrician! He's also picking up a foreign language with Rosetta Stone's $200 tapes, if you really know the scoop on 'Toy-Boy.'"

--"Needs a sex-talk with Madame Carla Sarkozzi for his anxiety after all those Starbucks Venti Mochas!"

"Is that Rick Steve's <u>Guide to Italy</u> on his coffee-table?!"

--"It's not Lydia Baldacci's <u>Neapolitan Recipes</u>!"

--"<u>Fate and Destiny</u>. Who penned that—Foucault, Derrida, Rousseau?"

--"It's the title Toy-Boy's Mall Chick has given to her journal!"

--"Yes, she began the writing exercise with,'A slow, creeping mist came up from the harbor…', so continue on your end as well!"

"'Cultural suicide? An expression from someone who has a hang-up about the dishes in the kitchen sink."

--"Yes, let's get the mitt touching the ground when they hit a line drive that grounds at the feet."

--"Slow day?"

--"They can't get the needle in the ball with all the spit in the can, so the whole stadium has to stop on a dime!"

--"I see why tensions are high on the field. It's a frustrating experience."

--"I see why L.A. is a higher bar—empathy like that'll get you the quarter for the loo in the Laundromat and a roll of toilet paper!"

Angels Or Upper Class Fools' Year?

So, in the end, one wonders if this is the awkward phase to shed from medication's side-effect, as well as religious conversion's ritual, or if indeed a demon got in that manipulates. Without going into the detail of the imaginary conversation I had with the Priesthood and class, I know Christ and his theoretical philosophy on religious growth has helped me overcome many of the speculations I entertain from others. It is though I am at impasse with angels in the forum on a critical subject, hoping for strength, illumination or transfiguration, and patiently waiting the miracles of celebrating both Hanukkah and the birth of the Messiah over 2,000 years ago.

****

"Well, you know, the problem is you guys are so busy competing when you're collective that you never can form a line, then you get problems, never-ending baggage, psychologist, and look now, no jewelry, no job!"

"No, Adam, look, your Valley perfection is showing, because apparently your Mom and Dad are aliens from a different tax bracket and you guys have loyalty chips stacking up to each other that are more than the San Manuel tables can take, max, because of the Common Bond, not to be confused with the Common money order. This is San Bernardino…different mountain, have buried more than Indians, why don't you get a piece of that rock, Real Estate- Boy, but that's the Highway 38, not a Red Carpet, try the 'pussy factory and helicopter' routine at these raves to see what you get and welcome to Downtown Redlands, were you thirsty? They can't even afford their own microbrewery's beer, but it's been solid America for awhile."

Faith and Politics(Chapter Five)

"The practice of religion is an affective antidote to the disease so apparent in our society. People who practice their beliefs will live according to moral and

ethical standards their religion teaches them. They will be <u>witnesses</u> against the tawdriness of the culture around them. They will be examples of the people God expects us to be. They will be that because they understand and live by the tenets of their traditions. That is the practice of religion. It is different altogether from the display of public religion."

To continue the dialogue, this <u>seventeenth of December, 2006,</u> on the first day of Hannukah this year we were reminded of what drugs and alcohol can do to the mind and body, for some strange, seemingly dated reason, as my view of such and those in that way of life should remain nearly a decade in the past. Yet the headlines of the San Bernardino Sun read, "Anatomy of a Drug Bust." This meditation, the first day of celebrating the conquest of a Syrio-Grecian emperor by Judah Maccabbee, as if a filibuster of anti-Semitic proportions on the subconscious, is indeed a dark cloud over the citizenry to keep gray area on economics thick as the winter cloud covering the sun. I got out to the center, a portion of it, and read from <u>Faith and Politics</u>, while wearing a cross that is a season's gift from my mother. You live, learn, celebrate, and witness.

On the second day, as we all light the candles, we become aware that we are in good shape nearing forty-four years via the report on our physical, that we have an enemy, and that others are not in as good of shape as us but we are forced to work them in, while fighting the enemy. We pray for freedom from this dialogue to have a smoother connection in the practice of longevity, which may have induced us to smoke in the first place. We certainly wish health and a long life on all, yet the stagnation of intelligent design in the apartment can be painful. <u>And</u> in the city as well. We are reminded that, in peace, the correct religious interpretation is to be equal to others, and treat them equally despite religious or political differences. We are reminded that Senator John Danforth is a model of Christian pathways to leadership, and that presuming we subscribe to the wisdom of his witness we can better understand how to subscribe to and benefit from our political leadership.

The Inner Voice: "You have to be polite and benevolent to be a Jew. I'm sorry I don't treat you equal. I have a different gift to protect." This is a thought that explains some of my subconscious feelings and pressure. So, off the subject of gifts all the same, benevolent and polite. *It is a separate voice addressing me.* Off the subject of snobbery, it's a good community. I don't worry about this home, I just muster up what peace I can this holiday during wartime. I wouldn't be welcome in many social circles that drink. And in the study and study alone, I have to stick to the subject. I'm chosen to move forward with my religious life. I am understanding more about Judaism in a Messianic context and like the Catholic Church. I keep in mind, as far as where to worship, there's always another church, but the same me.

It took me awhile to get the hint: I want to describe the demon for the reader, to see if he/ she can understand what I've gone through. First let me see if I may have the blessing to describe "angel."

A rather euphoric though healthy sensation will take over me, and my movements are in pacing stance, not really dance, in the presence of such. I will imagine them in my mind, attached to my thought, and my heart from the parallel realm of a celestial plane. Through the movement and cycle of celestial love brought into the home, I will be lifted from stagnate ideas to a higher psychology, which is to free me from the pressure of where I am at in the scheme of things. It is thusly the placing of a blessing to inform me in rhythmic process, what it is I am going through in my life, where I am called to bear testimony, if possible, somewhere. It is to excite me in a relaxing way and thorough manner about the future, myself a believer. They, the angels, may stay quietly a while, in fact, even escort me there, in a matter of speaking, with plenty of story and sights in the meantime while the story is being connected. It is demographic awareness. They do not speak to me per se': they may put an idea in my mind. It can be artistic, musical but silent, and can still me, not alone create my movement.

Fight Or Flight: The Angel's Whip and Mystic Ideas

Last night I fought through three psychological wrestling matches as abstract schizophrenic moments. They are images that are forced on my mind, I may be silent, though in one, words were forced out of me. I wonder if this is for Christ's sake I describe these things, if I endure them due to clinically defined, that is, potentially clinically defined, caffeine addiction, or if these are the demons I have avoided meeting in person leading to violence that I am hereby forewarned about in the guided imagery of a female angel I drew last night. I suffered temptation for drawing her, I try to express.

To start, I went for a brief evening walk, as angry people seemed to have taken the road, so I was discouraged from seeing the sights. A very young driving adult had parked in his truck, complete with his sunglasses on at nearing sunset and large cell-phone. His thoughts seemed to follow me on the phone as if mine, a man with friends who talked bad about police and had their beer money to do so. Not how I wanted to be reminded of my last twelve-pack of beer. As if he drinks right there in front, then throws the can out the window. I sometimes find beer cans there. In any event, his thought of drinking left me with an extra side-effect to deal with of a weakening nature in combination with the medication within. It was as if the spirit of our scenarios had momentarily joined.

As I continue to write for redemption from these moments, I hope an angel or saint will see my writing from Above and talk to God about my uncomfortability, when I could be creative writing on the fiction side. I'm now calm, though not well-rested. After this, I seemed to go through a controlled fit, as if a psychological force were moving against me. It is as if nicotine is pulling me into its arms again, though reminds me of what it would be like in discomfort to wrestle a high school delinquent, or twenty- something of that nature. Makes me see my nephew's mature side. It smacks of a community college roommate situation of tension, all going on in my mind. So some people we find comfort in, and others we don't. Living alone has been deep.

The second vignette of a mental nature seems to be this evil force that persecutes for not cosigning the argument of gays who want a prestige movement because 1) we're straight; and (2)

We have better things to talk about with our time. But still this force is eyeing me, while I become nude for the shower and bathe. This particular force is around a lot. All of this quickly goes to our last psychological and strangely demonic-style vignette, the passion of the choirmaster's son. Exhausting, dry, with richer alcohol involved and aiming for the goat in an aesthetic matter for pride and conquest's sake, as if he needs the whole floor to the Communion rail. Why me? Is this really an issue? Gall of an issue, and that's why it seems to be demonically persuaded. I've tried prayer and candles but, to tell the truth, these are all signs of a deep anti-Semitism that I am sensitive to in a somewhat telepathic manner, having nothing to do with coffee at night, as I light well my second ad-lib menorah's candle. Although, in addendum 01/08/07, the choirmaster's son is not like that. It came from somewhere. And now on my third, again on 12/08/06.

As Jewish: I'm reflected like that in conscience in my private prayer. In other words, the love of Christ as an active force in my life finds it much easier for me to manage the illogical with the logical if I entertain a Jewish facet, even though this facet currently awards me more than one struggle. Struggle for economic independence, well, the holidays do not have us dwell on it. Struggle for an identity that does not shake in the wind or is not unduly and unjustly challenged. All these things, even at wearing forty-four years old. As the doctor's spirit seemed to say, "Never grow up."

It is as if, not just a force that makes one think these things, but one that can invoke a slight, psychologically gray and artistically abstract sight response, as if I am connected by nicotine's power to these things I speak of, as if the gay hallucination is an actual geometric skeleton of cigarettes attached to each other that inhabits my veins, as if the passion of the choirmaster's son, a non-smoker, threatens not to release me from the symbolism of his mother's(a smoker) stroke on those beneath him, as if drunken youth fights me because I do no longer smoke(or had cessated for nine months straight when first written).

Macabre. Hanukkah is not Halloween but for me this evil side of Halloween in The Historical and Scenic District of Redlands, California continues. I encourage all to embrace the struggles of smoking cessation, as in the mirror I felt the staff of The Archbishop of Canterbury turn cancer from my path when I quit. It was a visual. Cancer from smoking is akin to insidious rotting of the flesh, all turned by His prayer and power. A parting shot in good faith as I attempt to embrace a more literary Catholic order to reach perhaps the Olde Country Italophile ever-present in the depths of my soul, which has now to view the power of eternity's blessings with these said chambers of thought to clean and purify, this time by Hanukkah's candle.

So I write furiously to fight its power, yes, convinced that through the rites and rituals of Holy Communion, the pen is mightier than the sword. No longer do we have clean tobacco, from Tolkien to Twain, to Einstein in elevation, not historical connection, to aid our neuroses. The Lord's Higher Order, the fall of tobacco, the feeling of the presence of angels, the victory through A.M.'s caffeine and love for Mary Most High, The Queen of the Empyrean, For the studies are to follow in older age of the history and current purpose of said location in The Christian Realm, yet at this moment Alighieri's poetry escapes me. Can I capture it, that level of cessation? Yes! His poetry remains castled by wooden and finished bookends in cowboy design on my shelves. One symbolizing my Uncle Armando, a saint who sold his slaughterhouse, complete with black bull to impregnate cows, before dying without seatbelt on, and one symbolizing Uncle Richard, a Colonel and jet flyer who had his life taken by nicotine. Alighieri would find protection in men such as these, bon-vivants of elders who ate well and were blessed husbands, providers and parents.

That is what Catholicism has me remember in its honoring of family tradition; I, the next generation, not to commit their mistakes as my sin. Inventory it is, as opposed to Twelve Steps and Traditions, and our pride fights on, called forth in the middle of the night as Christ's scribes. My muscles must not get sore from walking, so I take long, hot showers and bow in the tub, some-what stiff though well-fed from my parents' wisdom and trail of such through said city. This Pathos is apostolic, not political.

Even after all this, nicotine gets effeminate, attempting an angelic moment, and attempts to career itself back in with gentile, charismatic, and influencing whispers that leave a cutting exhaled smoke pocket of intoxication in the mind. The struggle for focus continues as a take-no-prisoners war, these sergeant-at-arms of complexes driven from the fields, while attack on the periphery for this point in time is stopped, as I read and have as much coffee as I want. I am holy, commanding my absolute adoration and reverence for the Trinity and Mary with my will and the blue of my eyes in the morn, preparing in disciplined fashion for Holy Communion, but not jumpy.

So, on this sixth day of Hanukkah, 2006, as the candles burn early in the morning, I am reminded of the heart-to-heart connection with my parents, how to stretch the definition of Kosher during the season, the pressures on the Jewish community, and the headache I don't have in this city. That's the positive. The negative is that I want more freedom this year if I'm going to practice Christianity in America, for this was a psychologically stifling, stuffy year save for moments in time. Yet maybe it is all the other side that is this only problem, their ugliness, their weight.

On this seventh Hanukkah day, as my candles are lit in the morning, I was reminded of a Rabbi's words, that it's a religion, not a race, and for some it can take years to convert. Yea, let's not make an action series out of it!

It became obvious that smoking led to (1) emotional entanglement, and (2) mistakes in relationships, but (3) guarded us in a sense from other deaths, supposedly, as well as is a guide on an adventure. So the problem becomes when there simply seems to be no other guide, as if Christ is occupied so much by the Sacred (protection of such) and healing that there is no justifiable, deserving leadership save some sort of angelic version of nicotine. Problem with such being it is not good with the environment or really the road: so many obstacles in tow attracted to the sight of such while one is waiting to arrive via said guide. Non-smoking guides cost money, hence the haves and the have nots on the subject.

Furthermore, it is for these above-state complexities of factors a shame more Christians are not marrying outright. For support. So psychologically I get advice but must support myself in my own dwelling. The problem becomes what to do in a legitimate sense, since legally there are attempts to outlaw smoking and cities are actually vehemently arguing with smokers. This is why all fight, whether smokers or not, and the old adage, "I'd rather fight than smoke," becomes obsolete, as one is not allowed to fight to quit though is consistently observed. The movement has not had anything other than an earthly psychology in the distance as incentive, and when quitting one gets weird feelings that accrue. Faith as a sword, the trail to enhance, must step in as metaphor if only for a little more time.

The temptation to smoke becomes fierce for a variety of reasons. Senior moments, the news, the news and the ride home, the popcorn on the nose, and more funny feelings despite nicotine patch and therapy. It's as if it goes against God's will for me to quit sometimes. Strange, but I had an observation about being asked to carry more than I should at the moment.

On the bright side: I saw a man who looked like Santa drive up the street from my parents: Thinking out loud, he said we would see what we could do about a new start.

Interesting. As I light the candles the evening of the eighth day, as I stand corrected, a phenomena that I will describe as the ghost of Freud is locked into the back of my mind, in a serious disposition at that. Seeming in part to be a figment of a serious thinking skull, in his shape, with deep sockets and a grip in said form in on my brain, for what reason may already lay apparent. Freedom or guidance I know not, yet ponder why in this smoking argument the tables keep getting turned, as if the devil himself is out to prove he has a superior strategy against my infinitesimal and damaged ego.

An impression is given, though traces of this feeling, in my experience, lead me to feel the voice behind such is that of Mary. Now Mary interjecting at this point leads me to a thought: that there is a riddle behind Paradiso's last Canto, which mentions her as I previously believe I noted. Her impression: "I'll leave it wide open to your imagination the life some men lead." Stunning words...Freud within me becomes a chess piece, as I remember, but not where or when, I was shown a tactic in using it, apparently taking him off that board for my own. As Christ is King I surely say to you in the psychological field, patient or doctor, that some are steered by said Yahweh the Father in this direction. Prophesy it is not what the quest is about. It appears to be a topic known as Art and Freedom.

To retrace my steps a bit, the local medical clinic becomes a multilevel paradigm where one can seek counseling on smoking cessation, though something is not adding up. Only the physician's office and perhaps the pharmacy offer serenity, yet outside the clinic is a different exam every day, a routine, complete with the engineering of a traffic light at the crosswalk and pedestrians.

I now wonder if indeed I am having clairvoyant sensations. On the news tonight is the story of a skull (human) placed in a bus stop garbage can. An illegal immigrant placed it there along with a four-page letter, saying that he found the skull in the desert. The letter was Spanish writing from Mexico. Is it my connection with the media producing this early effect on the news? Should I watch the news, or see if I develop some ability to predict? I do not want to bite the hook: though I have no explanation for the impression of Freud prior to the immigrant's findings, as of yet. A female Bishop handles the high school running joke enigma, a male Bishop claims, "I'm not a critic, but I have my limits, and the Archbishop concludes that '...once you handle a sexual knot dealing with the subconscious, via Sacrament and coverage by said blessing, you're not schizophrenic either.' A religious cure through daily ritual? Who knows, one's faith makes them well: Diagnosis during traumatic times correct, but then what? Not only the University but all of Christianity repeats itself in a 500-year cycle (?), and I am stuck, releasing myself from the captivity of change and movement in the course of new events. Rome and Canterbury have an accord. Does it stretch to intellectuals in said stymied circumstances globally? Do we really have to free ourselves from government to embrace the Church's power to guide? All important stuff on December 22nd, the eighth day of Hanukkah, 2006. It took me three years post-fascination with a discovery of Judaism to celebrate Hanukkah as Christ would; completion while candles lit tonight.

On that note I have religious leadership in this world that will not let me down, that is there regardless of the path I am allowed to take, my calling, the shuffling of a psychological card deck or the campaign to take certain dioceses and parishes. I am happy to be an Anglican at times, though at this moment Mary seems to be (1) either on hiatus; or (2) asking at the ripe old age of forty-four for more of my worship. That is an honor all its own, that many would say is my imagination, and that I am worthless.

Yet the candle is a yellow, cone-shaped and controlled light we have no power to grasp with our own hands, and its purpose is for us to see in the dark with symbolism of the servitude that arrives with Christ's love. Earthly themes throughout time reach for an understanding of the ever after: We are closer to Her at that point, in vicinity, and we must prepare, whether as described in Thessalonians we are in the ground, as souls for resurrection, or if, as Dante concludes, we have permission to come up, having prepared. I do not like to ask if Paul did not know all. I do not like to change the message of Scripture. Yet humility means to God's new decisions, what He will do with the new day, season, year, or soul saved by the love in a visionary's heart. Amidst a juxtaposition of nations, the candle burns, Mary's love above all. Apart from organization; in our homes. So, what I am writing about is vision from prayer and serenity of morning, on how God is helping me understand my place for peace amidst the confrontations of today. He invokes concepts, based on my faith, in symbolism. So, it is not always fact or fiction, it is faith in peace exercised. My place is that of humility. I seek not to suffer, though worship. Inventory is difficult to adhere to with all of this confusion. However it may be correct, corrected, or incorrect to say in a new beginning, we acknowledge our leadership's efforts. We wait patiently for Yahweh's decisions, ever-aware we have taken civic and medical, legal and parental counsel on where the place is we are to be satisfied with being at. It is Christ's blessing all His own that my parents' Sacrament of Marriage has over 40 years of anniversaries to celebrate and I play my part so that they may celebrate it with the abundance they desire. Not easy, yet highly rewarding.

Modern Episcopal Roots: The New Flask For Advent and Epiphany

With this prefaced, I still have resistance within myself for an Interfaith approach to Christianity. I am early in a dialogue that will continue for years, though not the earliest. I have some steps to take. Christmas celebration should help me sort some things out. I've been patient, increased this very quality. I've done well in seeking a better bridge with European thought through study of Dante, some Catholicism, and processing and praying for the Anglican Communion, which now has an agreement of a nature I need to understand well with Pope Benedict XVI.

A more serious devotion to Mary is the opportunity. I have kept up the pursuit of self-improvement, amidst it all. I have persevered on my duties, been responsible, prayed for the current and new world leadership though many have their own countries to worry about. It's no longer like carrying the world on my shoulders, though often I have a tricky path in carrying for my own American society, something very important to learn more of the "how to" factor. I wake early, I exercise, I formulate a new day on Earth, and I pray by lighting a candle, by waiting, with words as well, and by wearing a cross, with medallion of a saint. I pray to Saint Anthony this way in daily life, and St. Francis I seek as a guide. Two important men I should have a more historical understanding of. I ponder contemporary and private faith healing and volunteer work two years after assisting a member of The Order of Saint Luke perform his healing service. This is how inventory is to be seen. It is a shame that as I pursue these things, mental illness should pretend to be an obstacle in worship that I have to think of while trying to find a nurturing religious community, that triggers are present even in the pews.

Are the triggers the agenda of The Church? If so, we should be prepared, learn more and find a way to obtain our goals through worship without the pressure. After this study is done, I am faithfully positive that a whole new solution to many things will unfold before my eyes, and that I will be much more comfortable with the "place" I am at. The study that took place at U.C.L.A. is a long-term view to offer various insights, and I pray for my health in the meantime, having gotten a small stomach bug here the 23rd. My physical went great and my pneumonia shot is good for six years.

My long walks have built a new me who feels well for his new years in his forties. Prepared now to read Scripture through!? Father Caffrey, an Episcopalian under Bishop Bruno and Bishop Catherine Schorer, thought it is not about the spiritual (after all this time I spent pondering it). 'Christ came to honor the material, and part of the purpose of Christmas is the blessing we receive through the celebration of the season on our goods and gifts.' This was a day honoring Saint John the Evangelist (the name of the parish protecting us in Needles for nearing 100 years!). Yes, indeed, the author of Revelation may be honored on December 27th (also called in some translations of Scripture, 'Saint John the Divine.') We have done well to study prophesy this year, and there were insights right in Common Prayer teaching. God is deep, and thus am I, venturing now through a stage in eternity (thusly not to worry about marriage around the clock!).

I will agree with one woman that I have other things on my mind, in tune with yet another trip or two with my healthy parents to see U.C.L.A. At forty –four years old I return to a place of my mid-twenties. I am informed via the psychic depth of the Lord that it is not solely about memories as I honor the academic greats. I'm not to get overly excited and drink much water in the meantime.

Father Caffrey of the Episcopal Church also said that through religion with Christ, we may now look upon His face without worry of Old Testament rites of this issue. He commented about icons, including stained-glass windows. As not about art, though concerned with viewing into the depth of what God would want us to know. I have now joined Trinity in subconscious appreciation, and have noted, via my glance at Christ's window with Lamb in arm that I have some fear to deal with. Water will help. It was not about quitting smoking. Hard to believe, although I got word from one of the parish fathers that said subject can take two years to resolve with proper closure. We were reminded that the angels worship Christ consistently throughout the day. What a beautiful kingdom here. Want not to sing with angel most high when love captures your earthly heart in unconditional Episcopal fashion. This was letting go of the Episcopalian of the present.

To interject something important on a previous note, that being the healing levels of Sacred Mary, I am inclined to agree the human soul has various levels of interesting depth. To be forthright, I am precluded at this time from extrapolating on the nature of the levels. However, as Genesis explains, Heaven is a firmament dividing the waters above and the waters below. What divinely happens through the rites of the healing service and prayer appears to be this: Mary in maternal manner, connects us with what we have in store for us as Christians with the firmament, in belief in Her own Ascension post-terra to a throne of sorts, albeit not stated if this is actually on the right, or left of God the Father, or elsewhere. In this way, we become aware both psychologically, and often physically, with what things are creating blockage in our connection with Jesus Christ, the source of all healing.

I should restate: Christ is the source of many things, and after attending many healing services it appears both Christ and Mary are active agents in our healing process. Hence for the upcoming week I am joined to Christ's body (the Church) by Holy Communion to reestablish the connection, despite the elements of blockage, with the Divine Trinity. I become aware of my difficult areas in my health regimen through Christ's diving wisdom, present for me lastly in an icon of the Messiah, and a lamb, as I received insight by glancing a length of time at His visage. Yes, icons, as the priest stated, are for a window into said inspirations and others; not art as is. Sacred. As well, to continue, the protection of the sacred things to Christ and Mary on this earth, by obeying the commandments, is critical to an earthly foundation or firmament to connect us, at due time, with above for healing. Sin is a negative interjection though also can disturb the cycles of healing. Sin and its roots are also what we as Christians are here to heal and moderate in fashion, as responsible adult Christians. Faith received through worship (we often overlook that this is Christ's reward for worshipping—increased faith) is active, though rests. Not a passive religion, as often the humility of Christians is confused for such. I have to bear testimony that in an often-blessed country like America, I see many evidences of Christ's protection when confusion occurs because of humility.

I think mistakenly we look for conceptual checks and balances to the aforesaid statements. If taken as is, we can proverbially say,' Water your plants in the home.'" Literally water a real plant in response to needs for check and balances, not allegorically, and through conversation. Reiterate the checks and balances of Christian family life to enforce your healing connection. Albeit actually, two separate points dealing with healing.

There are two items now: the divine observance of a subconscious fear I had, having to do with smoking. I was worried it would interfere with my maintaining my abode, so I ventured outside to smoke. At the beginning I feared my studying would depreciate if I did move to smoke while trying to learn, down the stairs with two coats and out into a warm winter. Healing has taken place and I am even more aware of how to study without smoking on the spot. It is a new beginning in my contact, now reinforced with Christ. I begin not to just read excerpts, though the entire Scriptures in an entire year (in actuality you don't plan your life with the Lord, you lay it down for him—as a volunteer for the Church in 93-degree weather, I both smoke on the spot, study much less, and will never finish the Bible from cover-to-cover in my lifetime. But I'm a Cowboy and an Italian, a thinker and a sports enthusiast all in one. A chameleon hiding in a watered plant.)

A woman got confused by my appearance at a time when her relationship with another male needed troubleshooting. I was called to be an example again that one must internally resolve prior issues before taking a new road in companionship. This is how, while praying at healing service, we honor and support Christ's sacred side through obeying such commandments as, "Though shall not commit adultery." It was a psychological tug-o-war, as a transference of power she requested. Praying for a healing path amongst requests for sin occurrences requires effort, and is a fact of our Christian responsibilities to emulate the Lord, as Scripture calls us to do so. Get it? She had priors, wanted a new stepping stone with a male in town, by my previous experience (even though abstaining for economic reasons), I knew where the spiritual stepping stone was, I didn't cut in like a jack-ass holding his ground on said stone, I told her the placing, moved to allow her foot to come down, lost some of my cache in the ritual, but her new relationship took off by next Mass! The Priesthood the entire time, besides laying hands, declaring the benefits of 'water' in the healing process. A true London Conservative. Because after all that in Anglican pews, wouldn't you need a drink? I'm not Standish; Soltesz, Hungarian like that. It is not always simply God's test in resisting temptation. God has more projects in healing than the academic exam can sanctify, as exams will.

Insidiously comparisons are made, so it's difficult to merely compare myself to where I am now. Dennis Preggar says life has unsolvable problems, teamwork between religion and psychology is alluded to, and college at best will only give you a slight advantage economically.

The holidays were great. I laughed for the first time in a long time with my family when I informed my competitive nephew to swim for his life when he goes away to school, otherwise he could get drawn back into 'high school continued.' High School continued at home with his parents. He laughed.

There were some new familiar faces at the New Year's party. It was a refreshment I had half a glass of champagne the entire holiday season. I gave my Dad and Mom a large bottle of Italian wine for their hearts.

"We all have (a) different Episcopal experiences." Father Caffrey

Lo and behold I saw something in the back of my mind, after an otherwise beautiful though somewhat smog-filled day. The Devil was toying with a spot of fear in the depths of my soul's experience in life, but a memory he wished to enact on a plane of existence similar to Dante's Inferno. Then I saw through the murky, damp thought something stout and white appear, quickly donning said robe. At first I thought it was Apostle Paul. He flipped Satan onto the bottom floor of the plane of existence in severe punishment and verdict. Like a feather the subtle idea of the words, "Saint Anthony" flew through my mind. I finished my evening coffee and breathed easier, praying to Mother Theresa for continued protection on Earth preventing this plane from having further influence, in a manner of speaking. Like an unhealed wound this event had triggered anxiety leading the Devil to my conscience and now, as the Italian puts it, pace, or peace, and silenzio, or silence, came alone with the fan in my room circulating air on a quiet night outside and in.

For I love the Lord
He knows not failure
The weak know strength
Through all space and
Time
Using commandments
As cornerstones
Prayers as pillars
And thoughts
As candles
In the everlasting
Procession
Whether sun rule by day
Or hidden stars
By night
The saints must twinkle
With the grace of dominion
The firmament swifting earth
Of the dust from the day
As the Holy Spirit moved
To the sound of our decanters.

Money in the Collection Plates, by Usher Steve

Those that get to drink get the money and those that don't get the money don't get to drink

Tune them out if they don't get to drink
Because you need the money
A beer is two dollars, dollar-fifty
But that adds up over a year
If you tithed that
You'ld be firmer
In the pool
To get the money
Whether you drink or not
Now drive home
At night
Or in the day
And don't drink more than two cups of coffee
Because that's what they drink
When they don't get the money
After we tune them out
Tune in the music
And have another drink
Coffee does not tune them in
So don't argue
Just tune
It's not a toy
It's a conversation
About how not to drink
So they cannot play the game
For the work
To get the money
To tune us out
When we're tithing.

U.C.L.A. went well. Catherine from South Africa studies the brain, and commented she doesn't know of terms such as "schizophrenic pro-reactive" and "schizophrenic non-reactive." I describe the difference as to which pathway they take. In other words, "Schizophrenic Pro-reactive" is when a patient has criminal tendencies or past. "Schizophrenic Non-Reactive" is when they don't—when they are a good bet for treatment outside of the Corrections Dept. to rehabilitate and not rely on Social Security and S.S.I., by making at least

$1300 a month at a performable, paciently-paced job. That's what is needed to ease the burden on Social Security, to see the medication works, and that they are once again functional, not ever once a criminal worry: Non-reactive, not Pro-reactive.

An actual Muslim, complete with turban, walked in a nearby Coffee Bean and Tea Leaf in Westwood. He looked like a real-life version of the Turkish bad guy in James Bond's Octopussy. Wouldn't want his subjects of stress anytime soon. Many foreigners grace Westwood. Latinos in Burger King, Asians and Anglos at Coffee Bean and Tea Leaf, a Russian nanny strolling with small child, an underlying Jewish presence, and the sense that with war protesters, wage protesters, the medical community and student youth, it's a different and beautiful step to take to visit. As if Christmas with Cat-woman, the movie, was still in the air, we walked to the fifth floor of the parking structure and ventured home. Not much art was observed, though Mom mentioned the Ghetty (and smoking) plus the Farmer's Market on Olivera Street. I enjoyed the quicker-moving city at night, mindful of how not to test my own driving for the time being at night. The pay for collaboration on psychological research went in the bank this morning.

The Jews For Jesus newsletter was healthy as angel-food cake on my forty-forth birthday. I'm going to buy myself a gift. Makes me wonder if God has been guiding certain Gentiles to adopt an inherent Jewishness.

I corresponded today with the aforementioned organization to receive The Christian and the Pharisee, by R.T. Kendall and Rabbi David Rosen, this twelfth of January, 2007, for a donation of $35.00. I prayed for many.

The other night, while meditating in my parents' home, I was moved from the bed by the Virgin Mary's love and thoughts, as she reaches the artist in me, her words in my mind on the subject of romantic love being, "I'll think about it, Stephen." It was in good will, not condemnation in any way, shape, or form. She was sincere in letting me know that as I draw closer to her on my pathway to understanding Heaven, a relationship in my life is Her responsibility to protect and guide, as much as Y'Shua's. So there it is, an aspect of The Holy Mystery taught to me shortly after my forty-fourth birthday, and nearing my father's own day of birth celebration on January sixteenth, 2007. A year in the Lord it was. Taught me how exercise is a miracle in my life, a gift all its own, and the quality of mine in the open air of this city is a luxury I was blessed with for so strictly honoring the Sabbath by my accord with The Trinity.

Two other churches heard from me, as I researched and participated ecumenically, often with my mother, to see how other Christians in the area felt. I was enriched by these visits with both humility and wisdom in learning how I am perceived by my own. Much of the time this was under my mother's wing, though tired I often be of the memory of youth. I still feel young in many ways, and it has been a special three years of friendship with my mother, so I do not entirely complain about independence lost on her senior moments. She has mastered her cuisine and thus I know has cooked years onto my life despite my smoking.

I quit for three months this year. I see things conservatively piecing again together to quit in about six months. I may not have to smoke again this coming summer. I have started anew on the subject, ever- more aware only the Devil is keeping an historical perspective on this, and that with every walk I take with The Holy Spirit I push disease further from Mary's dominion. I walk and drive for Christ, candles burning low this good Friday the twelfth, awaiting this evening's news. Bishop Chester Talton came to mind as I sat with a cup of coffee, the thought of his possible angle on this imprinted on my mind, "Who in the world fed you to Jews For Jesus?! Wasn't one of our boys—couldn't be!"

I remember how the idea and follow-through on the subject of my thought that I have a Jewish identity now got started. I was living on the corner of Olive and Bond streets in Redlands, drinking whiskey, when on a hot summer day an overriding pressure came to my window. The window, large it was, had holding it up wooden design frames, something akin in shape to letters of the Hebrew alphabet. To let you know at that I will the extent of my imaginative side when pressure exudes to necessary transition through prayer, for though drinking I worshipped Episcopal-side on the Sabbath, having repentance in my heart. Nonetheless a Star of David I wore when voluntarily treated for alcohol consumption patterns in excess, and a comment I received on one occasion alone worried me. Treatment, with star of David now in hiding, once brought me face-to-face with a parolee having a swastika on his leg! Jail things, it was explained as, though I can never be too sure. Is this the price we pay for transition? The Lord God protected me and showed me all of Southern California on this end to that nature, during and actually proceeding the mark 9\11 left on this country's history. What a time frame to address the issue, which by no means was of the utmost of concern to tackle right then. I also struggled from home to home to find a sanity and serenity worth of such extent as to continue with my pen in hand.

In one home I was the only renter to a woman whose one daughter had just married a writer like me. I gave them the address of a nice place to rent so the couple could be alone, even if the marriage did not break the family, yet was against the mother's wishes. At this time I began to venture to Temple, the mother saying it was okay for me to think like that, despite her being a "conservative Christian." A surprise it was when an applicant for the other room, vacated by her, bore big black steelworker's boots, and frayed bell-bottom jeans, looking like a gang member with his shaved head. The landlady told me he had Tourette's Syndrome, but he didn't appear too shaken. I eventually moved on with my life, not one for an experiment in abnormal psychology by an older Christian female with a lesser degree than myself, simply over her race's way of handling sex. She had fed me well at times in exchange for work in the sun, around the home (the war had begun and jobs were drying up as the market opened), and yes, though her ex-husband perhaps was not, the daughter definitely was a believer in Christ. So, I'm glad little conflict occurred despite the brash nature of it all.

Shortly, after, I received my first newsletter from Jews For Jesus at my mother's home. It was now 2004. I know this leaves a gap, so in short I was in homes to stop drinking from Los Angeles to Loma Linda, and Redlands, in this time, being in my new rental since 2005, a studio with sunlight, cable, air, hot water, and a four-burner on which to cook Fettucini Alfredo alle Gamberette. As the song goes, "It's gonna be a long walk home…everybody has a reason to begin again." It is now, again, January 2007. I last worked in 2004. Drawing Christian art, 25 pieces in all, with pen, and writing, have passed the time, along with my attendance of the Healing Service on Wednesdays, as well an 8:00 A.M. Sunday Service, praying with candles for many and maintaining abstinence from sex since Olive and Bond streets' festive cycle. Seven long years. Quickly it has gone into my forty-forth year, from thirty-seven, without a relationship. What a war. Yes, I'm a retired Air Force officer's son, and that's me. I wrote, asking a recent correspondent from San Francisco on what Jewishness I was to perceive. I need an opinion and hope to hear from him soon.

Ever A Challenge Posed, But Never a Paycheck For
Our Two Bits: Cheap Ol' San Frannie

They did not write directly back, even after three correspondences. It's too bad. Not a real personal touch. I would say it's a mystery, but I know why Mary intervenes at times…I've spread the News in different ways a plenty, and I'm more a Catholic-to-be than an out-and-out Gentile. So they are interesting, though to be kept in perspective. They've made claims that are a bit unique, yet a newsletter is a newsletter. To let go of it now is a thought, or to keep it as a conversation piece is another, but is it a burden, is it baggage, or is it another piece to the puzzle on ecumenism? So I'll just leave it alone. Interesting. Almost possessive. And wanting a donation I cannot give. So no, not a Jew, not an Episcopalian, and war, with lots of books by both interest groups. Always the teachers, they are, and yes, you have to bear it. The mystery of that city leads to staying here.

One talks about spiritual crisis, whether from others' perspective or not, and this is the nature of it: Because high school cut and ran from the Inland Empire to L.A., I had problems in college. College roots have not translated, economically nor for a second opportunity. When the subject switches to community roots, I am constantly reminded of cutting and running. It is a slap in the face I should have never taken, because others away from it awarded me aplenty for my potential, but the Town cannot envision the new perspective without a high school cafeteria menu of professionals. I've made good religiously on all my faults, though Dr. Creflo Dollar is right: Lukewarm community response will prevent healing and so will unconfessed sin. Mother is also right: They can work until they are blue in the face and rich as hell, though they inherited the sowing of their own reap on the subject in spiritual terms. Better put psychological. Father Caffrey still has explaining to do if this is about the material. If you put it like that you are going to reap what you sow with hustling in this area. I think he has the seniors' best health in mind and does not mean to leave the gap for me. It's important. I'll bring it up to her. Material, but volunteer. Solve your own psychological issue because your roots, as your Mom said, are in the future. I'll try and be careful not to get their foul mouth, but it happens. I need a pragmatic, academic explanation for the roots.

What would Scripture or Dante say? Scripture would say I'm Christ's brother and God smites my enemies. It is a call to a renewed, practical faith. Trinity Episcopal is my roots. Some of their playbook is pattern. I should be careful not to bend over backward when they reminisce. It's an older institution now. To build a bridge with their conscience and yours is participation, purification through ritual, and renewment through the blood of the Lamb and your future. This parish is going forward. They have their material base. They are a Church. They are sustainable. Dante should give an example of what roots in the future off of which to draw while I attempt to strengthen community and parish ties. The Lord already explained in a divine manner of an intervention that they miss me as Americans and would accept me in my current situation, but that's not the point, I 'm moving on with studying, Catholicism, as this work, too, was on dry ground.

Paradiso Canto XXX 28-33

Dante talks of the inspiration the beauty of Beatrice gave him, in pursuing the writing of _The Divine Comedy_, at which work's completion he must let go of her.

It uses affection for the female in direct analogy for Beatrice, pointing to the two realms of Heaven, the first now being circular church roots. It talks that the raison d'etre is love and gives you a matter of fact belief in yourself, even at forty-four years old, if you don't get greedy! So that feeling of persecuting for being persecuted is the root of greed. As far as about the material, _Paradiso_ Canto XXXI, 22, talks of how God's incandescence can see through the surrounding darkness to see the value of each part in His plan, and nothing can cover an object or person being seen as the rays shoot forth. In a material sense, you're sitting on it; the value of your possessions, which is something to be somewhat proud of nonetheless. You've got nice items, furniture, plants, cookware, foodstuffs, electrical goods and dwelling. It's comfortable enough for art and culture to bear fruit. And move forward. It is a path you are on, clearly defined by Christ, pre-ordained, in other words already art, foreseen, and paved. And you have a guard on this path. Honor the guard. On the thoughts that God wanted me to be a Messianic Jew: "So did I journey, from rank to rank, along a path now up now down, now circling round. Same circle for all, with all involved, so it is your duty to learn them, but you are not them."

It's enough 'cutting and running,' so in the end what does the New American Bible say to that? 1 Corinthians 4:18-21: "*Some have become inflated with pride, as if I were not coming to you. But I will come to you soon, if the Lord is willing, and I shall ascertain not the talk of these inflated people but their power. For the kingdom of God is not a matter of talk but of power. Which do you prefer? Shall I come to you with a rod, or with love and a gentle spirit?*"

Why No Rebuttal? It's Not Their Field Of Medicine

--I notice people seem to bump into me psychologically when I have a nicotine craving, bringing a certain pain that renders me smoking more.
--The Europe of The Future
The California of The Past
The Christian of The Present
The Psychiatrist is an Ass!

The good news is he had a good impression of you. The bad news is he was the exorcist at whose church you attended before this, God is going by Hell, Purgatory, and Heaven on it, and is now up there with deceased psychiatrists' reviews of the City's sins that let in all the demons they had strived so hard to cast out. Before becoming an exorcist, he made money and met his wife of 62 years while bookkeeping in the military. So, this is why there is a depression in the region and a wine freeze. It was not from senior citizen apathy and poor management.

I think external perceptions change by seeing it anew, while helping moderate the internal experience when away from campus to manage things better through the process of release.

The Dawn Cracks For Some At 5:30 A.M.

SO, the roots of the neighborhood proper being assigned the judgment of the grace of the internal, at waking hours. A continuation with the reflections of the external environment, perhaps a different nature of root in entirety. To rise or to fall becomes the balance, or lack thereof, between these two factors and their issues when dealing with the individual God created. Legitimate advance, or Behavioral Schizophrenia with intervention by professionals.

The reflections of the morning:

(1) Five out of ten return the "Hello;"
(2) The Reverend stands sacred in white outside talking to two 40ish women:
(3) Catholicism as a proximity choice observed
(4) The nature of the other proximity factors nearby the historical setting of downtown:
(5) The mention of diamond in the rough historical homes amongst a sea of non-commission officer ambience;
(6) Memories of a dwelling I had in the area where I attempted to leave the past behind for a better future;
(7) A young woman who lived above me from my home parish, now a mother of three and a Vice-Principal of a new high school here;
(8) The throwing away of an '81 Redlands photo and two diplomas;
(9) The mention of ESRI, and the knowledge it began perhaps by studying this region itself and Earth here;
(10) The break of the sun at 9:00 A.M. upon returning home; and
(11) The knowledge of a) "Don't ask where I'm from;" b) "Stephen, you need your privacy;" c) "Like hell if that version of the external will get root in our City Council," and d) I have to move all this in order to become muscular again, stemming from the sight of the Episcopal church in the overcast morn. The large trees of pine are on Olive Ave, nearby the park flanking the Redlands Bowl on the opposite side of Smiley Library. I had never walked just to admire the deep rooting and size of them. Parkwood Drive. Are there two Parkwood Drives...two versions of Redlands proper? Interesting sub-context to the study of the subconscious where it meets the here and now.

* There is also a "Parkford Drive" by Ford Park off of Interstate 10.

For Dr. Murad 9:00 A.M.

The medication is now obviously not controlling symptoms. (1) While helping my father dig out a tree stump on a cool Winter day, I had hallucinations I was on acid and marijuana. I cannot remember another time on a cool day this has happened. (2) I am hearing voices doing the talking in my mind for me, though I do not speak out-loud. (3) The radio and C.D.'s have a voice, though be it Christian and counseling. Advice from music. I have a very low tolerance for radio commercials, though am patient to hear the classical music. Classical is all I can listen to without worrying about violence or sex, even alone as I listen. (4) Symptoms I must report, even when coffee is not a factor, are nervousness, increased urination, confusion, some weakness, numbness in the feet and hands at times, and unusual changes in my mental alertness. (5) Also, I report a spacey, gullible look on my face. I feel as if I am devoid of the experience I have in life. I have little identity associated with my image. (6) As far as connecting goes, I no longer have ability with social capital. I am nerdy in a state of being and responding. (7) This medication had me stuck in the past to the point I was having difficulty realizing some of what has gone on in my life in Redlands. We all are on the wrong page. I must say, as a result, some paranoia and over-suspicion have occurred.

I've never had serious problems with a medication change and am exercising fine. I don't imagine there will be complications.

On a positive note, when I watch T.V., I seem to have some sort of clairvoyance that summarizes my contemporary situation in life. It actually connects as if the spirit of the characters or people is addressing my issues, as if there are two realms, the actual dialogue, and them saying a completely different dialogue with all their personality and soul that seemed like God was using them to reach me. SO if this is hallucination and not clairvoyance, it has a personal efficacy. I know actual examples are rare, but Christ is resurrected and is a personal God. My beliefs haven't actually gotten me in a corner: This city does have some problems that I carry the burden of emotionally, not as a laboring, direct-action position in Christ's Kingdom at this time. I see what they see, the liberal element, difficulties with employment of others, Christians smiling—yet actual conservative over-socialization has grown sensitive overall, rather imbalanced in a certain sense. Or is it me?

Imagine a focus on the complex, not the healthy landscape of the soul, and over-conditioning taking place, or a stall game. Or worse, a judgment. So this risk is always present with certain stakes. Some risks simply become contrived and repetitive, so that's why it seems like a stall game, when actually, activity occurs. TO have activity with a vein of clairvoyance can be uncomfortable physically to the head because one cannot spell out what the real issue is. We say it's schizophrenia, or alcohol addiction\recession, but we can leave it at not wanting to call people on the nature of the complex that is shared. The respect of boundaries while perceiving issues is often a tightrope walk.

Q: Devise a strategy to avoid these clairvoyance cluster points that is flexible.

Alcoholics Anonymous counsels, "You cannot close the door on the past nor regret it", yet modern psychology would like to be on offense to heal wounds associated with our past. When it is not A.A.'s ball, where is what we left in the past in context with the traditionally fortifying nature of our roots? We certainly cannot lay stagnate in a root-bound psychology as the winds of the past blow into an unclosed door or window. Normally the door is shut and the window opened to see the future. On pure crap, a response I can hear them using is, "Don't always try to reconcile something from recovery to psychiatry." All this feels similar in philosophy to <u>my</u> version of academe, not others' that are driving this version of the boat now, though uprooted as an economic constant in context of esprit. Rehab did its damage uprooting the worker prior to war and recession. I've prolonged my poverty by doubting myself.

On upbringing… With my generation of military parent and their roots in their cohesive yet old-fashioned network, I face the challenges a tradition will have in California. I was informed by my father that I would not do well in the military. My mother recently admitted she raised my sister differently, <u>and</u>, in the extended context the family can be dysfunctional from this position. Only family in Los Angeles has gone the length of continuous psychological support for those with a modicum of talent. As someone who both as a child and adult was encouraged by the professional folk to write creatively, I have the time, yet my parents are staunch pragmatists. Due to challenges of youth and a dysfunctional family I became a writer who dealt with a mild alcohol problem at 37. It is purely behavioral. I was never fired over it. My extended family has their main drinking on special occasions and to relax in retirement this entire time. The denomination of my upbringing, from the top down, has had some alcohol and drug problems since the late 80's. My sister, a success, left secretly this faith at about 20 years of age. Jealousy, and not my own, can be a factor in conversation with this family, if it is not a closed system talk. No psychologist in near twenty years has solved the problem without working with me in a different location.

On the flip side, off the subject of a constant truth that won't change, I've grown into a person inwardly who cares, is appreciative, realizes he is in good health, has no sexual diseases (as some in my generation do), and is close to God. I've heard comments from the pulpit that Christ is God, Christ is authority, it is his plans for our life not our own, and within concordance with Scripture, in obeying His parents, we cannot begin to understand the nature of our riches in Heaven for the condition of our servitude on earth.

In that light, it would be wise to note that I had a Muslim friend at 18 while abroad who was also an exchange student, my family had Jewish in-laws by my 20th birthday, people from India feel I connect well with them at times, Asians have an amicable though quiet understanding of my servitude, and quite simply put, I have never had less than healthy instincts even before my diagnosis on subjects requiring sensitivity.

Life is not often fair and you cannot be a prophet in your own hometown, large or small. This last fact relates to money and well-being. Quite simply, as far as depression goes, we all have our days.

On attention span and symptoms management, anxiety and addiction, the dialogue proceeds as scheduled with morning being one and two and evening being a bit prone to negative thinking. Deep breathing has had benefit, and next month I will spend money on the patch again, smoking being the dirtier fighter of the two demons. Or is it my immediate environment—the other, a lummox or two puffing away in the monolingual left brain, machine-tilting realm?

An evangelist described a fragile contemporary framework to adult Christianity with a certain sense we've been given our marching orders, end days or allegorical. As all this comes off the top of my head to paper spontaneously, I think as far as framework goes I see trees out my window, old homes as I walk, though a sub-cultural diversity that can be a snag. I have a hygiene facet, a Naples facet (meaning I see stark ways to criticize with a humorous tone that I keep out of reach because of the fragileness), a rather unique Redlands facet that travelled well though can miss a cue at home, a serious work ethic in a generous family facet (despite miscues), an unfair clock I watch as a facet, a soap opera facet, and constantly calling me is a facet that needs an example of surrender in Christian example. Psychologically I have been charitable, dismissed demons, kept a connection with Earth, and fought hard to avoid danger through quite a course within my time, looking for a natural defense against the merry-go-round's ale mugs stealing the whole show because I will not ride. To end on a note other than the subject of tension or shock, it's thoroughly considered not only tradition but discipline for some to isolate in serenity, where Christ's critique is hard without too much interruption other than a baby bird nesting in the tree, until a new starting up of the carnival ride. So preparing for a better ride in the sun-dabbed theme park is good and fun with some fresh air(written while still unemployed, two years before passing a Federal exam and FBI background check to assist in the taking of the Census during the U.S. continuation of the Iraq War, cross torn from the badge of the Redlands Police Dept. that I've supported in prayer all my growing life, having neighbors who grew up to wear that uniform, while I grew to wear a Census Badge, a job of risk as well).

On the subject of less cognitive distortions to deal with on your trip to Europe (from your fellow man than "growing up"), America, <u>don't turn the thought inside out:</u> often it can come back. Again, as far as the extra-inning slow-pitch softball tournament in adult America for it to come back, <u>do not turn it inside out. Do not go over it again and again,</u> and odds are faith will conquer the field for you.

<u>The Room Is Never Left Like It Was When You Moved In</u>

In a room what you have is positive space and negative space, as this not only applies to art, though to architecture as well. So, in combination, given the planning you like your room to have, artwork is a control factor of negative space. To continue, the room is arranged to allow for inspiration, yet what is let into the room may create negative space on the subject. Not just in cleanliness yet in words left or actions done in the continuation of individual growth and learning in the room. The key therefore is to learn to dissipate the negative space, as much and all possible, for clear thinking.

(Don't worry about the baggage you've carried for other people, and never question God on your fate in relationships. You have a good fate.)

Closures and Apertures Withstanding

The Group psychologist's answer to the contradiction between non-monitoring of the behavioral side of mental illness and the genetic fact is simply I'm better than I think at times, and, as well as put, I am now out of The Devil's Triangle(my own words for it in notations months prior).

The Activity Center staff in the conference room, obviously with her weekend not having gone backward when she spoke, made an observation come clear. The soft, pin sharp voice I barely make audible in my room is not the classic symptom of something I am hearing that is not there. It is the inner voice I possess from my own improved mind due to exercise, rest, prayer, deep breathing, escape, and silence. I am thinking out loud very logically in a soft voice reflecting the agenda of my conscience. This was proven in the morning by a simple action: The voice created a new joke, and I did not lack the power of concentration to write it all down, a one- page long development of a conservative punch-line. This is most likely where, in response to this change in my self-awareness, I mistook the strong new development with being in the vicinity of a being from angelic orders. That's good humor as well. The idea that psychiatrists and therapists are equally as dry is my imagination. Instead of imagination and inner voice getting into their brilliant melody of "What Goes around Comes Around Again and Again, like Santa,!" I drew in a large part of Y' shua(or Jesus in Hebrew) on the canvas from top-hair to beard, His hair's length as well, in completion. The brows, lips, nose, eyes, and forehead are next.

Behavioral Science In The Now

1) What is one thing you are not doing that if you did it consistently well would make a huge difference in your personal life?
2) Making my influences over the course of my life move forward in a comprehensible, organized, efficient and respectable manner. Because

of my role in life, my triggers have been manipulated and I focus on peace of mind.

3) In combination with the above, my normally efficient discipline that I've always had instincts for is challenged. I slow and confuse.

4) It would create the boundaries that would give me the power of my rights so that it appears again as if the pitches are over the plate on all work, social, and privacy.

5) I'm aware of my agenda being spread out over the course of time and the influences are competing in the meantime.

Six months from now, look at it and see what you've done (Questions of leverage).

Next:

1) Explore the obstacles that prevent you from carrying out the group leader's idea.
("Your future is what you want to carry out,".....a shit load of war-time counseling for the tykes, isn't it?)
   a) Difficulties involved in obstacles
   b) Can be external in your environment
   c) You would not take the new pathway to work until it opened up.
   d) The obstacle is something you personally have to identify (Powerful question: Decide if you can do it and see how you are in six months).
   e) Internal obstacle.

2) The Group Leader's anxiety can take something already there and pump it up.

3) Leader: "Writing and reading go hand in hand."

So, on Jung and the collective unconscious, apparently prior experiences left in an unclosed state when the war occurred created a cycle of supposition, leaving a bit of detachment of the persona. It is like a cigarette that makes bad jokes, instead of inspiration. Call it the negative space of the mind. The Group Leader suggests one carefully investigate the layers of the unconscious, bounding to bring a little pain and further yet temporary disconnection, until the mind's layers are smoothed to flow in lieu of dysfunctional overlap. Not his own words, yet here this, as it should be, personal journey begins. Prayer was thought of, though Christianity that drives right into the point can stigmatize the detachment. Here continues what was meant earlier as Virgin Mary's levels of healing, so although slow and steady, it is a ritual of Christian maturity to investigate the mind in said manner. One will inevitably be better prepared for an occupation in society.

An age-old reason that I will insist on investigating in aforementioned manner is black and white, yet release is opaque: it is discrimination. I have often dealt with the role of the second-hand opinion holder in White society, while all the time carrying the baggage the Hispanic community has for itself while amongst them. I am a mixed ethnicity, not possessing a complex as often is believed, possessing a two-fold blindness—society's perspective of me in my grasp to observe. It is not as if I invited this, quite simply put, it is a well-structured trap of sorts, perhaps, and only perhaps, dealing with the multiethnic family dimension I possess as well.

So in total this is the character defect of hypocrisy in two different areas of the community, with a few hoops to jump through in between in order to ensure a less reactive and adaptable collective unconscious. The mill without water. In order to solve the enigma I get the feeling it is now a challenge to write out of a corner when actually, for the time being, God must be hoping I will not run out of aspirin so He does not have to change all them, but just me.

Yes, old-fashioned parents at times like rolling stones through society's decline. Am I looking, a product of an Episcopal marriage that has lasted 46 years, looking at the divorced White and Hispanic populations? No. The other old-fashioned people. Add a layer to the potential for schism. Dichotomy-trotters by day, the art world by night.

Crisis Situation

While going through a medication transition that otherwise was productive, with less side-effects and more exercise, the Activity Center had a very busy week, though this was not the whole issue: My parents', especially my Dad's, thinking was turned into a cage-rattling catch-22. On Mother's Day it was apparent: I get something out of Redlands, my privacy, and the new church. Though while worshipping, that's when I noticed how much pressure my father is putting on me in a competitive way, competing against me by over-managing, almost on purpose to create another problem, the same one for psychology, I've had in and out of offices for years. My parents have a busy life. They really don't seem to expect anything short of honoring them, then assign me a "space" of their impression of how I react, coupled with my sister's version of this. So this was later learned to be the old medication fighting back. The largest pain in the ass about this psychiatric treatment is the finger-pointing at my Dad and of my Dad, as well as other American men.

For some reason my movement was thrown off. I spilled some coffee. I looked up and from the other side of the room my Dad was staring straight at me, as if boxing in the whole worship experience to cause a mistake. At the table they talk about my sister's life in a manner that makes me feel like they have a plan with them they are carrying out and they don't want me to ruin it. My Mom once said my sister is manipulative, but this isn't the impression I get. So God, sounds like alcohol? My parents went to a party the night before. Dad gives me the feeling this is getting to be black and white. He likes the drink but is careful of moderation, and so are all the wealthy in Redlands. My brother-in-law is the same way. Drinkers', even <u>social</u>, cognitive distortions are now ruining it with a lot of information and forcing the agenda, even though their health is fine.(Still, only a medication, and perhaps a supernatural precursor to Aunt Hope's final struggle.)

Let's cut it down to size:

--They think out-loud. Right in the middle of church.

--My brother-in-law and rest of family think so loud I can hear them from another part of California. Trinity Episcopal had this problem. This is why, if you use a four-syllable word, it is assumed you want to be a priest or take a special vow. Everyone else is laying back after the last round of ESP, which, now, I think, you can see is a lesser power to clairvoyance. More common-place.

--So, rather than saying it is an insidious, over-competitive nature, or a curve ball that is not over the plate but <u>at</u> the batter, I am the unofficial "baggage carrier at LAX," so to speak.

Psychiatry wants me to carry the bags at Redlands Municipal Airport. Even a movie is more like monitoring demons of others in an assigned role.

--They don't want the old me back because he had ways around this.

--This is new since the war, <u>yet</u> eight years old.

--They don't even see the real me inside, and are overrunning the external. <u>They pick when they want psychiatry to be the issue, anywhere, any given time. I'm confidential. They act like holding this information is the same damn kid's confection as a cupcake or helado and their waistline isn't slimmin' down.</u>

(I don't want the old me back either).

We mentioned:

A) Med Change
B) Too much information
C) Discrimination
D) Religion as stabilization, intervention, and supernatural
E) Subconscious layers
F) They want a social drink
G) Many layers of healing, the pain of discovery, and a time factor

Q: What is the ritual, the process, the experiment?
A: Conversion to a new denomination of Christianity.
God's message is it is quite hard to free me, but He tries.
Next, you have to be careful where you work in this state.

Plus, Five Demons

So, Dr. Multani's inspiration seemed to say you've got a trail on the top going through your mind and a trail on the bottom going through the mouth, both in order to develop their own agenda. That's how health factors are affected. It's, yes, their version of psychology. Both need work for improvement with church and university to meet their standards for effective Christian output. It is a form of indenturing. How to snap out of this again, is not easy. It is an unfair foundation for competition and not personal, but the effects of the war and disasters and how it has been reacted to. No, it's not a dialogue based on the struggle your political party has had with the Democrats, and the psychological incongruities created. See, always present. It's symbolism. Some keep their jobs by reopening, lieu of closure.

So, this is the external condition we have based on the internal condition. One, by prayer, would be led to believe they are correlated, from breakdown to miracle in motion. Also, yes, different elements of diversity (ancient, traditional influences or contemporary, scientific influences) are dealt with in the same theme of empire's struggles and resolves. The Mediterranean vs. the Southwest U.S. So, if on the area of Christian existential moments and interpretation, we fail, don't worry. From Southern Italian mudslides to Hurricane Katrina, when 'ya wing it and nature kicks it out of bounds, the region gets a 'corner kick.'

Dante's Assistance Through Episcopal Inferno

Lighting two navy-blue candles and turning off the lights, I envision with the pin-drop silence the boundaries of Inquisition prisons. Vision, Romanticism, Materialism, or the bond between the two? For certainly through vision the divine comes down to earth in the fashion of said Materialism (to not only honor its timeframe, but also accept it as the most advanced phase of thought, under Introduction, p. 11, Allen Mandelbaum's translation of The Divine Comedy), yet when vision draws off of ancient roots its unfolding, we have Dante of time, space, platform and circumstance for the soul, not accepting of the era one reasons within. The status of an individual in a living Virgin Mary's eyes, as a Queen conjectured to have a throne, complete with Hebrew maidens at its side, lies prefaced through both reason and prayer in the candlelight vision of one's own Purgatory. Will she release me for an adherence to the guiding of The California Dante in its theological context, keeping an analogy in mind that *Paradiso* on this planet of beauty in God's hand awaits one who learns how to take the twists with said wisdom within as its own weapon of wits?

Will thorns of conscience picked up in this path of Romanticism and Materialism, tied and carried in our procession of saints, impede the very mind over matter Alighieri must have had in order to delve deep into the realm of poetry with the sacred as terms? And in continuance, does the verse provide a key to the healing on the levels necessary for our soul's incarnations from those purported worries? Was alleviation of pain on a pathway into literature his voice, to call echoing through the ages of a connection with Christ's Paradisoc chemistry, for all humans willing to not let freedom evaporate in the hellish experience of Earth's own demons? Is Holy Communion the only bond with the solvents of _Paradiso_'s planes possible to deem by our imagination's unstrained eye, looking into the petals of prose? Said solvents line the road divine, and don't they breathe through to cleanse our souls of the pain of the thorns? Assuredly one cannot equate the spears of an Inquisitional prison with the privilege and pleasure of any saint's well-deserved sabbatical. Yet through it's sating of our thirst for wisdom, the call echoing through the very chambers of the universe to all planes seems to save us here on Earth from the evil that may surface, and be from a Dante free of exile's boundaries, indeed a Dante awakening with Beatrice. Not a buried, dead and done Dante, not a Dante awaiting the confirmation of a resurrection of all the dead on an end day, but a man giving space for the dream of Romanticism to breathe, so that theology on any front not get cornered by further sins' seeds. Yes, institutions of religion make claims. We do not have a choice between Apostle Paul's letters and _Paradiso,_ yet as contemporary disciples who are no strangers to persecution's precepts, our absolution lies not with God's foreseen end to ancient text's tale. It begins with an often-authoritative Queen's verdict indeed alive from Above, that, as a resurrected Lord and Advocate sections and quarters, saves and sings, Her call instructing Him to let us dwell right here in Alighieri, right in the rhythm of the rhyme, is for her own pleasure in this fellowship. As the Bible tells of wars in the Heavens, I say Paul and Dante do not combat, yet one is esteemed guest at a banquet prepared now for many by the latter. A Heaven whose prerequisite is a servitude no stranger to us all, Christ's goal is healing. All worship will inaugurate such in His peripheries here and there, and fellowship after assuredly is not an absent idea.

Yes, we've seen it all, in our reflection from the glass frame containing a pen-drawing of the Crucifixion, though it is art, not icon: As if, to say, the Holy Gospel of the Anglican Church was pressed fully to the side of my head, and the ghastly image straight from a dwelling place in Hades of a Gollumesque devil drew my energy from me as I puffed my tobacco. Another mongrel of a male demon, with aesthetics similar to Satan's, scowls through at my own visage from behind The Cross of Golgotha drawn there, and yet as I shortly after get on the phone with my mother, I assuredly am all there upstairs in conversation, if nothing else at this time. Imagination run wild into religious, non-drug induced hallucination, or the macabre metaphor in sight of a problem's root? To say our mind is an arm in a sling is true: yet most definitely through perseverance it becomes all how you look at it in the spirit of the Dantesque vision. Sounds far from romantic, yet I believe it parallels through Romanticism a literary communion drawing on ancient vision. The truth.

So, as the truth unfolds, quite frankly put, the exhaustive escape from Infernesque Episcopal liberalism and its demons left billows in my bellows of smoke that Winter must expurgate in her months. On the upside a psychological barbed wire has hung on a corner of the Holy Communion's symbolism. When art is a sacrifice and not a sanctification, a Church makes an endless Purgatorial Island out of the creative side of the Sacred Virgin's meditations. Our Lady of the Way, Protect Me. To escape from this and many obstructions in holy healing arts is a winter where plants transplanted take an extra drink at night, while the day leaves a thick dust amidst the celebration of the Lord's benevolence to buffer against debt and death.

An Episcopal Church, whose dust is covered at the corners of the community, so much freedom for ministry in the community that the secular world, in its rebellion against war, now lies exhausted by same said dust it kicks up: Our marching orders in Messianic sense giving credit where errands are accomplished, and truth uncovered. It was finally that this hidden force of truth seemed somewhat satisfied with aperture of a New Year, as dialogue from the old left a trail the demons breathed upon, sending us finally to a cloud of fresh air where a barb-wire held back the undead, and life struggled forth with a more solid grounding to its previously- heralded, overconfident expectation of the current world. Avoid liberal coffee houses, be careful what you read, watch non-denominational Christian television and rest from this exploration's ritual of precarious information by the hordes. Such is the movement at this notice. The newspaper came out looking like a saint all its own.

So what we had, combining the influences of a <u>Divine Comedy</u> perspective with that of an ecumenical Gospel was a devil, not demon, of alcohol and a devil of what appears to be homosexuality that infiltrate the privacy of the thinking space. That of alcohol is horned and of a chemistry over a hundred per cent, appearing to some as a bull. This is predominantly why some abstain and some get a sanctuary where they can obstruct its influence, filtering below where it roams until Christ intervenes. At least, it appears to be the Christ. One must know that part of the ecumenical community is paid and part is not, regardless of the fact where the oxymoronic smorgasbord of issues shifts, the platter is the same. As universities shuffle for a new deal, and cities both rotate the guard and expand or contract spending/jobs, some in the region fell ill to a spirit of the law. It was over-bridled, simply put, with issues during congested demographic shifts. From sex in smog to shower and sunshine, a pneumonia vaccine sank in as offered, but the bottom on diet and hydration fell out despite exercise. In light of what California normally dines on, for this to happen seems abnormal, and compared to what troops can and must go through, it makes me a tinge bit curious of war's rations.

It's not always an easy-going mental environment for eggs to go down, California, from Spanish omelet to quiche. On a flip-side it's as good a guess as any for staying power, that and my cereal. Of course where wine grows in the heart is a different leverage on the devilish, where below the rapid, rolling steam of a caffeine river has its flotsam. Prefaced by winds that want to give a status report strangely synonymous in charting to the pathways of dialogue's agenda, the array of greens have some pillaring effects, or blockade-like power if anxiety 'over-spills the banks.'

You would think a two-mile walk despite a pot of coffee and pack of cigarettes would be enough to avoid need of antacid tablets, but no. It's why anxiety has, since youth in America, always seemed more dharma than diet, more karma than calories. The hand lost at the card table to a sister who won the sibling rivalry! Where freedom of speech became her platform issue to observe said characteristic, her tactics were enough for me to view any of her male courters as some sort of idiot in cahoots with the tempest regardless of brains, or wealth, religion, or race. The bright side of this state's generation born in the Fifties is not their child within's definition so much as their professional demeanor, not a tradition to overlook while those born in the Sixties make no gain with a culture shock dialogue unless it's a once-in-a-lifetime vacation. Same counter at McDonalds' all this time, yet eventually the Christian guard advanced. The siblings we are: an Air Force Major's daughter and son, and it's not a trite, "That's life!" Its war's stove pipes and engines embarking and landing. The point being that as Dante protects you, perhaps during all this, your safeguarding of your version is rewarded from Above in an unforeseen way. It's pleasant on the island mountain of California Purgatory, to finish a current version (of Revelation's Patmos as well) and know this Holy War is not the end.

There is thinking and there is doing, and there is healing and there is fixing. This was the premise, upon which noticing, I contemplated mental illness was not genetic, but behavioral and behavioral alone. And was healed in religious fashion. Out came Democrats, Crime Waves, Heat, Drinking issues, Arguments in the Ecumenical World, Deaths in the Family Due to Age, War Report After War Report, A Divided Congress, A Game of Survivor In the Southland, A Prodigal Son, A Candidacy For Rites of Catholic Initiation as an Adult, and Pope Benedict XVI. Infallible. No Joking. Behavioral. Traditional. Ancient. To Modern. In ambiguity. Yet a Church is a Gift, regardless of issue, calling, lot in life, deals done, story, or roots. The trail to establishing ecumenical unity. And its woes.

I have favor with God

God loves me

God is blessing me

God is answering my prayers.

God wants us to be active, luck is spelled work. God give us the strength to collect the bread you give to us.

Poor thoughts and difficulty with attitude increase disease.

When you read a Bible verse, put your name in it.

Read the Psalms and put your name in it.

Hour of Power helps with crossing culture and the shock, the choir is like a blessing.

33% of all deaths occur in hospitals, 10% of 1% of deaths occur in church or Bible study.

God answers our prayers by not saying "Yes" to everything.

Hell is not for people, it's for the devil and his angels. Some people are sent there for sins.

Three kinds of Hell: Hell in your heart, or behavioral persecutions that lower the instinctually-rising tide of emotions, like happiness and the three kinds of love; Agape, Fileo, and last but not least, Eros. Hell of the mind, such as when public commentary is bouncing sharply and the negativity affects our part in collective culture. Lastly, Hell outside of you, such as an endless chasing of the tail to make progress in an otherwise safe town, even as street-smarts have always been an adequate tool. People are like an army of ants trying to run out of their own anthill, spending a fortune to make a dead-end Shangrai-La…The world cannot give it to you, Christ only can.

The Cities Were Down: The Country Rose and Spread Wings over Them

On afternoon meditation, feeling the actual presence and power of the conscience of Christ in my abode, I read a passage from Scripture to use as a comparison to solve some current thinking that had been troubling me. I randomly turned to Zechariah 3:8: *"Listen, Joshua! You and your associates who sit before you are good men..."* Zechariah 3:9,... *one stone with seven facets."* Joshua asks the angel what these things are, as I asked the Lord why I do not have a better connection with two non-profit organizations in other cities that sought me out, past and present, seemingly through the power of The Holy Spirit.

Zechariah 4:6 *"...Zerubbabel...Not by an army, nor by night, but by my Spirit, says the Lord of Hosts."* In Zechariah 3:7, the organizations are, in this case the American Field Service, … the mountain. *"What are you, a great mountain?.... but a plain (before Zerubbabel."* Zechariah 4:9 *"The hands of Zerubbabel have laid the foundations of this house, and his hands shall finish it."* Zechariah 4:10: According to this, they won't always be scornful. The seven facets are the seven eyes of the Lord. The other non-profit originally was Jews For Jesus, though now stands as symbol of all going through issues in line with Scriptural interpretation: *"...his hand shall finish it."*

To continue, the Lord apparently is tired of this "action-oriented' path to God's will. Things to do, celebration of culture, accomplishment planning and seeking (all of it), is already under the guardian angel's wing and we need rest for our side, not 10,000 connections.

Enough Mud, Christ, For the Knights To Explore

And my patience is fine.

The sermon by Father Paul reminded us that Jesus gets in the mud to help, preserve and advocate our connection, as sinners, to God His Father and ours. We often do not understand, through example of the Homily on the Prodigal Son, the nature of where we are at, why we are assigned this role, and what will happen. The son returns to the father, yet the elder son who was more righteous has a jealousy issue, the difficulty ensues, and potential disconnection from God can occur. SO, we pray thanks to God for our return, ambiguously, walk with Christ, and reconnect outside of the mud we are in with the Father. Reminded not to be preoccupied with the vote by my angel. On the way home Mary, Queen of the Emporium marks my conscience with her heel in ownership. The war continues, all conscience, study delights and projects running right into *The Center Cannot Hold* (another book on schizophrenia), as we, sinners, shine to preserve, protect, and pray for Reconciliation with the Father both in small and great scale. Said Hello to Georgia and a Mr. John of Small Faith Groups. Noticed my bulletin was missing. There's a thief!

*

The Jewish World: They don't Use Interpreters, But Here's Their Translation (As Unique As Italians)

I have an older sister who has some other scenario than a jealousy issue. This is where we have an incongruity with the sermon that is additional pressure. "Mary with a heel:" This is vision I found. The Center Cannot Hold simply an added chapter to appease the "Ritual of Manhood" theory by David Schwimmer. This is where much time is taken of mine though as literature the story is unique. Speaking, that is, of where all the ensuing connect with current Jewish stuff, in defense of Israel. My sister married a real one. He does the Twist. No, not the dance step, my arm, at family gatherings. Taught my nephew too. Not like I asked. Good grip. Right to left interpretation stuff. The paycheck, ... Jewish brother—mine. And Italian—terrible film on life in America—D'ONT try again in the film world.

But no, Gerber's to New England to New York—a star is born again, but not in the Christian sense! Ha Ha Ha! Jesus will help you with this—those are Muslims. It's your baggage now for loading the card game and talking shit, until you compromise not with the Muslim, but with the American Gentile on laurels and cash. Believe me, your attitude is a burden to carry, and Christians have others.

For Christians In America, Where Is It Mr. President, Welfare Life Again?

"The perceptions and the realities are sometimes at odds with each other." This was something that rang home true about my pathway at this time, though was mentioned on air by a leader fighting abroad.

For myself, I am consistent in a healthy daily pattern, though 'a body with no bones' can often be the sensation. I learn this as I reflect on my evening walk. In making for a longer, more productive day, I sustained this 'no bones' feeling throughout shopping for food and household items, then at night clicked on the television. The update on a cable station was astounding. Much had happened. Unusually, another leader had talked about some sense of gathering together at the U.N. amongst those feeling the U.S.A. had done the right thing, the plan to have a stair-step reduction of military in the war for July 2008, and some of this starting in March. We've been there a long time. I'm no stranger throughout this second term of the President to a floating feeling of injury in me, and even though I'm well, it can perturb me, as many miles as I walk, as healthy as I eat, to feel this at the same time as hearing news of Americans living longer...to 79 years old. That's not necessarily long in my family, nor by what statistics I've observed in this home region of ours.

For the most part I'm grateful for a secure and safe location to constantly learn more, and it is a bit of early-month work to clean and stock it prior to study. Neither bored nor surprised, though tense from wanting resolution on prior prayer and meditation, I stumbled across a section of New Testament, in browsing, that made politely clear to wait for God's promise to be fulfilled. It's now twelve hours later, and a walk to Mass to hear more is ready to take place.

Does the Demon Have the Coin, Or Is the Angel Gambling Them?

Mass talked about praying for our enemies. After, I had a long talk with Vivian, the mother of a young teenager with autism, whose husband is a teacher. She talked to me about how everyone has a guardian angel and gave me, to keep, pamphlets on angels, explaining to me that angels can control your sense, as a higher intelligence, and that demons will be the ones to manipulate those.

I returned the coins and now have my refund from my order on T.V.

The Garden's Words, Symbolism and Breath

Father John's Gospel lesson and sermon had to do with honoring saints Damien and, if I got the name correct, Cosmo. They were doctors. Father John observed the area of Assisi as a young Franciscan, and, if I heard correct, this area iniciated the two saints' service to the Lord. The important element of their service included the dusting off of their sandals at places they were not accepted, the donation of all their money for healing to the poor, and the carrying of their Bible wherever they went. It coincides with the message for the today, that in not just a material way though a way of logic, we must not carry so many extra things that it has repercussions. It can slow us down, impede our progress, and hinder our service with the Lord. Our connection with the Sovereign.

Even after having thought about this issue of baggage over the summer in town, at home as well as my parents' garden, there have been many happenings that change the course of a day or week, so focus on exact issues can be delayed. Why, there seems to be issues in each location. Makes me wonder if the Lord just wants an endless dialogue on problems, as opposed to help, yet I know forward is the direction. The encouragement is I recognized it despite getting a bit bottled up.

In identifying the feelings I seek the garden for, it has to do with some sort of disenchanting air that has been present in the home. Strong stuff, making it difficult not to have a panicky sensation. There could be several explanations. Mother's pulled muscle, Dad's mind-set in leadership, and yet, between all of this, I believe an "if it's not broken, do not fix it" opinion is the best. The marriage has been together forty-six years. We continue to learn, educate, improve, and grow over the course of our lives, and what is important about our family is they make room for me. I learned to take my Dad's advice on my writing, as he is well-read, for the first time at the age of forty-four: whether or not he has put pen to paper since college, or reads Christian literature. It piqued his interest, my essay on my guardian angel, yet he feels I attempt to loop too much together with metaphor. My life has seemed like that for quite some time, as if motion and movement are the principle link in my Christian duty. Putting the foot down to rest at home has not been easy, and work has been out of the question. I should be honest, I should ask Frank to be my sponsor. I should confide in him about research. I do not have to go into extreme detail. I should also be aware of what my family went through with Episcopal thought and explain briefly why it has been a good decision to make this transition. I must pray we do not have endless issues in transition, and that the angels protect the dialogue of the fellowship.

Also, there are two ways of writing: right-side justified and left-side justified. My right-side justified handwriting, which is not the norm, aroused my father's curiosity. I pushed my words to the right side of the page.

I learned something important when watching my mother Lydia put the dry, clean clothes on the rocking chair for two: some of these questions answer themselves. Even with conversion. My father, Steve, said he would be back in an hour. Lydia is getting ready for Red Hats, a group of women over fifty who have decided on living as if they were still young, not with a senior outlook, and succeeding at it regardless of the loquacious details of time. I take a moment to remember the conversation my father and I had: he does not like church as an institution, believes in Christ and his teachings, yet Bishops and politics have taken away from the spiritual experience. He handed me an article on the Anglicans abroad chastising the Americans' Episcopate for the rift they started in 2003, and pushed them back in their approach from electing another Gay Bishop. As well, there will be no formal prayer for those same-gender couples. These are issues being handled after decades-long debate in The Anglican Communion over how to interpret the Bible. I am, to continue steering toward my new spiritual direction, glad to understand that since Vatican II in 1963(my birth year in Louisiana, and why, the location of the Episcopal Bishops' meeting!) The Roman Catholic Church has been decided enough to publish a beautiful Catechism. Emma described the Catechism as a heavy read...not something you look at cover-to-cover. Yes, just glimpsing in decidedly sets a different tone than an Episcopal sermon. In Mass today I had some sort of Christian existential thought from Above(although the term "contemplative" is more popular in U.S. circles: You would have to hear a native Italian say "Hans e-Kirkegaard-a to understand!).

In a serious manner, the words "Breathe through here" came to mind. I now am breathing outdoors, and a thought I first relished with a doctor four years ago subtly impacted my soul. Mind, body, and heart had removed an arrow of confusion. Rather, in avoiding the habitual metaphor, in practice, the question stirred from the beauty of a palm tree in noon sun. I seemed to be looking, indeed drawn, to discover what Christ's symbolism would be for the palm planted so purposefully on the side of the hill. The hill itself was a part of a mountain frontier, in appearance, from its incline. Yes, even though located in a green neighborhood far below the elevation of San Bernardino's mountain range.

Neither wind, sun, or rain this year, in its supernatural judgment of my family's life on this property, tainted the Vinca Minor plant in over-riding fashion. And to my surprise, behind where a dead palm had gone, was a springing tree! Many leafy green branches of a different variety gave a bush' circular formation, yet in there somewhere rooted, a trunk held them all together. Redlands' new Parish replacing Santa Barbara as a religious experience! The wind that brought the question of old rustled slightly. What is a religious, spiritual thought, and what is all in the mind? Would we agree on the symbolism of the palm, and be able to help The Holy Spirit carry on with a new theme He desperately seems us ready to start? Is this not only sacred, solemn, yet heart-warming Holy Mystery, that that kind of harmony, in agreement, brings strength, health and peace? Two lessons came to mind whether agreed upon or not, and at this, the wind rustles the left periphery, not only the right where the palm is located. The first is, God equates my father to a gentle king for handling the things in life he has. The palm is like an Indian chief in a general sense of identity who agrees with him and finds no conflicts in an abundant peace between the two, even if their blessings from God are for different means. The new tree grows by the palm that symbolizes all this, wherever that chief may be, and is always present in the yard for this connection in the Holy Mystery's work. It's not time to explain about either of these leaders' work, it's time to read a passage of Scripture to myself, and rest the pen.

High-Brow Catholic Writer? In Today's California Economy?

It's taken some time to see myself as a potentially-highbrow Catholic writer, after some of what mysteriously fell into my journal and essays over the past two years, and editing was my finalistic mark on time.

On today's agenda: (1) I looked at what I have not read, that is in my room; (2) Cleaned out the cupboards and discarded old, extra dishes before scrubbing the wooden silverware door; (3) Got my watch back from Redlands Jewelers, where I simply noticed it is the male in the store I associate with tension, and why, I do not know; (4) Talked to mother about my R.C.I.A. sponsor issue, seeing her divine ability in a sense, to process complex social situations; (5) wrote to Perla, realizing much is to be lost in research if I do not take my freedom to venture to U.C.L.A.

After all this I ventured to Gerrards'. The clientele has a heart though the overall draw has its issues.

Lastly, by God, my resume is a disaster drill similar to my short-term memorization of Scripture. On a positive note, regardless of World Hash-Me-Out Year and staunch, one-lover New York so-called expert literary critics, my angel is back. And not fallen.

Where the Holy Spirit Meets Street-Smarts Behind the Scenes

I looked back at what unfolded today. I walked past three separate sets of dog enthusiast on Highland Avenue today. When a distance came, toward where I cross on San Mateo to the other side of this two mile street, a compact car with two male passengers paused before driving off, a left from San Mateo onto Highland. I looked at homes differently then, for their side akin to remote forest dwellings, as Highland has a large variety of high, thick-trunked variations of trees. A female voice in my conscience, an American Petra Kelley, said to me in a revelation-like, overcast impression, "That's why we keep the book open."

"Is it that through Christ the personality has many facets and that in surrender to faith in God the Father, the soul has many exercises, by the uniqueness of the pathway of historically-set American miracles in Redlands?" The mist tells me anew for the same skinny-legged two miles, each walk there and back to my Acura Legend the same. I must grow to understand the mastery of control of the Holy Spirit over the spending of days' time.

Veterans' Day 2007

What a sermon yesterday by Reverend Jose Goopio. "Marriage on Earth is to avoid the loneliness, yet in Heaven, we do not have to be married…it is God who fills that spot." What a deep thinker, such as myself. I earnestly desire to refute the cluster of issues that stagnates my perception into, to continue, believing an evilness, to the proportion of character assassination in a contemporary dialogue on romance, long-believed the Church's aim. I believe, to fulfill my commitment, that they believe what they say.

What a revelation at Trader Joe's! They think I'm pretty good at the wheel in tough spots, on parking and parking lots…: $11-$13. I debated quite some time in looking at ads today on the new pan for the kitchen from Bed, Bath, and Beyond. I've got my coupon…it should be $15 plus tax.

Precursor To A Hernia?

Tuesday flew by in a blur. I had taken more of a five-hour nap than a sleep, woke, cooked chorizo con juevos and homemade tortillas, digested, and began a part of what they had previously referred to as the dance of the Holy Spirit.

It became clear that the shopping cart-Christian aspect of life stems from those other things that come up on those dance floors. "No, I don't want to speak Spanish to the customer today." Oh, really, did the automated service not react to credit, only when you punched in your pin number? "Next in line, please. (Grumble-grumble-grumble," to the effect of, "What are you doing in this store?!" "I was shopping here.") So shopping and rushing now hand-in-hand, and yes, what on Earth does this have to do with prayer? A clear message and lesson learned from the bank…no Visa payments over the phone and watch your pin in public. That's a necessary day if I ever saw one.

In my status report to Christ on the City of Redlands, between Hispanic construction from other parts of the County, and County leaning on the shopper who supports Hispanic retail, I guess we will not be needing the Humanities Department. The message of this morning seems to be we are going to pretend we did not see this imposter version of Latinos in Redlands while passing through, and hold tight to what we know about the intellectuals they are, through candidacy for Catholicism.

The letter to Char Burgess on art and the beginning, plus Episcopal closure sans fight, sent. Grappling with modern Los Angeles, the angel it can be, communicated in a letter to Perla Hallow at U.C.L.A. and sent on November 13th, 2007, 8:30 A.M. See? It's that drive past the historic library that starts this cycle.

Now, prayer, tension, reading, and breakfast. Then a reminder…no exercise for two weeks, sincerely, Dr. Osmani. Looks like a blood vessel of some sort. Use a lighter weight, later, and for the time being walk and wash cars but no heavy lifting. And you are looking younger, cleaner, and better. Now eat and iron what's left of the laundry. Is that Mother Mary's Eve side, all this?

Triple A Of SC Will Take The Photo

I get the distinct feeling God, Country, Community, and World would like to give me a new passport to life on Earth, once I start accepting Eucharist as a Catholic. Dad is deaf to Christ. He tries, but Mom blocks Dad's kick, then Dad blocks mine. Then Mr. Garcia and Mr. Paul act like the subject is offense, and Redlands asks for a whole new school. That's sports without the ball. We've played like that for eight years.

Images. Magazine photos. Jesus. What is his strategy for the Southland? Horse-pucky field to clean up while proliferating. Gambling referees, gambling neighbors, gambling in-laws, gambling Indians, and now back to the fact this was the 2nd half of the game with no ball. Both halves and a bucket of Texas health issues. Now, now, now…Homily, wafer, sing, and kneel before we do it all again.

Friday's Freedom to Multi-Task Mentally, With Movements' Metaphor

Cirque de Soleil was great! Wow! Get out of town free card from Dad and Mom, instant psychological relief: the acrobatic show with thin yet exceedingly strong women and girls, plus musclemen, midgets like angels, and a whole cast of cool-mist characters ended in a certain lesson: The possessiveness of certain persons in the academic community, off the subject of leadership, politics, and issues. Let us all blast out with full greed from the comfort zone for most American idealism and its results. That was the inefficacy I had in bipartisan communication when I was in high school, plus lessons in other things. If that produces predisposition to mental illness, I have an idea. Gabriel said, in prayer, it's an uphill climb to get out that way, with a facet of conscious feeling like a piece to stillness. Still, it is uncomfortable with the other pieces in resting, and gives poor internal feedback that is from the external while I know that others do not have their reputation and progress, social and scriptural, affected here and now. In other words, they rest without thorns. It was mentioned that by repetition that it should be melody, not disheveled foundation. So, the spiritual misjudgment of others is the source of the thorn, yet if God insists on another reminder, we must just say we're not old but tired, and a better foundation is found through ameliorating our influences' lines, like the ladder of Jacob, so we reach for the next rung. God reproaches the wrong in Christianity, even modernly, and we have our struggles with different religious groups. This is why unity is stressed. For strength, psychologically, the Angels (from Seraphim on down the celestial hierarchy) and the Saints explain in the spirit of prayer. Yet we must return to the moment and work to net love's stronger foundation. Others' insincerity and depression in the past have rocked the boat... must now explain the gap in work history. That's my father's word.

A Boss Like Dad?! You Mean This ISN'T A Recession?

My father, Stephen Arthur, and I got all the furniture moved to Apartment J from across the hall. I have an address change to make at the U.S. Post Office, cable to attach, and my head to keep on until the lease is received in the mail.

So guess what, Mary? They actually thought Jesus was not around, and came up with a huge new strategy, and then Jesus did sneak off, the whole Holy Spirit and all, and apparently has gone to a place where they have no traces of collegiate Steve in the working world whatsoever. So, at the Knights of The Roundtable meeting Stephen Arthur took me to this day, the men who are members said, in unison in the end, "Those who seek to serve others serve themselves." That's me to a T, the opposite of Dad. I just pray, as Christ promises salvation, that I am redeemed from such and do not become a complainer. It's great to be on good terms with a big guy like Dad (compared to bad). We deserve more when we're together, though I'm happy he's had space with fellow, retired U.S. Military Officers.

The Tools and the Tool

To continue, what a milieu of religious discourse. This education is vast. The Holy Spirit seems to want me to play a greater part. The reading must be done early and done well. As far as conceptual discourse, it is being provided for me, not extrapolated upon through me...the switch requires a transition. Yes, I receive some ideas for new essays and art works, yet pen to paper in the room of J, well it has not come to full fruition in transition. Lots to do. A new pattern has not arisen, yet earlier rising and an idea for work have resulted. Brass tacks are the exigency and yes, I am now accepted formally as a candidate for, and noted member of the Catholic world. What all needs to be said is not easy, yet, politely put, it is the car.

Walking Up A Porch's Stairs For Chai

This morning's walk was polite, like a Christian discussion on what my life looks like in the New Millennium. First off, there is the controversial start at Lynn Case's, where her daughter wanted to marry a young man who is not a believer, and they took an apartment at Dr. Zelinski's (the veterinarian) across from my home of worship on Fern Avenue. There is a one-bedroom with a den now renting there, for the sum of nine hundred and twenty dollars a month. A lesson on home ownership was given after passing the place on my walk…it's not easy to make a house payment, so the beauty of the place when coming home just gives one the feeling of having the magic touch(!).

The houses are (1) Continental; and (2) mysteriously empty (This is the next stretch of Fern Avenue). Halfway on to Center Street's crossing point I seem to picture something Catholic about a dwelling or two, and the lungs start to air out well. I start feeling myself again circulation-wise after Thanksgiving '07's feeding. I never had tryptophan last that long. I carried on a length, noting a home that said in front through a small sign, "Some of my friends are flakes." I am confused now with which home Achel Frederick, the Indian intellectual, took me to in order to say hello in 2003. It looked like this one, yet also the one on the corner, two doors down, is the same, and my initial thought. Only problem is an old landscaping truck is in front, not what writer Steve Sabel would own. I do not recollect his wife's name, though thought I saw her at a Footlighter's banquet with a different man. Achel and Doreena now divorced for two years or more, and word was he either returned to India or had posted up in the Greater Bay Area, never to be heard from again.

Ah, only if something more polite could have happened to the Episcopal intellectuals we were. I'll give us that…we may not have had a warm welcome in the Redlands economy, though amidst the arguments we lost not our spirituality (Medical treatment, unfortunately, became a habit of the community, not an absolute truth, so it was often repeated).

The Anvil of the Age

This brings us to my noticing of a conservative dale of a side-street named Michigan Street. I first completed the sidewalk on the topographically-higher part of Fern to Center Street, having started at the corner of United Property Management's other building, a University-image building, of liberal and rugged ilk, smack at the four-way light. Today I had the extra calories to walk down Michigan Street, which will stretch to the back lawn of McKinley Elementary in a quiet, still and green way. I notice in the backyard of one of the homes, where a dog was heard barking, the longish-haired fellow who walks the two small dogs with his shorter female friend. They usually go in the evening. The neighborhood is great. It was after this the story began of my historical yet current dilemma, and now that said, my mind tires of its weight.

Before I get into that I observed Michigan Street's dwelling and spirit as the politest definition of traditional Redlands life, that time and money could afford to bless individuals with, provided they were true to God and wrestled free of troubles. It is a place where, after all is said and done, a down-to-earth aspect of health and proper aging does not let one dwell on what dilemma they were given in their time: A country that replaced culture with issues. It seemed like a humble road for prayer, provided critics, whether internal or real, do not get the upper hand.

Inside Our Next Great Depression

The rain has begun to dot the kitchen window, while the arboriculture in all parts of the seen sphere sways slowly, not as if battered. As I say a prayer to Our Lady and Saint Joseph, I am sorry to say my first day after praying, I was rudely interrupted by my mother, who pretended she came in peace though, in fact, parodies as a Bush, Jr. spy for her own economic purposes. Her mission: to get a wine bottle removed from my fridge. That's got to be about as much Jack-ass ball as any Jew put up with in South Beach Florida. And yes, for arguing sentiments like that, in lieu of scooting the definition of said moment into psychological terms, a woman who uttered the expression "Ugly American" recently lost her vocation. And Now I See… put it another way, that God refuses to disclose Himself, without a price, "without the ulterior motive of commissioning the visionary for service to the whole community."

In a way I believe it, though the only problem is I feel sick when it is put like that. We have emptied the sacks containing previous grains of symbolism to gain a hold on a high mountain climbing, and though secure at foot's crevice standing we must reawaken to the new balance necessary amidst contemporary mission's strong, eternal social current, continuing a cycle conducive to the production of my own aging mother's inner depression. Others, many, falling prey to the said omniscient temptation to ward off said blues by eating out everywhere. Some claim work is exercise, others own the working world and take expensive exercise vacations. It remains amazing amidst all this that a mustard seed of faith, for romance, despite poverty's own chemistry, is protected, if for no other reason than an almost sullen, museum-like beauty. "The man who had several European female companions in his thirties." Hardly the same face it was in the course of American rage against terrorism, yet way downstream to a Roman Catholic conversion the face will smile, though all other prospects seem to wrestle with something invisible at my mere utterance of said and such. By this stagnating scenario's exuding effect for chronology's desires, I feel hollow, only hoping all of the future does not reflect, moment by moment, on a soul gutted by the mentality of the quest for victory in said parts by all, evil and lesser of that, while I sacrifice a bit of privacy's excitement for a choir leader's number and my mother's desire for an honesty I have never trusted my family with ever since certain actions were taken. So, it's not just myself who wakes to this: it's my Lord Jesus Christ, the Above encamping with other Americans like me.

The Evil Spirit Passes

It's always the drinks, and now that's way in the past. She led me past my father to have wine at a recent wedding.

Sifting Flour, Sifting Brandy

I think over-defining the bitterness of rejection on issues weakens. Yes, it's like people in the same boat as mine firing accidently at their own. On the subject of wine, none of this family sees worth a damn since the days I drank in a foreign part. It's more like a Revolutionary War analogy with Franklin departing to France, and no, it does not have to be the meat and potatoes of holiday celebration. Alcohol and medication to Christ may be an excuse for their own mistakes of past they prefer not to process at their age. Faith, and one field we fire in at a time, knowing with Christ at our heart and God's psychic wisdom symbolically sealed on our heads by infant baptism, as such we carry on a reflection of the Eucharistic Lord arisen. I continue in my prayer, for Mary Above to dictate my social agenda, despite this misfiring of both intent and ego (A year later, 09\25\08: It's a war! And we're together while it's fought--the important point, not where the finger points).

The Doctor's Party Game: "Armistice or Escalation!"

Wow! It's amazing, how a few negatives, and the right to play the negative cards, have to stay in place for the rich to enjoy a Christian party, year, holiday, and meal. I think the challenge is to remain true to Christ amidst people's loving to force the negative on our plates, like a life-long plan! They have all the excuses to make faith a game. Church services do no good for some people. So, yes, the thought came in my mind...dirty Christians, not just sinners and saved, formulate the terrain in attempting to manipulate the flow of the river. Yes, all this seems like observations removed from the arena of what will have efficacy in strengthening the Church. The motto seems to be, "If I have the wheel, I direct the dialogue." There's very little room for interpretation of personality. I believe that's the exact reason why I think of old friends at this time, both the American and European, Mexican and otherwise...this time in America's history has sought some incongruent things. Personality for profit, profit for personality, personality and competition, the economic agenda of those with good humor. This ideal has brought more diseased reindeer into a fight or flight display of skills, to get character-assassinated by geneticists subletting from a psychiatrist planning to win the big one on the tennis court this Spring. This is the crux of unemployment, off the subject of the entertainment field. This is what Americans want from politicians-- what they did not get for the past eight years. It became a contest to moderate the extra stipulations of advancement and modernity. Some are now competing against their own.

I believe I struck upon what may seem to be a simple idea, yet a process of which to understand can take quite a long time. Initial discouragement and deconstruction of confidence, to get to the root of imperfections in a person, are rewards received <u>while</u> the flurry of society and one's role in such ensue. When society's often repetitive techniques and\or issues are accentuated in the advancement of life and intellect, a use of extra energy results often in a stagnation of tension reduction in the home. The atmosphere inside won't hold, i.e., the interior is challenged, as opposed to given survival-like chemical, literary, or spiritual rites of passage through time's agenda, whoever is setting it at the time. The dormancy of such factors leads some to reach for aspirin, or otherwise, on a regular basis.

Parable Unfolded

One evening at Rites of Catholic Initiation as Adults, I could have sworn I heard Emma mention something about lots in life (even though she said no, I must have misheard, that I do not have a lot). I have had that mentioned to me before in this area. I believe after a three-mile, post-Mass exercise routine around Highland Avenue and Chestnut Street, I felt the urge to have coffee out for a change due to growth. I have arrived in the third week of Advent season, celebrated as a Catholic candidate, which pronouncedly is a progressive pace filled with events, feedback, and lessons from said deity, Yahweh Sabaoth, at all hours around the clock. Issues to solve in a theological nature come from the inspiration of prayer and meditation's socialization, modern medicine, the classroom and the community.

So it is interesting, that as I ventured into Augie's Café', during an otherwise serene airing out of mind's activities with my previous walking pace at Prospect Park, that the exact nature on one lot would come to me in a subtle air. It lingered for quite some time, and came now today in focus through its desired Christian metaphysical definition. The desire to smoke again not the issue, that is, of the voice of Mother Mary sensed amidst the birds of nature's mist, I continued this subconsciously-guided morning agenda available to my eyes and ears. The subject at the urn for a Coffee of the Day was an Ethiopian blend, slowly sipped with cream added for an intellectual backbone (black coffee is a union member's drink). I started at the beginning with the placards by each large photograph (framed) surrounding the inside seating room. Nigeria, Peru, and India were seen. Muslim-Christian disputing amidst oil-rich, yet corrupt, Nigerian areas had sidetracked the subject of medical facility upkeep throughout the country. In India, a Muslim praying in the morning is the first thing heard who separated by language, spells out in other ways such as his visual image what is felt to be the simple net of our misunderstanding. A Peruvian lady in colloquial, yet somewhat street-smart and fancy attire leads us into the vision of a Catholic country, home of Lake Titicaca, which still involves much Inca ritual in its projection of said faith if I understand correctly. Off the subject of the Church's presence the Incas there today have held to still-fertile roots.

On this note, before summating the brief metaphysical idea expressed to me, I opened Holy Scriptures to Luke, Chapter Eight, verse Four, and read onto the page. The Parable of the Sower describes seeds on the path, or my big city life, which was trampled. It describes seeds amongst the thorns, my former living area in Redlands and my connection with people of my past, choked when the thorns grew. Then the other seed, that on rocky ground such as my relationship with the working world, withered away due to lack of moisture. That seed which fell on good soil, listed in order(1) prayer and its network (2) Catholic life (3) Psychology's conclusions and (4) Talent's survival mechanisms that all come together to support the exuding of a simple key insight; the casting out of the demon which went into the spirit of the pigs in downtown life. This is similar to the lake pigs ran into away from Christ as he healed the sick (universal, not analogous, for all cities). So Emma also mentioned that not only the lot of the pigs (apparently where her discourse led me) yet the scripture which says we, at times, are "fed only the pork" coincides equally. The parable of the sower reminds us to bear fruits through perseverance. That's contemporary Catholic art in a nutshell. I will prepare this way the best I can.

Knowledge Is Power--Normally

I look at how negative I had gotten about the roots I shared in common with the area now, and now I know it has not been easy on all, though other than that, I had not been focusing with patience on the catharsis. Oh, I know the year built up tensions, yet I have to understand that I must be steadfast in my desire to see my family meet the challenge. The homily today reiterated my Catholic duty with a reading from Sirach 3: 2-6 and 12-14. "Whoever reveres his father will have a long life…" SO out of all the potential God gave me for thoughts both captured in ink and free in Spirit, I have to forgive all others to keep negativity from entering the relationship I have with my father and mother, "…and, when he prays, is heard."

The holiday at Felice and Bob's reminded me of the healing from the wounds of the past by the Holy Spirit proxy that if I have a lot to learn, past stigma may linger to prevent such, or…I must do conservative research so that when I reach for more, the facts protect me, along with my will, from traps. All of this is reflection on the evening.

My car is definitely not ready for the commute some are, yet certain things I can still do. If the public is tense as I do them I have to pray, for certainly Saint Anthony could get exhausted on the Southland. I believe I now know why the avenue has been bump and go on my career, to change the subject.

The Riverside Renaissance has begun, and though no complaints have been registered, the city has spent too much. At a time when Governor Schwarzeneggar may declare a fiscal emergency, this city had more projects than Gilberto Gil and John Harrison did for Redlands in the recent past. To find another job is not all… one out of 150 children are now being born with autism in Southern California. U.C. Riverside, in their Fall journal, also announced the selection of an Arab to head their Creative Writing Department. It has been one strange, continuous tradition: it seems of some semblance of a liberal efficacy to draw much of the area's energy for life, and a rolling tide of curves and new connections leaves me still wanting more Scripture for news on where to continue the use of knowledge and of acts. I have to do what Man told me…get a business card. Italian\English as it is, again after all these years.

"Non Si Puo` Tradurre"—It Doesn't Translate

At this point, I feel a lot of people have a better relationship with God than I do after all my devotion, and I'm inwardly upset and do not always feel like socializing. I heard Mary's voice say, "Believe me, I do not want anti-American sentiment to come out." So it is arguably something else that is tedious, dull, painful, and wrong to me. I've been more true to God than the majority. I've had to demonstrate a huge amount of patience for poor taste and manners. War again, and my language's (the Italian) thorn.

"Si Parla O Non Si Parla"—You Speak It or You Don't

A thought on the Communion of Saints came to me: "They would like to welcome you to the travel dialogue." See if there's a book or journal by this title.

Christ With Myself Through 9/11

It starts off one of those days where the ceiling cap is on my inspiration, yet everyone else save the poor are working just fine. A goldenly illumined feeling in the well-lit, tiny studio bathroom seems to impress, "This is perhaps where Christ would not want you to be a complainer."

The drive beforehand to deliver the rent this first month of the new year, at $775.00, impressed upon me that God would like us in the area to start over, in all our diversity, and He does not really mind where, as long as we are just and well-collected. The immediate question, at my birthday approaching of 45 years this seventh of the first month, is, "Does God favor me looking up, to an academic orientation for work, what with my background, or down, as I am starting anew?" At this point, as a writer, I seem to picture God smiling, as He may remember this exact situational question arising when I wanted to fund my first project. The memory of the stress scene tore at me while I fought for spare time to produce story. The tangent Christ and I went on was Judaism, alcohol treatment, identity searching, psychoanalysis, and most-importantly, a modern-day re-enactment of parable, The Prodigal Son. I now have no job to tear at me, though have my parents' renewed blessing, in safety and fairly free, amidst a moment in American history where anything that could difficultly falter made its attempt to, from natural disaster to the stock market, to moral fiber and back to medicine. Indeed, an incredibly challenging time to earn the forgiveness of a man of retired rank as an officer, along with other lifetime achievements, and his wife of 46 years, an equal by the status well-earned of similar career achievement.

To exactly recollect the most interesting of timing in the swiftly-enacted detachments in culture, faith, and hence social milieu is an achievement all its own. The summer prior to September 11th, 2001, I had my one and only private talk with a female of The Cloth on religious issues facing me in my life. An actual authority on ascertainment to receive closure it is to me, the haunting denial of demons trailing me by those ecclesiastical (Episcopal-side) that the wind naturally bends for, or used to. At this notion of an issue arising in a panoramic sense, eventually action-oriented prayer by the disciple amending his nature is received, and followed upon, by the very Savior in us whose Father is all around us, so that a remotely-chambered discourse does not echo through a pardoned mind while trying to "Buck up!"

This...this was the gravity moving in my abdomen like a cannonball on morning walks prior to asking for an M.D.'s opinion on the spirits of Travel Dialogue Past. The M.D.'s strategy: to show me what alcoholics with different backgrounds in the region looked like, and to introduce me to the concept of humility. The other M.D.'s option was a Judaic rental-sharing experience in an otherwise uninterested Los Angeles, and yet a third showed me the men my age who had been jailed for narcotics in a S.F. Valley half-way home, and did we see the two towers fall on television in New York City, all after somewhere in here I had managed, if prior truth not be twisted from root of belief and incarnation I am, to place my first vote for George W. Bush, Jr. I believe it a most difficult circumstance reflecting myself, and often, my region of Southern California, to decide on this leadership. God save me from the controversies it caused. Yes, amidst this grey area of exactly what The President would do with his leadership and the effects it has, I backed out of first the Republicans, then the Democrats, then the Greens, yet lo and behold, a reminder Catholic Cardinals in the adjacent city's press acknowledged better not to vote than even such a candidate for highly-held office allow abortion. What Christ is this, scribbled on the mind of a retired non-commissioned officer I draw on 18"x 24" tablet paper, cross and all? Invoked is a covenant with me often apart from all those Federal, State, and Local, while showing me that their terms, with His guidance, need not be excruciatingly self-defeating for me and others. This is whether the politicians in them have a bond with the human in us often or not. It is the Christ who had sooner told us a dirty joke through another than get us in a beer brawl, that has a humor not just a sacred side, that honors my dialogue in four languages(English, Italian, Spanish, and French), yet honors the doctor as well and must reconnoiter healthier connections at present, solitude or not. See if they can do it while reading an L.A. Times article in the morning about the "Ghost of Hitler" in Germany when Iraq broke out for one and all. Is L.A. free of this too, or are we duped at times systematically by a vindictive sanctimoniousness? The kidney is healthy, the health at time of physical very good, and the blood qualifying to donate, if you want to know the truth, and "he who cast the first stone" may debate health with the next President and all Patriots.

"It's Not Our Life Alone"

What an abrasive though structured 45th birthday. Father Gunningham stated, in a certain way, "Next you'll be a Cardinal!" (It was funny). Yes, use Italian to your advantage. So the only stars you are able to interpret seem to imply.

Anyway, the only check to balance this dialogue is <u>Diary of a Retired Officer's Son.</u>

> "Rather than fall into a trap of depression, keep pushing forward and pushing yourself."
>
> M. Lydia Soltesz

That's what we say to a three-step plan of flu-carriers at Morning Mass, low-man on the totem pole politics in psychiatry and a talk on the Holy Eucharist, absorbed with as much energy left in me to get the cake home without dropping it. A good night sleep, a cleaning of eye's clutter on my desk after a bagel and a persimmon, and with the dishes washed I headed to the Southside: once thought my bane for my behaviorally-challenged past, now my refuge, where I heard the good news. This very woman my mother had chosen and pushed to be might appear on television! She had participation in a senior woman's group that drew interest from a local writer/show hostess. That was serum that diluted the aggression when I received the news.

Cut Off The Rebels At Suit and Tie Pass

(She had taken the front. She missed "Chammy," the family dachshund. Stephen took Chammy's place. For the entire Breakfast Club. What once was a morning through 3500 lbs. of Redlands Underground a week was now a guest-actor training arduously for a private, behind-the-scenes soap opera role as the son who never fully recuperated from the varicocelectomy at Loma Linda University, and walked as if he did not understand the female. Yet for his efforts in courtesy and affection, the Major's wife bought him her version of a male wardrobe. The community did not want to confront the powerful Breakfast Club, not even the Police Chief, so they just occasionally looked in Steve's direction even though they would not lay claim to having a job they could line him up with. As far as food goes Steve's alter ego 'Chammy' was allowed human meals and bathing quarters. Lydia, who would be appearing on t.v., studio schedule permitting, had dated-a-plenty in her youth and worked many years for the Air Force herself. TO her, you did not have respect in Redlands unless you were successfully divorced by the time you were 40. That apparently is where Lucida was stepping in to remind him, 'Chammy,' that there were near-50 in the area from 25 to 50 who did not work and had an apartment his size. Next: <u>Lydia of the University</u>

On this note:

--"My Alma Mater? After over 15 years? What good are you now?"

--"How about a second chance?"
--"I'll take it."

Let's Not Forget It's a Blood-Drive

So, in the homily, actually before, we prayed for those caught in a web of deceit and lies. Now, in the homily, through John, we are reminded that just as he, sometimes we have to get smaller, so he can get larger. On that note I was reminded I must push on, outward, into the city's life, and keep what I know, have faith in what I know and have faith in what is there, as well as how I react.

Then we went to the coffee shop looking for the one ex-girl-friend who would let me off the hook, after the milkmaid ran me to the Community Hospital for a blood-drive.

Environmental Science Near the San Andreas Fault

So God pointed out to me today how nice he wanted me to be to have children. And you've apparently got to give people a second chance. I think if that is the problem, out of some of the things that have impacted life at this age there is a chance we need to find where the breaking point is in the misperception of others and "win the front."

Are there not a lot of people in research?
15% doers, 85% thinkers, as opposed to 38%-52% before Iraq (empirically speaking).
82% of the doers are in the Media.
(note: post-Iraq commence date)
6 professional sports
2 politics
7 medicine
3 accounting, real estate
100% Law and Order

The IHOP Omelet is $9.23
Science Fiction Book Thought: HYPOTHESIS: Yes, empirical feeling of over-all agenda shuffle, psychodynamics affected due to division, instability. Plus, creation of an all new dynamic in hallucination, something crawling in your head:
Alive a thought that is not realistic with a body. The mutated hallucination due to genetically-engineered corn and not <u>actually</u> the fly-bite, but the mutation

reacting to the fly's bacteria when the actual immune system is fine: yet, still the mutated hallucination reacts, crawls, sticks its praying mantis-like tongue into the vacant brain tissue, restructuring from coffee cessation. Surrealistically alcohol-compatible, preferring 75 Proof for spiritual connection to outside the 1st galaxy within the Black Hole's extended, exponential periphery, even in 100* heat, while considering to platform experimentation on planets both much colder and warmer in surface temperature than Earth. Test procedures for formal case studies performed around the clock, with daily symposium for hygiene and carbohydrate intake, including calisthenics and lock-down security measures for Land-Rover functioning to supply the corporal lab. I took the crawler inside me for a walk in the rain. Contrary to contemporary research in environmental studies and veterinary science, crawler crosses cultures with domesticated animals as if raised with them in Assisi a long time ago, during the epoch of St. Francis' more prolific years. On nighttime terms with undomesticated animals and their litters under three feet in length, a separate neighboring colony both on the ground and in the air, filling in to support Patriot armies diminished by the Leftists (For further topic-related research, look under Founders of East Valley Common Law, 1985-2005, as well as Post-Demyrjin Downtown Redlands Circa 1981, in the card catalogue at Armacost Library!)

--The tension in the library
--The cactus bulb, bloomed, of the nopal
--Community Chest High
--Community Chest Low
--Repetition of said dynamic in another community post-grad level

(Your premise as defined in terms of American, not necessarily patriot, is that demons of liberal culture, in their own primitive theology hunting for humor, leave unchecked, yet in the balance, enigmatic equations for the Conservative and challenged. A similar struggle to the American farmer with a cause who is outspent in the business enterprise before the spirit of the land yields its lesson— the same feeling.)

--"Why couldn't the undercover cop figure out the kid was schizophrenic?"
--"Because the kid's psychiatrist ran faster!"

--"What did the kid's Mom say at the Policeman's Ball?"
--"Please stop throwing the curve to do your job until his father loses a few pounds too!"

--"What did the psychiatrist's son study in college?"
--"Astronomy."

--"What would the pigeon on the wire say to the kid if he could speak?"
--"I'm back from flying your complaint to Big Brother."

"The Lord is my light and my salvation: whom should I fear? The Lord is my life's refuge: Of whom should I be afraid?"

Perhaps not the cults, perhaps not the Muslims, perhaps not those that bear false witness against their neighbor, perhaps neither militant female nor ultra-conservative, perhaps not the shady, the shadow, nor the shallow of said gray space. Perhaps none of these enemies can destroy God's plan for me and someone else. Perhaps critics when I shop actually have no control over how I am perceived in the public eye as regards my reputation to work, write, and worship.

"What About the Feast?" "It's About the Lights Now"

And in the gray-skied Winter afternoon, I realize there has to be more hope and options for the devout in contemporary America, especially when they are continuously and conscientiously supportive of so many who have set lofty goals in the global community. They should not be forced to absorb the impact of the patterns to depression, apathy, unproductive comportment and poor psychological hygiene; This all before they alarm the walk, the shoulder, the breakfast, The Mass, and then the job, for the lunch, the news, the prayer, the heart. The soul, the song, and the nightlight.

"Are You Going To A Dentist In California, George?"

This was a Monday morning beginning with a 9:00 A.M. appointment with the dentist, a ground floor office holder, in a modern building two half blocks away from my second-story studio. It is a much longer trip mentally, after all is said and done viewing American politics, Hallmark's latest in The Russell Girl, contemporary University of California and Los Angeles theatre in The Biacciarelli Issue, and last but not least the in-Mass viewing of what will be the design constructed of the merging of two Catholic parishes into one church edifice on San Bernardino Avenue. Was it I who moved the dentist and his staff, his other clientele, or Christ, to change the discussion to daily pattern's priorities for the rest of what has now been 45 years of life? Fight some of this label of having problems in the well-moving trenches. So to quote this anonymously of things like ivory powder and some sort of gluey substance into a paste, to keep a temporary filling in place.

"I have an alarm clock in my mind that goes off at 5:00 A.M. and am up after that. Also, come 7:00, "I'm out." It was a bit earlier than an O'Doul's and meditation evening, whatever time shut-eye is for my neighbor in the business community. It would be a long trip into a most painful modern Redlands conspiracy role for the U.S. Air Force Major's son (in what author Joel Rosenberg contemplates is The End Days, to tell you that the inspirational thought coming to mind in the dentist's own voice after the appointment, young and attractive assistant and all, was, "Don't count yourself out on bearing a child that will be a doctor or dentist and save your drawings until you know much more about art, without rushing). All that cache' for not flossing this temporary filling for two weeks! I think I'm instinctually trying to tell God I'm the right man for the job with my enthusiasm, attitude and potential to iron out the weak spots in my faith argument.

Another voice said, "You win a convert like that dentist and I'll knight you!," yet these are more random thoughts that bounce around and off me, from a brain storm that is traditionally as ecumenical as Mother Nature, and today, it is a gray sky. I had the sound of Mary and Jesus weeping, the rain on my window His tears why God the Father set me aside from normal, everyday fare. So if drops fallen from above were what woke me, before I felt a little nauseated to address the dentist's ambience, and rested another hour, moving on schedule through choppy, psychological braking and accelerations, That's truth. That's hard work. That's Redlands, not American or the University. Tie or not, that's the Christian I hope I become.

As far as ancient thought goes, the message was, "We are going to climb until we see what we want, and we've got money. And it's here for the time being." This, this, this is where the S. Holmes in us all leads a silent protest of those maybe not criminal, yet organized all the same, taking inventory when it's time to support the Househusband Olympics. Good title for a movie.

On the other part of it…My resilience in the present studio ambience is the weather change. A Capricorn in denial of the world's end piths a blind ignorance to failure, as if the entire wool on the Lamb of God sacrificed Sunday was a fable, and he had to see the strategy in the new stars to conquer demons risen from the new excavations in the depths below sea and land both. It may come as a migraine of temptational personality in the dark of the night when the temple it was blessed as at birth looks from within to the what? Friendly wind in the tree on the outer landscape, despite wars in corners still digging for the extra tip of silver I gave a Seventh Day Adventist today for working on a hidden black cavity. Yes, Eucharist, taken, or candidate's blessing received, is the art of The Word to live on after Mass(The art. Actually not rich, thought it's art. I won't treat you like you're disabled if you get up.).

--"It's a bottomless pit."

--"Well, no, not actually. BUT, it is deep down there. So the strategy is to drop one of those large ham sandwiches down there, so that whatever is causing the internal turmoil spreads out."

--"Curiosity killed the cat."

--"Then hire someone to watch the new cats, or get a volunteer."

--Southern California: An oral tradition on the subject of cache' that works like liposuction for some and Bulimia for others. Here are USC and the new generation of Lakers to provide the balance.

--"O.K., George, welcome to civilian California. The guys driving trucks in the suburbs during the day are siding with the Mexicans for propers, whereas I'm siding with the Church and Mexico, for religious reasons. So, it's not about business, least of all oil, agriculture, banking, politics, or the media. It's a daytime American something or other. So, war. Good. I cannot comment. The something or other keeps us busy. God's move. He said some are not looking at the ball when they swing." (Or, I would guess, this is a devil that followed the trail from your delusion's attitude and words. Calm. Quiet.).

"Valentine Call You A Serious Lover Too?"

So, Lent is not about purifying yourself medically: It's about purifying your conscience. Then you will be able to read better and more fluidly, your coffee will create the chemistry you desire for writing, and your day will not expand in anxiety.

And then I saw the devil's hoof imprinted on my mind, as if stepping on my head while galloping by the faun with the goatee' who was in search of Valentine and his devotees. Satan, from yet a different and still not Earthly realm, tried to rutter his power by steering an up-turned rib, reminding me in the wake of his failed attempt not to let him steal mine. My left side ached from the onslaught of
the temptation, the smoke in Lucifer's anti-incarnate being clouding for a moment the impression, the sun's love high above lightened on the kitchen seat where I lay. My intellect had snagged like a bone fissure on the tense drama as the images began to fade, The Prince of Demons' voice telling the faun, "Leave him alone." And only thus did I breathe anew.

"We Know Why You Like It Quiet"

Quite a decent, well-breathing mystery it is as to why the public me seems to have all the bases covered for a clear conscience amongst his own Americans, regardless of race, yet the private mind in the dwelling abode...without the sun

110

and a fresh breeze, it's almost as if eminent danger faces me before the breakfast meal unless I escape out the door by 10:30 A.M. I've been cautioned repeatedly previously from seeing this as the fault of the establishment, <u>Maryvale</u>, itself. I've always had the caution of my parents from shifting the blame to the residents in other abodes as well. God, according to the by-laws of belief, is present both in public and private, from whom to seek counsel. Holes in my head in which beings of undetermined affiliation dwell seem not to have much to say when the subject is the park, yet as usual, in private, the very thick of the Redlands conspiracy against a few of us comes alive, evening for O'Doul's or no. It is not possible with the naked eye employing its empirical observation experience, as a people watcher who has crossed two other cultures, to come back home and deduce who is in the know on this issue, so for yet more time I live alone. It also is a touch-and-go scene as to whether or not my parents have details on this secret network that keeps some Christians two tent flaps away from being Saturday's sacrificial roast being for the economic future of the community, in a vain attempt to right even the abundant and historical of wrongs with the Messiah!

A scout asked me why there are holes in the tree. I replied, "There probably were limbs there when it was shorter." This was curiosity's most far-reaching attempt at information for the day.

02-14-08 Valentine's Day: Progress Goes In The Pidegon Hole:
The Day Is For You To Observe For Others, Not You

The Holy Spirit whispers thoughts ever so beneath the surface of the connection between what image the eye sends, and what the heart absorbs. The heart is not weak, yet distant, and to the eye the details have been immaterial, according to what dialogue the soul originally had with the heart and Beyond.

"You're stubborn about this behavioral cure for mental illness, as well as for art's potential and truth within to set you free."

"Every time the subject comes back when others repeat their actions with a new presentation, go back to this very spot, internal anxiety based on a trigger or two that connect in the same time frame of a day's slot, even though I do my part, to return to the era in the sun that there was the initial confusion. It's as if you want me to repeat the down, the 'extra test I took for your social pleasure.'"

"So maybe that's what the doctor means by what you know worries you more than others. You're afraid that even if you test differently, we'll still react indifferently. This fear...wisdom, guilt of sin prefacing the exact weakness noticed, or God's humility? Yes, all this must be digested before a lesson on the Beyond for <u>you</u>, and not for those we moved along. Call it what the Church did...a transition, and new common ground."

"Let's hope this new common ground is able to rid us of these feelings of martyrdom, and not exhaust in the process. What a turbulent process from one passage to the next. None of these explanations deal with social capital."

"Well, that's a lock. A Deadbolt."

--"Train yourself, you too, to let go of everything you fear to lose."

In meditation on Holy Mary, her voice came to me and stated, "No, they don't want their cake and to eat it too," speaking of people from my Redlands youth in athletics and education. This includes family and the strategy exacted to move on from misdirection. They call it a trigger when a childhood neurosis is re-enacted by a photo and an article of childhood acquaintances' recent successes. There seems to be, in accompaniment of current anxiety factors, a psychologically internal, which coincides with a physiologically preempted, occurrence resulting in a sequence of thoughts, which perhaps produce audible sentences with the semblance of no realistic correlation to moment's scenarios. What I have noted after several attempts, over a roughly 20-year time span of psychotherapy with seemingly well-tuned professionals of both genders, is that once placed back in the "off-the-couch" environment, the pattern for preempting occurrences remains the challenge (and seemingly a psychobiological reference point in the mind is the very raison d'etre of the latently humorous prayer time, to reduce the anxiety late in the morning). The gene for mental illness looks up, not just down(09\26\08). This is where the right mood scrutinizes the foods and beverages that go in the body, in enactment of a culture's mentality and faith's ritual, rendering itself paramount.

--"Married?!"

--"I dated a foreigner a few years back, but no."

--"Well tell the whole town!"

--"That's what they pay my barber's daughter to do!"

--President Obama's wife made another faut pas of sorts yesterday. She said those seeing a psychiatrist with no legal problems, that are old enough to vote, will be considered her husband's "Lucky Vote," like the last cigarette in the pack.

And WILL Keep Going Around—In A Good Way

....Voted "Most Ignorant Concept of Romance Leading to Children With Behavioral Problems"(Pick a winner in your life, any winner!).

--The Grinch's resolve was to come clear with me on his raison d'etre for moving to Redlands: "I'm sorry," he says, "there were so many errors on behalf of those who have known you over the course of your life, I thought I would offer you my opinion."

112

So, Choir Practice at 8:00 A.M. Lisa is 32. Her birthday Mass afterward at 9:00 A.M. The homily dealt with Christ using mud and saliva to heal a man who had been blind since birth. Father Jose` brought up how animals lick their wounds, and people do that to their burns and cuts. Saliva has been believed to have antibodies.

Also, I got the humorous feeling that some of the mentality behind this city stems from Law Enforcement and their acting ability in being problem solvers! (Maybe that's part of my problem with unemployment too…professional jealousy). Enough, the town had an amazingly bloomed feel to it as I walked out of Mass toward my car, driving through much of the West Side and noticing how it, too, has both a historic and modern pattern to its living. I am hoping to be awarded for progress in the pace needed for maintaining both characteristics, by (1) the healing of my mother Lydia; (2) the good and clear flight pattern to and from Texas; (3) the joyous get away to see my relatives; and (4) the return to a good 75% chance at the Library Page job without any psychodrama involving Eric Weck of The Quality of Life Dept., in order to get the beginning of a new writing project off to a great start.

The idea for such is a love triangle between two Junior Law Partners, one who eventually marries the physician's assistant when she's clean from drugs. Her brother is bound to catch on if they don't trick her parents, during the season of their normal arguing (Fall), into believing schizophrenia has set in on the now twenty-year-old lad. He is hallucinating, fallen for this reason from the religious path and no other, and you cannot believe a word he says. The away field from the city of their upbringing, for the wedding and reception, is where his signs of social difficulty begin. Bio-medical puzzle, schizophrenia? That way every voice that's in the set-up gets a word in when the deed is done…and it's not dirt cheap.

--On the subject of mudslinging between the Parties and the brilliance therein, after 'refudiation' got into the New Oxford Dictionary, I wouldn't be surprised if this turned in to a discussion of "Space Poop" launched by the rival!

--"O.K., Crew, the Bad News is the priest said we are strongly urged not to go out with people on the work staff…the Good News, besides Jesus, is that we caught the priest giving a pro-Prop-9 sermon on the other side of town, and he's in deep trouble with the Bishop!"

One And All Putting Feet Down

I need to more effectively learn to use my sixth sense, so that when it goes into application or is triggered, I don't nervously lose strength. The adaptation is the fruit of experiences I had growing up in the neighborhood. Phenomena, through prayer, is a contemporary situation and definition. IT has to do with the present.

More understanding of eco-psychology could be a bridge to further adaptation, i.e., strength retention. It is men of said nature who have been strongly marked by maternal instinct, yet are not the category of feminine. To further adapt the connection while continuously cultivating and controlling the masculine gender is a foundation topic in sixth sense control.

Due to nature it is either abused horribly, or used with a distinct nobility. It is the man on God's side of the war in good vs. evil summarizing his use of collective culture in an exterior, social application for promoting many esteemed qualities in society in an efficacious manner. Men through marriage learn to master this at its source to create a harmonious family unit, and single men learn to be elastic amidst a combination of immediate family and environmental issues, an equally important role in society only lesser repute for its sake of singular productivity. To better help the cog function is an esteemable task of married men and women. Through endless bearing of a testimony of maturation to youth, and while in many capacities, volunteering is not necessarily a timed designation, it is God's pathway, step by step. Christ as mediator and advocate in conjunction with the tutelage of established parents and grandparents may allude to the fact that America has not so much as lost appreciation for all things familial, yet is threading a needle amidst a national call for unity.

In said manner tensions ease without imbalance of the scale for haves and have-nots to render more in accordance with their own through patience, love, respect, humility, and new-found fortuitousness in the faith aligned with God's roads. As is pointed out in Pope John Paul II's <u>Humanae Vitae</u>, said instinct or "dominion"(p. 6) is not always man's to control, for its grace comes from God, who is "love."

Recreation of Patriots

There is much vindictiveness in the world these days. The Armed Forces have an expression, "Freedom is not free." With all this, what is freedom, if one paid the price in American society, barring the worldview issues we see our leadership intervening on for us on when the news is cast in the media?

(1) Safety and security in one's abode; (2) an esprit in such that is uplifting for our thoughts and worship, without the Devils' tension interfering with our very saying of the rosary; (3) Unobsruction from a healthy life, and concentration powers enabling both our pragmatic and creative instincts. This is freedom: Love is when we share these precepts with others, free of boundaries, such as at a pot-luck or bar-b-que (Do you cook for both? In other words, has your leadership been expected in both quarters?)

When boundaries are respected healthily, both freedom and love are due for arrival to be fostered with American craft, one sure in its old traditions and historical fiber's culminations. To go on is to notice loving thy neighbor as

THE CROSS AND THE CANTINA

thyself is often sacrificial during tough times, as reciprocation may be evaded or neglected, condescended or conned period. To lay to bed and rise through this day in and out is the modern martyr-like light of the contemporary American Catholic, as Christ's fondness of The Church's attributes that have survived grows, more and more. They come to the Catholic altar to begin their walk in debating whether these are intellectual or moral issues, issues of money or issues of repentance, scrutiny and inventory or fest and the Arts, divinity in the latter or shouldering the burden of the mind to yet be educated and converted in priority of said time frame.

Can one angel take these ideas under his\her wing for a Diocese in entirety to manifest altogether through regular obedience to rites and rituals, Holy Days and Seasonal adhering? Or should we kneel to seek intervention from an entire order not of monks, nuns, and friars, but of angels themselves to guide our mast through the eye of the times to again discover our Plymouth Rock at Thanksgiving time, to share and survive, showing our thanks by the very act of our private gatherings, greetings and service?

Angels indeed. I often get the feeling there's one in my room, who has in her power to free me week after week from obsession with my sin and weakness with her blessing, to guide me through the toughest of times of month in and out, yet cunningly tease and then giggle uncontrollably when my face shows my want to be taken with a new seriousness.

Who the voice is in the back of my mind at Mass when the priesthood talks of the corrections in behavior we must make for fear of God's grave wrath, I do not know, but it is a strong, enveloping, not heard yet loud in mind, as if I have not lost connection with said angel and she is in an array of disbelief as we go through this again. She lurks in invisibility of sight, perhaps yet present to the senses, back home in my somewhat tidy abode. I dare not say to the wind, "What might I call you?" Whatever it be, someone said it, and she found me, now tucking herself in as I turn on the heater, preparing this week to set up my Christmas tree.

Europe Never Changes Either

--"Guess what? The 'Ghost of Christmas Past!'"
--"Yes, got it. 'Closure' in California comes with an early Christmas gift of Pepperidge Farm Cotto Salami passed to the closers."

--"And the University this year? They hired an architect to build a leaning dorm for Family Student Masters' Candidates on the Foreign Language Track!

(The Café San Bernardino Conspiracy Papers and the Long Arm of the Hungarian Right)

--The Longest Discussion in the First Week of the Exchange Student's Year Abroad (that got cut short eleven months because the student got sent home)

Student: "May I go to town tonight?'
American Mom: "I don't see why not."
Student: "Why not see?"

--Did you hear the one about the guy who holds three jobs so he can have homemade <u>menudo</u> every Saturday, at the same restaurant he's had it in since he was a <u>baroncito</u> (young heir)? He's not connected—he's <u>muy popular</u>`!

--"Childhood Sweetheart?! What does this have to do with it, Doctor, I haven't seen her for thirty-five years?! This is a dinosaur of a psychological theory—Einstein complained about these Freudian bourgeoisie tricks!"

--"Because I'm the girl's father, when it comes to dinosaurs. I'm the real McCoy, and I'm going to eat you, Caveman, apartment dweller! Grown-up? Back in your cave, infidel!"

--"And now, Hallmark returns to the conclusion of <u>The Boy Who Would Be Dude,</u> inviting you back next week for the season finale, <u>The Legend of Pancho's Peev</u>!"

--"I believe I was in the middle of a more appropriate intellectual conclusion when I inadvertently got hooked by the Democrats' new one-hour series, <u>Colonoscopy Squad</u>."

--"What else are women looking for besides an animal in the sack?"

--"A Martian in the sack."

--"Be careful not to wake the Dead this New Year's 2011."

--"I don't give a damn. Got news for ya'. If they're that strong a spirit, get a uniform on'em and have'em go protect the country!"

<u>Prayers to Saints and Products of Such</u>

I seem to remember an actor, Keanu Reeves, talking in an interview about his study of Buddhism, expressing it's fascinating, though actual conversion is what he call "The Golden Noose." This is applicable to aspects of our culture as Americans. The practice of medicine in healing mental wounds, (this week's battle scars) through Modern Psychiatry, as practiced methodically and institutionally in America, can have social and economic disconnections that make the expression resonate. Many diagnoses can be avoided, yet we have a high degree of naivite' on the subject of the abstract, or the abnormal psychological field, especially in the field (if this isn't a straight line drawn, Hollywood action movie plot-point style), and many Americans are on a much different face of the mountain climb than female abuse when diagnosed, both in a domestically-defined and Freudian (or childhood-wise), worldview-defined cultural context.

Many practitioners keep it simple on this topic, that the Gods have interjected with a world religious paradigm for their patientry. Psychiatry is our class on the topic in a coated medical vocabulary. That is what many feel is "over their head." I refuse to surrender my flexibility with cultures from distant places as an exchange student, I do not have a mental block on adapting my pronunciation nonetheless, English is now not defeating, yet potent: A strong stout of a tongue, and America I am proud to be a patriot of in this superpower tactical struggle.

"Major" as my father was in the U.S.A.F., is not a word denoting rank in the military: it is twenty years of a family whose innate loyalty and discipline may be foreboding to other cultures or professions in the same domestic paradigm. When we have a disconnect at retirement, followed by inadequate suggestions and behavior in laissez-faire civilian cloth and companies, we take hard measures such as psychiatry to maintain our balance. However, psychiatry is not the temple in which we were baptized with the power of the Holy Spirit. The world religious discussion is not about a political leader's trip or discussion alone: it is about reflecting bedside-manner in a way reflective of the greater whole that the entire patientry represents as community microcosm. I think, in what going around coming around, what one psychologist irked me with in describing this as "our own culture" has now a more luminous connotation. Research scientifically is done at the University on mental illness: it only took fifteen years, not twenty, and I have handled the ball, not doctor or Dad, on democracy. Some feel we shouldn't vote: my sister was a more polite woman when on her period, so this fact too, takes adaptation, and to answer, no I am not gay. So, in memory of those American military and their remaining families slain at Camp Hood.*

*(There is no correlation between schizo-affective disorder and PTS Disorder, let alone other psychiatric issues facing those who have seen action, in a medical context, other than the environmental factors in an American upbringing and return from abroad). I honor them, here by letting their descendants know I have medical help of other faiths as well, that is quite healing on occasion, loyal and in

117

a different state of Democracy and medicine as they continue to abhor terrorism. Thank you for serving our country.

*I have yet to revisit Shreveport, Louisiana, where I was born, since my early childhood, and culturally, as a schizophrenic, apply what domestic culture is appropriate, flexibly, at any given moment, though yes, predominantly, am considered to be Californian. Many Californians don't know how to sheath it, at times, once they've drawn it, though, as you know, the mentally ill are not allowed to support the NRA—those incidences are the system's mistakes. It doesn't mean we're not Republican despite this law, though want to encourage some recognizance of how to let the dove fly when we walk at night downtown. Therefore, we can take this, finally in the true nature of a Catholic convert, out of the soap opera of the moral discussion, and embrace a more adequate academic tone to ourselves in the American Catholic constituency, a flight away from the Spanish Steps with credit card provided by an American bank, or we can keep this in a simple ecumenical context of Satan's disgust for the chemistry of slow-release tranquilizers provided by an ethnically-diverse psychiatric field right in the heart of Southern California, only forty-five minutes from sacred San Manuel Indian burial grounds (not tourist attractions), located in the mountains of the Diocese of San Bernardino.

That's my tongue, not my temper, and that's my blood. That's Christ's blood. That's my Master is resurrected and directing me (I don't have an overriding temper, though yes, am sensitive and feel psychotropic shots destabilize more aspects of mental problems than simply covering up buttons people push—they're an oil change for a college-educated mind that is prescribed by a licensed professional of conservative nature, and this particular medication is slow-release, not affecting motor-ccoordination, in fact, due to anxiety control, aiding it with everyday bounds).

Sagrada Communion. Holy Communion. Congratulations, Spain, from California, for moving the entire contemporary world on with your victory in South Africa. !Que Dios vos bendiga (May God bless you)! And were serving for at Camp Hood? Did you have carte blanche to embrace your Battle of San Jacinto, as my uncle did, killing Nazis in World War II? Do you prefer Texas—is California off the mark to you, even though she, too is alive with your pride, and can be trained? Did I hear correctly, the traitor had been penalized?

My mother will sacrifice me, the Isaac, to keep the Muslim physician I have now got giving me my shots as the true friend he is—he's married to a Christian woman and is a man of peace. Mom—her olive-green eyes have a Tex-Mex temper in the sun when I am given instructions on clearing the patio so the plants can breathe: She served the Air Force for thirty-four years as a bilingual civilian secretary—Que Dios te bendiga tambien, soldado(May God bless you too, since you shot a round, and we will take the information given after!)

So, on December 30th, 2010, as I head to sleep after writing and drinking the coffee in California, I have asked my patron saint, since my guardian angel and I are tired from a long walk into town this evening at the store, to take over on the prayers to reach the Heavenly Father. This is what San Antonio came up with: *En el nombre del Padre, y el Hijo, y el Espiritu` Santo, con la bendicion` de la Santa Maria, Amen. Padre nuestro en el cielo, que das de las bendiciones al Presidente de los Estados Unidos, nuestros soldados en Guerra y vueltos, y sus familias, de proteccion` y salud, que puedan pensar claramente. Tambien queremos bendiciones para aquellos de otros religiones en America, que queri`an fumar tobacco con los Indios, hasta que termine en paz nuestro vicio. En el nombre del Padre, y el Hijo, y el Espiritu` Santo, Amen(In the name of the Father, and the Son, and the Holy Spirit, with the blessing of the Virgin Mary, Amen. Our Father in Heaven, may you give blessings to the President of the United States, our soldiers at war and returned, and their families, of protection and health, that they may think clearly. Also, we want blessings for those of other religions in America, that would like to smoke tobacco with the Indians, until they end in peace their vice. In the name of the Father, and the Son, and the Holy Spirit, with the blessing of the Virgin Mary, Amen).*

--"Who is it?! I'm tired!"

--"It's San Gennaro, Esteban, the Patron Saint of Naples!" He's visiting me on surprise! He has a prayer to say, on my behalf, so that the Bishop who helps Pope Benedict understands the time zone!

"Eh-em, Nel nome del Padre, ed il Figlio, e Lo Spirito Santo, con la Bendizione della Santa Maria, Amen. Padre Nostro, fateci il favore di communicare che Steve e credente fino a questo giorno, senza parlare delle tentazioni ai suoi amici Italiani, che 'ha visto Napoli e sia morto,' come va l'espressione che conoscono in questa citta` sud di Roma, e che questo diologo non e' finito per l'eternita`, per lo scambetto che Steve ha vinto contra San Manuel per le vostre alme Napuletan'! (EH-em, In the name of the Father, and the Son, and the Holy Spirit, with the Blessing of the Virgin Mary, Amen. Father of ours, do us the favor of communicating that Steve is a believer in Christ even to this day, without speaking of the temptations to his Italian friends, that he 'saw Naples and died,' as the expression goes that we know from the city south of Rome, and that this dialogue isn't finished for eternity, because of the bet Steve has won with San Manuel of San Bernardino, CA over our Neapolitan souls!").

--"Oh, I don't know Cousin, I think our cousin on yonder was right—maybe a girl with a little more SPICE."

--"Oh yeah, I believe we have a consensus for the time. I'm glad I have your support on this decision."

--"All break-ups are F-I-N-A-L, as we say in the South."

--"Uh, yes, that's the Last Call For Alcohol, that's for sure!"

--"On our end, yours truly is waiting for her to finish her first beer, so we can get on the road to the Family Reunion in California. Get some rest."

--"Are you kiddin'? The only job out this ways is the changin' of the seasons, and I think Jesus got that one."

--"Well Jesus is the reason for the season! That's why, I understand and I'm sure He does too. Volunteer for Him and we've got an idea we'll tell you about promptly on arrival!"

--"You mean it's a SURPRISE?!"

--"You won't even hear us comin' in. By the way, what seems to be the main environmental factor in California that is so testing?"

--"The plate of thirds with the early bird's worm on the side."

--"Why don't you rise a little earlier if you're hungry?"

--"There's an old Indian expression if these parts—though I'm not sure which tribe: "The first one to take a morning poop is the last one in the buffet line."

--"Flaco! Be nice!"

--"It's a Brand Moo Day."

--"All right, you two!"

--"That's a Freudian slip."

--:"Jesus designed it. Like it?"

--"Out here they wear'em with a little higher of a hempline."

--"Are you sure she was that bad?"

--"Do you want to change that to, 'Live the American Drink'?"

-Why don't you make a cup of coffee?"

--"Because she hasn't left and she's usin' the coffee pot to brew some Evan Williams."

--"Does she sing?"

--"75% of the town has a gifted voice."

--"Are you trying to tell me you're the skinniest man in town?"

--"Damn near. No seriously, how much do you think I weigh now?"

--"I already know. I got the information off the web." At the U.S. Census Bureau's site.

Thank you for cooperating this time."

--"They really haven't stopped talkin' about it in L.A. for some reason."

--"That's because you said the nasty word."

---"Census?"

---"No, U.S."

---"Oh, they're not bad. Aren't you thinkin' of the United Nations in New York City?"

---"I changed my mind on New York City."

---"Why?"

---"They won't grant me amnesty."

----"I thought you liked the South."

---"I'll leave it to your imagination."

---"Those beers…Seems like you're trying harder to see how long you can make them last than it is to actually drink them."

---"If it's in the fridge any longer, I want you to sit it on the wine rack with the vintage bottles."

-_--"Now that we've got, 'Lead us not into temptation,' how long does the crucifixion that followed this abstinence go on?"

---"Oh, really? Those beer terms? I see you're kissing for the bucks, but you really need a hug!"

--"Oh, was that a couple different movements and a few faculties holding court before the game? No, that was the L.A. Riot before I gave the password in Italian and received probation from Uncle State Bear!"

--"Are those the Lakers fans now? My advice? Wait a year after getting your tattoo so you can be one of the 37 percent of Americans who have the genes to donate blood to the military locally."

--"The difference between a blood donation on the Cost'Amalfitana and the one in San Bernardino: The Italians said your change is a croissant, S.B. gives you a sub sandwich. I'm American!"

--"Yes, if you don't mind dinner and drinks at your place has been great, but I don't think I should drive. Do you have an extra blanket and pillow, even though I don't want to spoil this by rushing into sex?"

--"Sleep in the tub."

--"The tub? Why?"

--"It's safer."

--"There's a bolt on the door and no one but you and I. What's the worry?"

--"I'm about to drop a 5-gallon hormone bomb on you."

--"What's in the well?"

--"Oil! Black Gold!"

--"Oh. We'll try the pub with more ties to American companies."

--*"Oh, Gran Britannica, si fatica? Si lavora? Bella, questa economia. Ciao!"* (Oh, Great Britain, workin' hard? Labor? Beautiful, this economy. Bye!").

--"Yes, it takes more than a good tub-bath after complete immersion culturally to snap out of it, but thanks for the advice, 'Nan'!"

Well, I believe I see it, clearly in the dark of the night yet. At Mass, while the Celebrant watches the center of the aisle with the bread of Heaven, server #1 goes to the left with the chalice. Server #2 goes to the right with another chalice, and as the young married couples with children cheer them on against the sluggish defense of the sinners, the entire inner city team scampers down the right side of the periphery to the psychiatrist's, whereby they fill out the paperwork to get a reality check once a month to compliment their reward for honesty, and a second check from Social Security! Thereby via proxy the largest county of the Continental U.S.A.

This particular aspect of Pancho Villa's legendary strategy, uncovered yet not in modern times again, is actually quite dependent as secret held only by those who buy their frijoles at an undisclosed location! For after properly removing stones from beans, a good long soak followed by a change of water for boiling (bacon, fresh-chopped onion and garlic salt added if so desired), the pinto beans will talk! Apparently, not tooting by modern agricultural methods, as in the days of yore: Apparently, now a method of covering one's back that is capable of throwing the fight either way it pleases, much more important than eyes in back of the head to the street-savvy.

Villa enthusiasts of our time can sense one such as I approaching, though now that the principal allies of good are keen on new goals, a saga for new legends superseding such likes exists. With this is a distaste for the American doctor's influence (in this call for repentance) is held in the legal binding as a definition of the term "confidentiality." Hence, no, despite the calling of those influenced by the writings of Octavio Paz, those intent on ending the Villaesque Era will not "take off the mask." Holy Communion, Fart Knockers et al., yes, Rock 'N Roll in fact 'Knocking at Heaven's Door':

Let's see, Haiku…
> Crazy Man
> Smokes mota
> To travel planes
> Learns lesson from Don Juan

> Returns to Earth:
> "Where's home?"
> Misses nature,
> Hikes far
> Drinks beer
> Packs out can
> Recycles at home
> New hike lasts seven peaks.

--"Father, I have sinned."

--"What is the nature of your sin?"

--"Original sin. Wasn't going to mince words about a card table in the sky with Adam, I probably would've taken a bite of the apple myself."

--"I absolve you with these conditions for repentance."

--"Yes, Father, what might those be?"

--"Eat another apple from the market and get out of my town."

--So the senorita won't date you?"

--"No, she says next week is going to be a long one so she's staying home this weekend—just going to watch a little Saturday Night Live and get ready for afternoon Mass when she awakes."

--"I see, that's how high the social I.Q. goes in this town because of the translation."

--"Think I should try again?"

--"I don't know, what's the last time you could fit in your ballet tights? I think it's a safer bet to go back to professional dancing!"

--"I think she had a word with my sister."

--"Your sister doesn't live here anymore!"

--"She says she loves her new car and only has to drive through L.A. to say hello!"

--"Is that when it's personal?"

--"Personalized. No More Men."

--"Sounds like you're up against the odds."

--"I fare well at Bingo but that Pechanga curse won't quit!"

--"Now don't drag your tail!"

--"That's my shirt. It got a little warm when I came in at lunch, remember?"

--"General Custer never said this country was one tribe at a time!"

--"Don't make it look like I'm related to him!"

--"So you're takin' it pretty hard?"

--"I'll have a double of some of this harder than this beer—it's last call!"

--"Okay, Whiskey Pete, don't bite off more than you can chew!"

--"That's comin' from you?! The President of Overeaters Anonymous in these parts?! "Un-fair match-up."

--"Okay, I'll watch the rumors going around and maybe you can have a second chance."

--"Never mind. I can't stand Saturday Night Live!"

--"You recognize this guy, Captain, in the camera?" What about this guy? I think they're a group of men hidin' out in these parts plannin' a bank robbery!"

--"Each of these guys is drivin' the same car!"

--"Yeah, that's the get-away car—they're bein' sly, sneaky, and all those other names mothers use for scouts not livin' up to the code!"

--"It's the same guy in each clip! He's just changing his clothes each time, meathead! It's a 110 degrees out there and he had several trips into town to make!"

----"Yeah, I'll say, I think I'll go home and get a shower. I've been on him since the early A.M. Patrol!"

--"I think I've got an idea who's planning a robbery during tight times that you haven't seen in a film clip this month—your dry cleaners! You stink to high hell and back!"

--"Did you play college ball?"

--"Are you kidding?! I didn't get away for a single sporting event!"

--"Don't complain—they didn't get us to a single play-off until we got out of college."

--"Walk-on?"

--"I think I walked a little farther than the guy they gave the free car keys to at that!"

--"How did they miss you in the draft?"

Catholic Life In the Ground War

Men and women do not support the Catholic Church in America because other of this nation's interests (quite possibly on all three levels, Local, State and Federal) have discouraged their employment in a realistic and viable capacity. They then award their time, thoughts, and passion to that search, often carrying the burden of conscience and anxiety of more than one side of the coin in all its metaphors, while living an upright Catholic life without volunteering their lives, spreading Christ's message, or allowing their home to become a happy conduit of God's love for self and others. Solutions? Common sense, direct answers? Or a share of The Holy Mystery on all three levels, regardless of cash corner or economic anchors?

When the anchor reaches its designated location, why does a certain type of woman (type or two) keep showing up in their life? Some are part of The Gospel's analogy, and other are a part of The Devil's Paradigm. From video game to attention-span training with Java T.V., keep the mind sharp while they slip through the side door of modern entertainment's stage and get a few words in to the world at large. "NO more Guantanamo Bay!" An esoteric subject not necessarily on the mind of those representing a role choice in all three levels of America's culture, and knowledge thrust when the new week is more between God, County,

Mountain, and Man. Now, key #1: If you cannot move to another county, do not want to change God's, and are stuck to the set at the sight of mountain-climbing alone, just be a man on the move. This way critics at least do not lose interest in themselves, as yes, depression is fluctuating upward while the economy fluctuates downward, in the pattern of the rise America's real enemies get when they hear, read or *see* what we are going through now.

As Psychiatric Chambers Keep Their Own Score

The 9:00 A.M. appointment this morning was a clear, fantastically balanced view inside the potential stability of modern psychiatry and the resolve of issues on a client's plate as well as such to actually search for the solution to symptoms' triggers.

This was a professorial meeting of the minds where two key issues, completely off the subject of the vast religious, behavioral, and\or environmental arena previously discussed, were brought into the light.

On the positive notes, to squeeze honesty into the anxious tossed salad of regional culture and political airwaves, attention span had won a front in the overall pro-con argument involving psychotropic medication. Returning 75-85 % for rapid-dialogue media viewing, and showing little moss on the stone for in-person conversation memorization of facts, while only moments of the build-up of anxiousness and fatigue had poor reception, the injections obviously paid off. This is the best of my recollection.

Often, the piling up of issues gave a course, a seemingly physical jutting in to the brain that I describe as a cat claw or untrimmed human toenail. In turn, as a combination of genetic and environmental factors that merge in time when they want, a distraction from reading concentration at home is left seen by the aforementioned light. From prioritizing to catch-up, a supernatural explanation on the morning suburban walk noted that yes, there were strong factors tugging for exercise energy at the spot, of the stretching ankle. Prognosis: "Kinks." It's the patient's kinks, this is the root of all disclosed and included exacerbation—mind, body, soul, earth-to-sky and land to sea.

With this in mind, what can one say if, at waking up by 7:00 A.M. and being ready to go by near 8:15, a strange feeling of having no muscles whatsoever in the body occurs, similar to the immobility of a groin injury minus the sharp, immediately-apparent effect. At this point, all jest aside, I wondered what on earth certain environmental issues were mine to carry, every Monday, as if I had been in a driving accident, or the victim of a fall was I. My center had an inherent confidence in both prior knowledge\experience and current integrity\ religious identity. My strategy thusly, despite esoteric difficulties, blossomed into an expanded view of my surroundings from my conscience. Treatment with over-the-counter aspirins had proved environmentally non-habitual and genetically non-addictive. I had energy for what I needed to continue with at my age. Middle-age was not a closed book with no view of youth, warm regards for my past-times during it. Blessings on my forehead given by lay ministers at Catholic Eucharist were actual, alive events involving well-prayed for seeds of healing and illumination, to my relationship with my fellow Catholic, man\woman of American descent, and city citizen.

Now penance for the body of the anxiety I normally sated by my appetite for cigarettes. The doctor had asked again, at a check point of who knows how many more months than five, after our detailed conversation on triggers. These things were not, by clinical experiment standards, associated with either nicotine use or withdrawal symptoms, issues, demons or dirt. A breakthrough in breaking ground for monitoring improvement was found (At Dr. Murad's on the third of March in 2008, four days before the birthday of my mother, though one event at a time). I celebrated by ordering my first coffee at Jitters Café in the downstairs of the medical building and watching Reese Witherspoon play a ghost for a while in the apartment of an old building similar to mine. I celebrated the season, all events lifting me for a view at the new mobile me.

Dr. Murad:

> Genetic or behavioral?
> Both.

Is there really a law that says clients, say at a rehab facility, cannot, even on the outside, date the employees of the facility? Is that universal in terms of medicine? It depends on the policies of the establishment.

Lifetime Puppets Due To The Challenge They Offer Freely, Without Obligation

Concentration for writing: optimum listening and task-performing of simple cognitive task: Quick and efficient. Concentration: good. Anxiety: Minor tension noted that is not interfering with tasks (I saw a mosquito. I wonder if I was bit. A warm, sticky feeling, not the cool air of the day, came to my skin, after a sandwich).

One environmental trigger I noticed in my forties stems from early work of a psychological practitioner dealing with the image therapy for my child within issues. Memories, which in my pragmatic stride to reenter the job market were not the subject, recurred. Her technique which I followed was to write letters to old friends, who at my last recollection I had parted from for one reason or the other. This sealed off some grey area in an attempt to let more light be reflected and connect internal with external feedback. Then we simply were to throw the letters away.

I found that some in the area of city history actually needed the knowledge of the documented portion of memories from those years for their research as private writers (archivists). Others were connected well in relationships both social and business through the project of the local high school Alma Mater's Alumni Directory. Is there closure on the ritual of young adulthood beginning in this time frame? The psychologist did not mention this in her work, yet stated human beings are in growth mentally through all stages of their physical bodies' physiology. In the end, Biblical Study may prompt us in a way we wonder the meaning of when we recognize names from our own lives in those pages. Only consistent Biblical Study will get under this layer of coincidence that may lead to unwarranted or aforementioned triggers of a combination between genetic and environmental factors.

Do The "Fuzz" Clean Themselves Out At 4:30 A.M.?

You set the alarm for 6:30 A.M., you wake up feeling like you just had 50 kg of MSG dumped in your corn flakes. You go back to bed, you take your walk at 9:00 A.M., and all the new people who have moved here from other parts of California and the U.S.A. have moved back over night. Only pedestrians for the most part are out, save for near five cars passing intermittently, until...

I walked back to Fern in decision to stretch this walk out another length, as a tall and well-dressed woman came across the pedestrian walk on Fern heading last for the high school. When I reached the corner, there it was. "The FUZZ," as my mother calls it to this day, waiting for a green light, yes, Redlands' finest going in for the 9:00-9:30 A.M. shift. After walking down and back westward on Fern, through the door and up the stairs into my apartment, it appears I'm not so different from others in that inherent difficulty of being awake and about 6:30 A.M. sharp. I'll keep trying.

On the doctor's end, I jogged for the first time in quite awhile on the fourth, yesterday, as well as recycled, got my teeth cleaned, visited my mother resting up from a flu, and went to Choir Practice. All this time save for the dentist and the jog, my attention span was being moderated by angels assisting me in the process of realizing what to do with your life is an answer that may not come from God with as much clarity as others. How to clear up difficulties in your life at the same time is not an easy shot to live up to all in the same day. The jog felt great. The teeth are cleaner than ever. Mom sounded better than I had heard before. I arranged for an extra choir rehearsal. I was resting by 9:00 A.M. And today, the attention span...I'll take notes if it's an issue.

Later...10:48 A.M. Reading seems to be at a less obstructed level of concentration. I am reading an article on facts regarding energy sources. I should research this well.

Graduate Study My Ass, Mr. President!

A thought that the good doctor would say," Take it easy on yourself. Do not overdo it..." came to mind. Yes, even I, what with the City of Redlands Quarterlys, Westways from the Auto Club, UCR's Alumni articles\news\events, the daily paper, my personal work either to extrapolate or edit, the edition of Internet website reading, requirements in At Home With The Word for R.C.I.A., and my priority reading made clear at Morning Mass, this day I rose at 7:00 A.M. to shower and fold up my futon, to start the most important Gospel of John. That's what all came to mind that was headache worthy, though not from sin. Did I miss something? Imprimis sent me an article from award-winning journalist William Tucker on terrestrial energy or nuclear energy and why. Indeed, the hook for graduate study makes me feel like the Catholic in me fought free of the line with bait in the mouth on the line, to face a rough current. The good news: The Colorado River is much less muddy for fish than it was in the 1940's, according to Channel 2.

These news anchors seem like they are next door in making the viewer feel important. Well, moving on for the day, I must pray about what to say to the Personnel Department on my errand vehicle, though the interviewer, Estelle, has stated "Page" is only going to be expected to handle errands afoot, such as to the Post Office. As far as why she feels Personnel might want to know for some reason, in a circuitous way, if at all, I will not ask, though it will come to me what they need to know.

Raison D'Etre Behind The Times? Do It For Yourself, As We Did

The Lord's lesson, on a stale Lenten Saturday morning in the County, was the effect of the laziness that is often displayed in a passing secular culture of early morning motorists. The lack of religiousness stiffens the air with what I would deem a repugnant arrogance, had I not gotten the feeling that boorishness and poor excuses were the reason Christ simply did not introduce himself to them in some way, shape, or form. Certainly I, as one who bears Christopher the saint as a middle name, have carried on the subjects of etiquette and law not only for the region's secular community, but for all of international film and television as well, quite far enough. You can take a three-mile walk with a brisk pace, and come back in to find their baggage not necessarily at your door, nonetheless parked behind you! By all means, yes, I see, they see, it is red, white, and blue's rut, not theirs. So, the key word, with all that "guard up," is not lazy…it is spoiled.

Make a list, if it has to be, of the things that are bothering you, and give them to the Lord.

After Texas, Just One Los Angelina With A Hair Up There Threw The Field

To continue on 03\08\08, Soltesz' Theory of Abstract Cognition alludes to the idea that normally, what is viewed as the pressure of the exact social stigma endured by the mentally ill, has an equal and adversely reciprocal, chemical component in the environment as a counterpoint. This, coupled with the latent effect of the dissipating medication's path over the course of the day, all weather factors aside for the moment, induces a potentially alarming arrest of reading and retention, regardless of the accuracy with which a religious concentration can produce a marriage between sanity and serenity.

In turn, all of the aforementioned criteria, coupled with a moderate change of external checks and balances with nature by an additional 5* F, displays as one walks, the streets both urban and suburban, after breakfast and two cups of coffee plus vitamin, what will here-to-for be referred to as "The Greenhead Affect."

Upon the carefully-performed research by a team of the University's finest, budding psychological instructors, of a group of 15-20 people who had volunteered for further scrutiny of the Greenhead Affect, it was found that indeed: Behavioral abstractions are not induce forth-rightly, yet a perception of the external will be augmented in a quasi-phobic way, where volunteers described the, "You should talk" phenomena. To pinpoint this phobic awareness as offspring of the chemical component as counterpoint, it simply appears to their sense as if all others look at them with a, "You should talk…you're an Ugly American too," sentiment in mind. Some noted this is regardless of time of day (Ad Infinitum).

Vocabulary List #1

THE PAROUSIA: The Second Coming of Christ.
Benign: gentleness, kindness, favorable and\or mild, for tumors.
Kinetics: a branch of science that deals with effects of forces upon motions of material bodies or with changes in a physical or chemical system. Also, kinetic energy.

Diffuse: to scatter; to spread thinly and wastefully. Also, to break up and distribute by reflection.

Photovoltaic: of, relating to, or utilizing the generation of an electromotive force when radiant energy falls on the boundary between dissimilar services (The lights used in the stadium to shine on the Cheerleaders in the Stanford Cardinals\ LSU Tigers game that never happened in President Bush' first term, because we all were suitin' up for war and needed the lightbulb on the tan. So, college football has been playin' catch-up since. And pros?! Apparently, we never saw the real quarterback hit the field, since Staubach, who served. I think. What, what, what was that? He served in the cafeteria diner at the YMCA compared to the action now? Well, well, well, on chattin', since we're capitalists, and if talk is cheap, this country is not—and we're all makin' up to Marguerita to get the weeds pulled. His rank isn't the subject, apparently he's a prune, I mean, how many of you been goin' to an out-o-state game and sock-hop or square dance if you're not a trucker getting his Imodium AD out of the truck-stop vending machine? I get it—Real Estate's best and brightest only go until the doctor gets his state of choice. Congratulations, fans, you've now gotten more blow-jobs than the L.A. newscasters and the Clinton family combined. Now you know why the Youth made Animal House and Porkies'—so you could learn your roots with a campus pan-scrubbin' minimum wage job to work your way up if you were too dissed to honor Recruiter.

The Revolution? In my generation we reminded each other that it was only in our pants since the Jewish banker's son had agreed to be our roommate if we didn't sleep around. After all, Moshi does the talkin' to the Landlord in the college scene before Koby gets in Mamba mode. The tacos aren't stale: the professor's got a secretary that has a twenty-five word vocab in French, a 'Guia, and a film and television class schedule, though.

"Oh, how narcissistic! Honey, you should have a raise! The Italians never show up to class in shorts and tank-tops when we're on Sabbatical in Parma! Does he think he can even get the Lakers Walk-on try-out without getting a parking ticket in that Hundai? He's on pot, isn't he? He wouldn't understand the professor's English in Northern California—he doesn't even get along with our theatre families…SO what do you think…volunteer work until he decides where I want him to shove it? Give him a "D": He's trying to talk that street-wisdom from one of those other countries with those Republican girls who lie about the fact that their Dad's an Air Force officer! I mean, really Honey? She thinks she's so smart with that All-State ass! She's no Los Angelino! She's no Trojan! And she's sure no Britney Shields, if she thinks he's a painful date!

Why, I remember the fish tacos we found at the Pier to beat the MCAT superstition when you became "Born-Again" and then passed. You were a loser until I taught you the surfers don't mind if you get extra paper napkins out of the dispenser! Now, what time do I set the clock for so you can get to the surgery center tomorrow in case the patient needs a Caesarean? Let's make it before 10:00—I've got an Amway appointment in P.V. at 11:30. The doughnuts are on top of the fridge: (In case y'all were wonderin' why the Brothas got together and grew Hip-Hop on our ass instead of getting a radio station in honor of Scott Coltrane). Thus goes the caveat of Republican in-fighting dumped on the Bruin field to pick up at Roses time by one honorable LAPD and Southwestern School of Law, as the Magic Claw rises in the sky to demonstrate that even a Chicano can own a $750,000 home in this bastion of an economy under a Democratic U.S. flag. Don't break it down with self-medicating. Enjoy the line to buy the tickets to the game and don't throw down. Amen(Now, back to the doctor's private practice...)

Nuclide: A species of atom characterized by the constitution of its nucleus and hence number of protons, the number of neutrons, and the energy content.

Isotope: Any of two or more species of atoms of a chemical element, with the same atomic number, and position in the periodic table and nearly identical chemical behavior, but with differing atomic mass or mass number and different physical properties. URANIUM ISOTOPE.

Half-life: the time required for half of the atoms of a radioactive substance present to become disintegrated.

Thought on Extensive Research

So, if in a combination of defining inherent cognitive distortions between Jewish and Gentile religious tradition and culture, you prove a fallacy in the definition of the particular factor, attribute, or act of a case study of schizophrenia, you have grounds for a platform to restructure the founding premise for life after a misdiagnosis, observant of post-relief support needs.

What We Go Through While They Try To Create Jobs

The problems the mentally ill had in this country during the two terms of George Bush, Jr., other than parody laws, transitions in treatment plans (including logistics), and the calluses of another decade gone by, were simply the feeling that although supportive of American democracy on a deeper level than benefits alone, it was as if we were paying our yearly taxes to be the superficial people popping in and out of the dialogue.

Yes, there is both an academic explanation and a *simulation explanation for the inheritance of what,* at first, appears to be a genetic mental illness, which in fact is a near trilogy of behavioral psychology. The key factor to note is the absence of all but a bad religion dialogue to conduce repeated falls. The <u>raison d'etre</u> on the simulation explanation is the mental illness\caveat-natured, and powerful discrimination of a *socially deterring persona, similar to the Riddler* or a *Britney Spears, a Scooter Libby, an Elliot Spitzer. All sides are pulling themselves forward in positive religious practice against a counter-energy that can have deeper than a psychological nature. As Dr. Burns comments on patterns conducive to depression, one sees the enigma of yet other devil's advocates historical or new, who play in respectively with their knowledge of modern society's rituals. Christ in here?* <u>Hebrews 3:1-6</u>*: "Therefore, 'holy brothers', sharing in a heavenly calling, reflecting on Jesus, the apostle and high priest of our confession, who was faithful in (all) his house."* But he is worthy of more "glory" than Moses, as the founder of a house has more "honor" than the house itself. Every house is founded by someone, but founder of all is God. Moses was "faithful in all his house" as a "servant" to testify to what would be spoken, but Christ was faithful as a son placed over his house. We are his house, if (only) we hold fast to our confidence and pride in our hope!"

Now, as speculation occurs following M.D.'s orders, what factors are contributing to the imagination's over-suspicious cycles, in address of the (mentioned above) 'counterargument'? This is done to keep the studio apartment, as psychological laboratory, intact.

Don't be a hybrid...take shorter showers...turn off the lights (live in a cave), buy fresh but not frozen vegetables...don't buy so may air fresheners...use scented candles after airing out...walk to work...try working at home...(e-mail, teleconference)...select products that are sold with less packaging...bring your own coffee mug...try clothes that don't need dry cleaning...clothesline...keep your blinds open for heat..leave shoes outside, or in the closet...bring used bags to the store...protect your image and get a job.

<u>No, California, Honor the Tex-Mex, Academically,
Socially, and Physically. We Own Too.</u>

Let me try to be concise. I want to know if you agree on how we mistakenly diagnosed me in to a less functional corner than I was originally in when I was a drinker and an Honorarium-receiving writer.

We had me part-coerced and part-voluntarily treated over three times for alcohol cessation alone. We've taken that many notes on the Subject though never once in rehab or treatment after opened a window for college, psychologically fit for adult Steve. We've in addition now made several new claims. When I was a drinker, they hired me with my college background. Now they don't. Without drinking in spite, it seems as if we only wanted revenge on the patient, not advancement of him as a formerly high-achieving patient with better manners. We reprogrammed him, but now he has problems reading and headaches, whereas before he read fairly well in yet another language. It's not nicotine. Psychologists fail to do anything with the patient other than a Prodigal Son that is supported, but uncomfortable in his own skin. Why does the baggage resurface at any moment as if it's a separate, competitive being? The patient had problems in this area with aggressive boys and sarcastic girls since his youth. Is not the label schizophrenic and the administering of medication merely the verdict of a vindictive peanut gallery? If not, why cannot psychiatry exorcise the subconscious demons when the individual cooperates? That is, its own slant on modern American religious persecution.

--"I know where the headaches on the subject of local economy are coming from."

--"Where? What on earth?! I've never had migraines before."

--"The mayor's hard-boiled egg timer is in the shop for repairs."

--"Dr., it's my eye to ear sense when T.V. is the issue. This is the challenge to the short-term memory functioning. There is some semblance of both internal and external resistance. The internal manifests similar to a web and small sensitive spot the size of a coin, like having something on the eye in distraction. I don't think you understand the plans that were already producing results for International Psychiatry. When Washington catches up with itself they'll be calling. See? When I crack my knuckles, my hands don't make that noise because I'm flexible and smooth."

--"Later on CNN, why the show has been named 'Fiscal and Ebert': after that last god-shot on the Academy Awards hit the Conservative Right square in the crack of the ass."

So, on this Sunday, the sixth of April, 2008, we have unexpected inserts of songs unrehearsed at the beginning of our ministry in the choir. It seems to be an ad-hoc exam on how supportive we can be in following the leader in an alerting situation. It went fine, save I forgot my overcoat that I left on the pew. The amount of tension-causing thoughts going through my mind surprised me. It appears whatever psychological or sociological dynamic that Dr. Francisco Vasquez started in 1984 could predestine a failure: a misinterpretation of contemporary Catholicism's efforts, communications, and destiny. All this, plus a milieu of current Redlands difficulties with positive impressions, fortitude, and windows to a more amiable festival.

I pray Christ sees me qualified to troubleshoot this intellectual dichotomy once and for all, while staying the course and keeping the pace, if for not other accord with the Communication of Saints past and present than the resolution of enigmatic feelings for Mass' progress. Father Beto presided, again leaning into why difficulties in clarity and translation adopt themselves, amidst other presences, by his celebration this morning.

"In continuum, do you, Steve, have a better picture? What are you going to do with your life? Do you want to support the field of psychology by seeing if they can solve your two problems(the Redlands Enigma and, as well, what should indeed you do about your romantic instincts given your past)? Do you want to support a Matriarchy? Is this a misinterpretation of such? Would it not be of historical accuracy to cite the challenges she has faced and what her reactions have been? Does this exercise detract from the Holy Spirit's truest desires? And, again borrow some of their time to honor your father by reading The Hour Game with it, by a leader in contemporary global literature. He deserves a little bit of expert, fire-tested honoring for this latest coup de moits:"

"Just take your time with it: you are the master of it, not the other way around," speaking at first of the computer. Later, like the morning fog through the window before the paper takes home plate, it occurred to me that the subtext of incredible weight is this applies to my studio's tension as well! He solved it, a psychological loose string as flopping and tangent-causing as the new mailman's drop-box service.

Cuando llego a tener razon`
Con uno de estos que escojo
Dejame saber,
Porque siempre empieza a decir,
"!Has equivocado! ! Equivacacones!"
Pero despues` de exclamarlo,'
Hablas demasiado,
Todo el dia, y toda la noche,
No hay mas color
En el equipo sin lonche.
Viene la cena
Y todavia estas hablando
Sin darme una galleta
Escondido en la mano

Te perdono para eso
Y esa y esto
Pero en el futuro
Pues en el futuro, por supuesto
Tu tienes mi fe, mi coraje
Mi apoyo,

La laguna es nuestra
Agua y todo.

When I arrive at being correct
With one of these that I choose
Let me know
Because always you begin to say,
"You're wrong! Mistakes!"
But after exclaiming it,
You speak excessively,
All day and all night,
There's not much color
In the team without lunch,
Here comes dinner
And always you are talking
Without giving me a cookie
Hidden in the hand.

I forgive you for that
And that and this
But in the future
Well, in the future, of course
You have my faith, my courage,
My support,
The lagoon is ours
With water and all.

What Is In Play When Italian Comes Out

Let's recap. Mother. Surgery. Aldrete gathering in San Antonio.

(I have a side of me that is speechless as it experiences beauty of family, after the long haul of this timeframe in America). Problems in Texas. Stagnancy(and germs) in San Bernardino. The Catholic Church uses the example of Christ, a rejected stone who became the cornerstone of the faith. I've been unemployed so long, it's as if I've never worked—and I've had over twenty different jobs up and down the Golden State. I could write a scathing complaint letter, or...I could let this version of Nature have its elusive side, build my faith, try new options and complete my esoteric University research. Now, back to the neighborhood.

How do you change a turkey into a ham? How do you sell the goat and the hen for the house?

How do you get the librarian to move the mountain, herd and all? Currently they say on the Catholic side following Christ has its tough holes, and we must move on from those. Also, the shepherd that is good like God brings the lost sheep back even if the rest of the herd waits. That's why I ask about the turkey into the ham, because Christ keeps bringing me back, to Helena Lane. Mom talked to me last time about moving to another place, yet I think this house has potential if two things would give: the area's interpretation of vision and Dad's ego. So turkey has two wings. The laughs the ham is known for? We might as well try and find Ponce De Leon's old maps of The City of Gold.

So, yes, Dad's habit of making me the short end of some jokes, when the atmosphere is Redlands life and moving along, is where schizophrenia and nervous breakdowns happened between the two of us at the beginning. This is why if some things don't change, stubborn does not equal difficult gene makeup, just a mix-up, and a Jewish opinion on "Survivor" is a backstab, not an objective topic, with its inherent raise the bar junk. Mom and the Aldretes minus Diane(my sister in all of this). What can Dr. Murad do about that? If I answer that, the rest of the country may give me a Nobel Prize despite an MRI's conclusions…for Psychotherapy. And an Honorary Doctorate. Now the hen is the outside family path to a healthy relationship, the goat is the City's business, and the horse is the future of Christ's mission with me involved.

So, Mom can have the same issues with Diane, and this is where I got the impression that the wine drinker in both Dad and Robert(my brother-in-law) had some bad influences. This is where vocabulary-building became the corn that was chewed by something like a Dog and stuck in the middle like an axle, for me to get wrapped around. Imagine translating this into Spanish! That's Aunt Flossie's 91st! My sister's house is not paid off yet, is that right?! Wow, and I'm eating mystery chorizo from Vons on Sunday Brunch-time.

****

"Be still and know I am God." Those were my instincts' silent orders following Mass, the day after Pope Benedict XVI's 81st birthday. I was, pragmatically-put, brought forth by the movement of the Holy Spirit to still my own pond, yes, calm my own nest, after a series of muffled lamentations had caught the Divine's attention. The very first thought, as I heard the 1920's home emit a cracking sound, was, "It hurt," and the voice reaching me was my mother's to describe her recent back surgery. I sat on the black futon sofa, feeling the sensation of limpness attached to my own being, for my recognizance place from Above of what this ordeal had been for the family to hurdle through.

My head tautened now, as this pattern of stand, move, and sit had been observed to be a new norm of dorsal treatment adopted by Mom that I wanted to understand, and I then stillfully searched for what caught the Spirit's attention, off the subject.

In my right side, as if pierced long ago by an object, upon a surrendering subjugation of the head(a genuflect to the Cross at the Altar brought home), lay the bitter and nerves of a complaint habit developed in circa 1982. I for many years had smoothed over Neapolitans' seeds of distrust for Americans, due to land usage and the U.S. military base's supposed local reach, having processed much further beyond this, yet the paradigm lay in what indeed is a current trap God only knows how many American-born citizens are caught in, for their cross-cultural sojourns.

The humanistic bond of using customs, acquiring a language, and hence assimilating mentality-wise to walk softly, while three hours South of Rome, had created a time-tested backdrop of mind's psychological corner with which to view the American political discussion. On the Italian side during the early 1980's, discouragement of sympathy with Fascism by poignant Italians landscaped Naples proper, and in the communities lying between her port and Sorrento. *I Cristiani Democratici*, the Christian Democrats(not to be confused across the board with Catholic Democrats in America, agenda-wise, let alone mentality), were introduced in amicable fashion. The exact party providing the corner in the mind stapled one concept by name to associate with, whether discussing politics or religion, sports or film, and that was *Napolitano*(Neapolitan), the name of the dialect and the citizen speaking it in the region therein.

These lamentations, noted by my Afghan-American physician to be the physical effect of coffee in abundance, are the muffled disgruntlements of a ruffled feather in the cap, the Neapolitan in me, because this cussing was not Catholic me and, whether weak or not by European standards, Folgers will not irritate me to this point of actually smelling the air thick with blown smoke (without cigarette near me for months now): This dogmatic Bush move into American thinking, that produced highly-numbered tangents of knowledge and a well-fought war in exchange for a new Christian Church, coupled with several amateur art works for my results, had been a scepter's slymieing and stigmatizing of a once-valued maturity, so it appeared to the Neapolitan within. In addition, branch two of this mode had Democrats challenging us so cautiously because of our Bush vote that the momentum of the day shuffled Christ's Mass into a makeshift tardy bell on the subject of reproduction.

The Italians never used the expression, "facendo l'Americano" with me, or "playing the American," yet these two examples were the side-poles of a jungle-gym I pained to climb across to find God's version of American Catholicism in actualization, off the subject of surrender to obedience. Now I sensed Pope Benedict the XVI had new ideas on what to pray for, as Naples is normally a source of joy and laughter, if ever callouses are noticed, so airing this out means taking responsibility, one American way or another, but enough migraines on that: the side's wound must pass.

Before we get hasty and point-blank about the wound on the left side of the head being anything to do with the headache from Catholic sex issues and beliefs, after departure from "The Episcopal Community," I might want to look one more time at where I am located: The entire state of Sinaloa has relocated to the North Side and is getting their degree in these parts, The South Side, beyond the Emerald Necklace, has connections with old money involved with The Utah Prune Company, and that down there is the infamous Downtown Redlands.

Cattolicesimo ed il Rinascimento della Balancia Genetica(Catholicism and the Renaissance-or Rebirth—of the Genetic Balance)

Quando si parla, in questi giorni, dell'inaugurazione d'una eta` psychiatrica e il ragionamento delle malattie mentali loro stesse, si sente un poco a una distanza. La scienza, per investigare, ricercare, e risolvere o adottare delle situazioni, non parla di un modello statale e il suo protocollo, perro' existe. Sotto questo "soffitto," varie idée (alcune molte creative) esistono per la sostanza filosofica o religiosa dell' paziente dell' dottore. Per trovare la piscina di amore celestiale e proverbiale dell'ammalato e` lo scopo della partecipazione, ricerca e trattamento psicotropico. Qui in punto puo` cominciare un dialogo che ha varie difficulta` da risolvere quando la sfida sia esperienzata, avvanzata, e progressiva. In altre parole, il paziente puo` pregare e ricevere dell' intervenzione, ma una societa` nondimeno tante, non solamente un movimento, esistono e che dirigono tutti i mezzogiorni nel sole. E` come se fosse la domanda per il partecipante e` di non essere terreno in radici a mantenere una balancia meno abrasiva, ma di crescere un' ala o due e navigare gli ostacoli con la faccia di un Cattolico con delle risposte. La societa` cambia velocemente ed e` maneggevole procedere nell'atto continuo di onorare Dio per la Domenica, e la settimana che segue dello Spirito Santo, col Credente. Allora in questo passaggio e ritmo societale, il creare di una soluzione per miglioramento delle facolta` mentali e` pregata, chiesta, studiata, e incorporata per delle scoperte dell' togliere dei blocchi mentali ed emotivi, nella cadena delle idée pensate, in una funzione anche medica. Cos'altro, che solamente la psicologia fa la raffinatezza dell'abilita` ascoltare e imparare?! Gli studenti sono stanchi di prendere risultati inferiori dopo che abbiano lavorato in ore ai compiti: La memoria nelle quarente per il participante, dov'e`?! Qui a distanza nell'dialogo riposa la risposta che la confidenza rimane con un lavoro. L'intervento di Dio vuole l'organizzazione concreta dell' participante, non solamente la passeggiata dopo di aver smesso il tabacco una volta per sempre, o si prova.

Catholicism of the Renaissance of Genetic Balance

When one speaks, in these days, of the inauguration of a psychiatric age and reasoning of the mental illnesses themselves, one feels a little distanced. Science, for investigation, research, and resolution or adoption of situations, doesn't speak of a state paradigm and his protocol, but it exists. Under this "roof," various ideas (some very creative) exist for the philosophical or religious sustenance of the doctor's patient. Finding the pool of proverbial celestial love of the sickened is the goal of the participation, the research, and psychotropic treatment. Here at this point we can begin a dialogue that has various difficulties to resolve, if the challenge is experienced, advanced, and progressive. In other words, the patient can pray and receive intervention, but a society nonetheless many, not only a movement, runs all the afternoons under the sun. It is as if the question for the participant is of not being terrestrial in roots and maintaining a less abrasive balance, but of growing a wing or two and navigating the obstacles with the face of a Catholic with answers. The society changes with velocity and reacts to progress in the continuous act, of honoring The Almighty on the Sabbath, as well as the week following of the Holy Spirit with the believer. Now in this passage and societal rhythm, the creation of a solution for the enhancement of mental faculties, is prayed for, asked for, studied for, and incorporated for the discovery of mental and emotional block removal in the chain of thought in a function that is medical as well. What else, does only psychology refine the ability to listen and learn?! The students are tired of getting inferior results after having worked hours on homework. The memory for the participant in their forties, where is it?! Here at a distance in the dialogue rests the answer of confidence to remain with a work or job. The intervention of God wants the concrete organization of the participant, not only the stroll after having quit tobacco once and for all.

"It's Not Our City"—We Only Shop Here

So Matthew apparently points out to forgive people from the heart, not from the wallet. Meaning? Meaning true forgiveness soothes the tension centers that recalibrate the dopamine flow for the brain's extra psychosomatic campaign, and a cup of coffee and a Danish for a friend keep them floating, semi-forgiven, but not prayed for above the low-harbored fault your conscience holds them in observance for, while they stumble on. On that note, on this subject, I have arranged myself into a corner to fortify my survival path, a 'prophet' in an eschatological Community House of Horrors who laments poverty over the waste of his lips, on the service to the Queen Ant somewhere hidden in a garden high on the hill that the water runs down!

On a more serious note, one could write down a list of people they are having a hard time forgiving and for what, then scrutinize and self-scrutinize to see if there is a link of some sort that kinks the chain of events in (a) forgiving; and (b) the continuance of such sin. In other words, do not just absorb Matthew: get homemade P.I. license on what deep down is getting you the headache, as well the call for extra aspirin.

The scene: You are a year older in the town you went to grade school in, though now in your mid-forties. The mayor for only the past five years of your life has been winning popularity by the tritest of election speels as "subtle change." At the same time people are either calling you long-winded, interrupting you, tailgating you, or dumping their extra baggage that keeps community attitude balanced right at the step of the proverbial Church door, all the time while the Haves are reaping incredible results and progress for the Community Bright-Siders with this "subtle change." The windows on balance stay open about as long as it takes the cupboard to take its morning coffee can back in, once the brew begins. In that time frame you briefly fire your one good 'bullet' to make the subtle, new-improved impression, and then drive home. Safely, the mayor will throw in.

--"Hell, Steve, your Inner Critic sees what you're saying about Catholicism, and is going to comment, then see about teamwork to help you get a job!"

The Inner Voice

"Run The Ball Up The Middle?!"
There came along, with what I always recognize now as a beautiful drive to Redlands Community Hospital, an idea or two from the ride's daily lesson. (1) Yes, some people in this area have a beautiful life following country, not classical, music; yet (2) more succinctly, not all people in this County are positives, team players, flow-continuers; and (3) God knows me, and despite grave errors, when the full story is told, I'm not a failure or let-down who went so astray at one point that when it comes to push over shove, He understands why by the Laws of Nature those were my challenges, and medicine may be painting something much more holistic. As I'm a blood donor at times.

I did not lie nor have to get embarrassed by adult questions about sex and needles. No, after all that tomfoolery with one eye out, I do not have a sexually-transmittable disease, and I do not need breathing therapy. I won this front against the Devil pretty well for Christ and his Court, so it seems when the light from that side of the San Bernardino Valley comes down on us all; and (4) I must follow this light as a writer who is better being well-thought of and safe than owner of a very impressive domicile!

This part of the American-viewing audience influenced by what they watch on television has been sold to <u>Frazier</u> for $300.00, whether they are working or not! On with 2008! In this passage, viewer #1 takes refuge in his Southern California apartment from the winds crossing over because of the weather in the rest of the nation, which seems a bit upset that at age 45, he still has the right to swim in the family pool despite the controversy over his stream of consciousness. Apparently his Italian words, "Pace! Pesca! Perfecto!" did not ring out well with either Democrat or Republican ("Peace! That's a peach! Perfect!").

So an independent he is who has now scratched Tagalog and Portugese off the list of what he will learn under Schwarzeneggar Rule. When General Powell and Company get back from Indonesia, please tell them the Major's son has now quit smoking clove and regular cigarettes for 11 months, no turning back. Tonight, he sleeps on the Hill.

"We have a cross-fire at Sweetroll Pass! Load'em up and out and go left."

Some Divorce The Economy To Protect Their Own

It is interesting, apparently, to the Holy Spirit, when looking at my life from the angle of "Law of Nature" in the surrounding area of my parents' permanent residency. To me it has always seemed a steep, demanding side of defining such, and my ability to adapt. On the pro, so far, it has withstood everything the world had thrown at it, almost with mastery. What it seems to have is an angelic side covered by the shadow of its own mission, as well a fierce split into a domain with a bed of problems adequately defined as diabolic, and is the aspect I name The Social Arena. Unmistakably it is the almost secret attribute of modeling for the others like it is a silent, underlying effect. It pinpoints painfully, in a way, the desire to define an illusion in the belief that relocations offer such more fortified branches of freedom, from which to return to one's center of solitude. A core belief is what with the nature of America's challenge as a, said, universal whole, this jagged rock with a few nests at overlooking points is a climb that is indeed psychologically arduous, yet the underlying effect mentioned earlier is partially connected to the external's handling of fear as a subject here present.

Indeed, yes, this issue is at once the place attempting to manipulate said split into the diabolic, which circles around to grapple onto the symbolic core of a Holy Matrimony that conquers this very split. Work and life ideas wrestle, as their own separate platforms, with the very same configuration which stealthily observes and eludes until it can again, jealously, practices its ritual of the Challenge. This configuration is not the devil, in my opinion, yet a side of the face of Redlands to save that is hard to protect. To collect impressions all week of said area with this facet doing the looking and return to quietly wash it, is a feat of sainthood that is done by many, though apparently when done with a marriage partner problems of tension plant on the inside, despite proper dining. Climbing to these platforms of the Social Arena for exercise is an act all its own. Where and what is the working world, and Christ within, up to?

The Mayor's Role (Roll, As In Jelly)

Besides his once-a-year temp assignment delivering on the presents route, Santa acquired a position helping Americans process their long-term memory, as it relates to contemporary times, in the religious fashion particular to individual need. Mrs. Claus was quick to point out that one elf, though not overly religious, had facilitated Santa's reception office by handling clients in a dialect of Double-Speak, as Santa often forewent extravagant explanations in English.

Meanwhile, the Major's son, taking on his next heavy read, is perusing the preface of the first volume in the trilogy, Echoes of Scenic and Preservation District Past. He squeezed by the Mayor in the Smiley Library Silent Auction after said leader laid down a few collector's Canadian coins (and an old pocket watch) for the 1911 edition of Redlands High's yearbook.

--"Where are you going?"

--"I'm going to City Hall to get this straight."

--"What will you take for the road?"

--"Um, yes, you handle the register, I presume?"

--"Yes."

--"I would like three bags with approximately two pounds in each bag. Economics of the Day, that is."

--"Coming up right away. Don't forget to stop by the pharmacy. Sun glasses sale. Two for one."

--"Yes, it's getting a bit bright out."

See all that news and show-business? That's how many things I'm tired of settling to have a thinking cap and a plate of food they will not peck apart. After that interpretation myself, based upon words and the faces, I've got a good sense of humor, though am somewhat consciously disjointed. I'm trying to be flexible, but end up feeling broken and used, and so I become slow-moving. The truth is they have a strategy to share you into control with the way the problems and issues are balanced in every day life, given when they chose to present or cause them. They need intellectual enablers for ratings and profits. That's not what my injection is for or about.

Southern Uncle's Wisdom

If the gateway to Heaven is narrow, do not worry about 'tunnel-vision.'

Can't Get Enough Of That Professorialship South O' The Beau-dy

Bella: "I don't like plans."

Savoir: "You don't like plans?! I'll show you what plans are good for. I plan to make it big! And then we'll have a party. But we're going to have a party now so the plan is full proof!"

--"Please, let's regroup the people who have now all participated in a group and group on. Before the clubs come up with another version of this fiscal emergency!"

--"Good, we'll regroup, in a theistic and philosophical sense, but not in a tangible locale. The last time we did that, a visiting professor from south of the border caught wind of our artwork, and told all of Latin America that a cult had cropped up in the American Medical field, right in the middle of martian territory!"

Next: Legend of Senior Moment Canal

The Good News

My humor thwarted me into a new dimension, post-exercise increase where my parents noticed my muscles were enlargening. After a fright of the chills during the hail of the tornado in Riverside, I reawakened without the stomach cramps of Excedrin Migraine and O'Douls' Non-Alcoholic beer to this new influence of Christ's own halakah. That was a week in the past, it is quite a bloomed season with ample sunshine at this time, and though the working world escapes me through yet another stately season and Church season both, all I can muster is God blesses America!

Meanwhile, back to the pragmatic exigencies in demand of a reformation of Sophistic nature (debates being handled with angel and demon alike in said solitude), I have been hereby informed through the Old Testament lesson at Morning Mass of an axiom: "If you are not right with God, it will be like a curse, and one that continues until your sins are righted." As well, a reminder through the fable of the Three Little Pigs that my apartment in Redlands, combined with visits to the parents' garden a mile or so yonder south, yet at an incline, is, unlike Riverside, Los Angeles, Santa Barbara, and the fourth pig, San Francisco, my apparent sedentary, mundane, highly old-fashioned yet abnormally paradoxical life within, given said tornado's exit, and is stone and stone enough to keep the wolf away. The pro, the plus, the positive, the perk. And now for my n's, my negatives. Never to now, neither ner nor the remainder, being assed with a verdict of narcissism by the Lord for here-to-forth what shall be known as your brazen wake of sins' recourse. I think, after all, it was not a misdiagnosis, falsely proclaiming me schizophrenic. D'accord, docteur. I will see if there is any news on the status of how the curse medically has affected me this week. After all—he said not to worry about my attention span via the nurse's phone-call. Wednesday with Wadsworth, it is, and that's Dr. Murad to the gallery of angel and demon alike.

A thought through Maria Callas' voice playing is, "'I have no time for speculation, than interruption." At once a doctor's office time, a career choice, and a summarization of the chain of command's next job, as well being undoubtedly the stone house despite the mood swings of the wolf. What I do that others won't. On the subject of languages, if there's an egg here to hatch indeed. And that, that as it looks, is my corner, my second-story room to enjoy an NBA Championship this year.

--"On the subject of the period, did you know there's now a pill that can bankrupt the nation across the board?"

--"Yes, there's now a pill that can bankrupt the nation, across the entire board."

The "New Script"

At night venturing to the cinema with my parents some important links in Redlands mentality where noted. The theatre itself is a psychological well that has the potential to return life one-hundred fold for her support. The options are avenues which help fortify the desire to guarantee the viewer, eventually, all out of California life one can get.

At the current time, for some unforgiveable error, we seem landlocked, despite ample exercise, to a rather sedentary apartment life of simple thought, anxiety, and processing. It is an incalculable amount of energy locked up in the persistent mental blocks and rules of others, who in turn have earned breathing room for a pattern-like committal of social error, leading to my current economic predicament without a calling. The trouble started when we rejected an approach of reaching for transcendence.

In effect, several pragmatic bases emphasized our overreliance, in error, on said pragmatism, as opposed to protecting the situation. The grouping of several to study and foster different culture, science, religion...

Play your part in building the social order.

Economic Dead-End, Playing along the path.

Fear that schizophrenia is merely deep-rooted character assassination.

Also, despite healing and faith, father's resilience, to anything but a competitive and financial debate ending in continued treatment, persists. I have not even mentioned alternatives to treatment. Can see the writing on the wall.

Medications <u>can</u> slow learning ability, according to some, despite their mood stabilization. To me, that was not the subject. The initial circumstances of the diagnosis are involving growth pains due to familial denial.

Barbara Shlemon Ryan, author, <u>Healing Prayer</u>, Healing the Hidden Self. You want to call it a Portrait in Healing, but be better along.

This Presidential election post-Bush is intellectuals vs. snobs. The snobs are all acting like athletes…The intellectuals are being quiet about the medals they've won, before the subject turned to the pros and a rousing hand of "Favorites" began again. The minority communities are going with the intellectuals, if they've got their senses, after turning pro in sports and getting a taste of the sportsmanship.

I used to smoke. I used to read too, because at first, smoking helped with my alertness and my attention span. Then the long-term side effects of the other chemical in the cigarettes, beside the nicotine, built up in my system and I developed dyslexia. To make matters worse, I had to go to the hospital and have my left ear removed from my butt because of the other demons associated with smoking. But I quit a while back. Now I'm trying to quit again.

Because the social order has affected the economic order by pushing a dialogue of selectivity into the realm of regulations for order, as a historical antecedent to a contemporary formula, the layman's peaceful process is manipulated almost systematically, like a hand on a clock and its movement. How indeed does this translate to a comparison with the Ancient Church times, looking to Christ's own obstacles, and responses, for adaptive reason?

--"Caught between the corners-cutting of Black Americans and the ever-present threat of another Mexican stand-off, Whitey decided to take a few years off from nursing the country's economy."

Room At The Inn

I'm beginning to wake up somehow in my new Parish, as I seem to experience the same dot-connecting on card-playing in the choir I sing in as my family I come from. When I got back to the outside world for a double-shot playing in like a double-decaffeinated, I brought some tension home, despite worship and going out, that wrestled for the stagnate me, until some kernel of faith dislodged an apparent theory that Christ had been walking me forward to learn. There are, through music, literature, and art, updates on Modern Catholicism that, though secular in objectivity, can act as boundaries or 'new' murals on the walls. These keep symbols of reflection in place where the death of sin used to sit along mind's trail of memory. For example, the darkness of flashing back to a morbid collegiate experience in a neck of wood otherwise reflective of personal growth with God and Earth: It can be made new with the music of a group that is familiar in sound without carrying the stigma of the media's center, and what we bear cross for, even for our distance from it, is felt by the weight others see but not say.

Art is more of a 'sealant' we can put in front of us to give our prior acts the conviction they need as important steps we took.

It is, to continue, interesting to note that our proper reflections of Traditionalism are what prodded the Devil and his co-thwarts to trap us, while attempting to study and produce. On the day commemorating the martyrdom of Saint Peter (by crucifixion) and St. Paul (by beheading) we see both these elements of Catholicism, traditional and modern, normally contrasted, aligning for a strategy in unity for Western logic's sake and promotion. Those in the pews reflecting tradition remind us of the support needed to continue a legacy of local youth travelling abroad to spread a healthy Gospel message. Those in the pew keeping us ever-current, with the modern bookmarks in our faith, remind us as well that many helping others catch up in life need breaks to change and grow. We all worship for the choice of becoming one Body and Blood of the mystical creation the Holy Spirit used as a vessel, to instill wisdom in each and every one of us along the road we peruse for signs of our hearts' next steps to leap. It's a warm and sweat-filled ritual the week becomes for Eucharistic love.

Father Jose Goopio today reminded us to listen for the mission and missionary work that, overtly or privately, the Church in the Almighty's name would find our deeds in as a most shining way, showing God's love when there. This has been a great challenge, and discouragement can settle in when I cannot get off my butt to hear more. A strange blending of new assignments often decided on by my own conviction could lead to a freedom not before in sight, even locally, which is hard to admit.

"We got it! There are some Americans who cannot go back under the 'Once you have Black, you never go back' clause, nonetheless, have hit a stall in the program! I believe credit goes to both parties. Wow, America, your Allies must feel this is a breath-taker of a realization. Definitely don't move, and that goes for the economy too."

"I cannot believe it. This is the longest-lasting glass of red wine I have ever seen. He is not nursing it, people, he is applying to make the hospital staff!"

So on the Dead Sea Scrolls, the trick was that all the demons of sin and tragedy, in a scholastic sense, had been dealt with, though demons present within the study of scripture remain. Nothing to do with Redlands roots.

--Note: This video was a polite way of advertising for the Scripture-oriented foreign universities. That's where I was after Naples, before problems...planning to go to Bible College.

The prayer was polite, reminding us that there are some less fortunate, that don't know any joy of a relationship, nor even their mobility.

Note: make room for ghosts of Reaction Past in the story of the game. All of them, all the people in this country you remember, and the argument about loose connections the Dr. presented is to let you know he is a tough ref. This is the Obama Decree on how psychiatry as a field is now handled from the top down. McCain, in a less sympathetic tone according to intelligence gatherers in the European travel industry, has implied more than once to Thatcherites that the American Psychiatric Community will continue its sanctions of top intellectuals, in exchange for more support on the warfront. This is where we can't get over the "Who's On First?" Video we ordered, because it's Saturday afternoon and the car race on T.V. is over!

So he read it from start to finish, and then he began to pick it apart with a fishbone.

"Wait?! No! Who's cutting in?! Why is the subject my calling?!"

"They're going over the notes Grandpa took of Robert talking in his sleep. Even at the age of 53 with two children in college, he's been cut off the sauce indefinitely by Lord Grandpa until this habit is settled at a psychologist's."

Meanwhile, on college war-bride front: "Thank you, Shannon, I was afraid I wasn't going to get the engagement ring back even though we only participated in mutual masturbation."

So, really concisely now, we seem to be having the New York mystery on the Italian world sucking the wind out of the room when a Naples-America dialogue is opened, while L.A. is a sorority\fraternity-compatible version of the Italian world. People are being taught Italian language in San Francisco elementary schools, the most hyper-defensive city in California. Off the subject of Christian members of the American Democratic Party and their lives, why is it an egg-shell walk not to turn over D.C.'s apple cart, going from Naples-to-smaller city realm? Why does this rebound always go back to New York, regardless of the language spoke? This is economical if you play a hand that poorly, save for the Internet.

Dr. Burgess and Nathan Gonzalez in <u>The Case of the Touchy-Feely Holy Mystery.</u>"

--"I got the message. You want me to kneel here, then walk back to the car and drive home. And you're going to take a cab....Waiter I'll have another TiramiSu, for the left nut that got blown on that conversation!"

There was a down-slope, wet grass ride in my surreal dream. A lot of familiar personality types from around America in my youth got in an all-men's event that I somehow found myself in at the last minute. I did great with the frontrunner on not falling <u>and</u> having a blast, and was at the bottom, not in a puddle, at the hill's base. I noticed an Episcopal office building. I was carrying something upstairs, that had a familiarity like my studio apartment home. I put it from one end of the kitchen counter to the next, ducked, came out in the main room, heard a "Pop!," and looked back. The coffee pot I know I saw in the dark was not plugged in. I got out the door and someone was coming up the stairs and into the hall. I surprised them but they were silent and chagrin. I said, "Father Saville, I'm just putting a toboggan in the office. I'm going home now. It's me, Stephen Soltesz. Good night." He half-smiled in a "Lots of priest work at this hour," way and I pressed on. He went in.

Last night at the Vigil I got the sense the Christ wants me to take my time. He wants me to keep journaling my progress over time. On that note, even though I have a clearer idea of my goals, it's sad I did not come to accept my diagnosis, its territory and history, and my family for their role in it more. There are <u>some</u> reasons for that though as a Catholic, God would have me move around, not dwell on, and as my Mom said, "If people aren't responding as friends, you're entitled to just let it go."

So, I felt some presence of my Catholic cousins in Los Angeles counseling me on how to handle the high school "brainwashing" and how it affected my balance, in other words the snobbery I experience. Apparently God <u>does</u> give us the wholeness to move through this, to Christ's <u>real</u> point of our work and lives here, without any sacrifice. I saw much of it all day, at the Laundromat and the store, the desire of some not to leave that social construct behind for growing up. I'll be fine. I prayed for them and their children.

Father John Gunningham was in Yugoslavia when the war broke out. What a story he told, of the Virgin Mary appearing in a mountain village, known for its utter peace, the same day.

I have to say this district has a certain peacefulness after Mass in the evening. I'm not going to complain.. Wow, it has been warm.

I guess I want to get art off my chest, for something to do, though I miss my inspiration in Italian, and translating as a career, and a good writing idea. I wrote my brother-in-law about it. I think on my own, I should carry through with them just fine.

You Think That's Funny?

--Why would the 'center not hold', as mentioned in the title to Dr. Ellen Saks' novel?
--Because the team had an idiot on the pitcher's mound who kept stepping off of the bag to throw to first for the runner trying to steal!

--"Stephen?! Hello, this is Col. Brown. Your father said to tell you you're off the hook for the thorn-patch joke you cracked about the Johnston College Writing Program, when the subject of Redlands history and The American Catholic Movement came up around the time of The Orton Center Awards Banquet."
--"Thanks for the news, Colonel! I'm glad he was finally able to see where these were Gospel seeds planted amongst the thorns, and the apartment complex is not necessarily the rocks."
--"The apartment building, Steve. Don't trip these people out, especially if they're U of R students or otherwise. Oh and it's not time for a philosophical discussion, Steve. I'm not Pastor Creflo Dollar!"
--"Okay, Col. Brown."
--"Keep in touch, Steve."

--"Hello Steve?! Col. Brown again. I checked with U of R Admin. After receiving your note, I swear these people with the license plate placards are not paid by the U of R to drive around and keep an eye on City of Redlands' 'what comes around goes around' Tournament."
--"That's good news, Colonel! I'll tell my folks you send your regards. Bye."
--"Steve, wait…I need you to be more low key this time around. I'm glad you sent me your perspective, as it does lend weight to my wife's theory that Uof R Admin. is connected with Big Brother!"
--"Oh My Gawd, Col. Brown, city furloughs, economic rough patch, and now this—what do we do now?!"(To Be Continued)

--"And the winner for Best Comedic Film is…Dude-ing Europe." And the winner for the Best Actress Award goes to…Michelle Tuzzi, because we know she doesn't love all of the other counties!"

--"Steve, Col. Brown. My wife is reading Porky's, The Novel, after seeing the movie. She says it's quite prolific. She's on page 262."

--"Really? O. K., Col. Brown, tell her not to spoil it for me. I'm only on page 179. I set it down for awhile. I'm getting ready for the 5K Walk."

--"Good luck, Steve, my money's on you to finish it this time without smokin' a cigar after.'

--"Thank! I'll let you get back to your wife now."

--"Col. Brown, this is Steve. I can't figure out what in the Hell is going on in this place. I thought we had our freedom to be Christians in this country."

--"I was ruing this over with the Mrs. after we saw you pass by the store when we were on our own errands, Steve, and what we figured out is, you have a bunch of old friends sneaking' around pulling on your ex-girlfriends' string! Ha Ha Ha!"

--"What's so funny about that?!"

--"Think about it, Steve—some of these folks are that hard up to earn points with Big Brother! I like the way you're building on your own for awhile. The Mrs. said you should try making Hungarian soup with pork for a change instead of chicken, that cooking it will handle the evil spirits. Get a new smell in the house on this pig concept."

--"No kidding. They race home from work and yell, 'Honey, where's that string? Might need a new idea for that old spiel."

--"Bad Pig, that's the bumper sticker I saw on some lady's vehicle, and that's about it, Steve."

--"Ha Ha Ha!"

"Make some corn bread, too, Steve, while you're at it."

--"Which hand is it in? Which hand is the key to Alice's Wonderland in, Prime Minister Brown?"

--"Hey Steve, this is Col. Brown, how've you been?"

--"Well, Col. Brown, good to hear from you…I don't think it's the same problem anymore. The university is playing appearances ball. So, it's 'My other brother besides Big Brother'…the one who can hear, or tries to, between the walls in the apartment building."

--"Hmmm. Steve. That's peculiar. Isn't this a private rental situation?"

--"Yes."

"I'll give that some thought. It's almost like a bad joke from Verizon on surround sound loud speaker stuff. You're not a loud person, Steve, you're from California."

--"Let's just get back to work, Col. Brown."

--"Steve, good call. In the meantime, until I get back to you, don't have a dementia reaction. Be cool."

--"Si, Senor."

--"And that's enough of that."

--"Bye."

--"Oh, bye, Steve."

--What do James Hennessy and Jody Foster have in common?

--They are both considered 'sloppy seconds' on the Continent compared to Bono's track record on A.I.D.S. awareness. (In other words, 'silence of the lambs' for those who get their Irish feedback on this side of the Pond).

--"Father John, how does that song go, 'I am the Lord of the Mexican Hat Dance, said He?'"

--"Don't bother me with that right now, I heard the Pope and the Youth watched Romper Stompers this summer in Australia and I can't find it in the video store."

--"Don't go by me, Father, but the Pope and Youth don't really need to know that's what they watch at the potlucks when the Protestants 'make the soup'. Because they won't let go. So I think I'll save the chicken soup recipe."

--"That reminds me-- they did not show the plans for the kitchen for the new church."

--"I have a small kitchen in my apartment on the second floor. Sometimes my upstairs room is confusing and I forget things, too, so I fast after walking home from Morning Mass."

--"You're killing me!"

--"I can recommend a doctor if you don't feel well."

--"No, it means you probably owe me a Confession."

--"Why? I haven't done anything wrong recently."

--"Then it must be time for your shot."

--"That's what the Lord told me—if I get my shot, I don't have to go to Confession."

--"Since I get a shot too, for diabetes. For symbolism's sake, IT'S A DEAL!"

--"Oh, by the way, try to be confidential about the shot the doctor gives me so the other Catholic psychics don't disconnect my attention span when you roll the Sermon."

--"So otherwise you're good with converting to Roman Catholicism?"

--"Yes, you held the fort while I completed my temporary assignment with The U.S. Census Bureau. That's the good news. And now I can attend Morning Mass again once in awhile."

--"Yes, more prayer, less repeats."

--"Right, it's not about 'repetition' like Jeff Rickard said in The Daily Facts when he retired from the U of R—the other schools have to get on with it now."

--"Wow, that's all the wisdom in the world. I see you <u>have</u> travelled."

--"I would travel to Rome again, now that I'm Catholic, except I need the shot."

--"Yes, well, Europe also has current spiritual problems."

--"My Dad said the same thing."

--"I'm sure he doesn't owe me a Confession, with Father's Day here and all."

--"Yes, that's true, the doctor is an Episcopalian who thinks the same way on certain issues—if the children get their shots, the parents don't have to say Confession."

--"By the way, how long does it take for the shot to actually take affect?"

--"Between Obama's ground game during the election and walking home from Mass after singing 'I Am The Lord of The Dance', it's hard to tell, but I'm determined to keep up with the pack."

--"The Pack did not do to well last year as Pittsburg won."

--"It's a good thing the Catholic Church is a church of history, and we get to keep the good memories."

--"You can't be old enough to remember those day!"

--"No, Father, I'm born on the cusp of being a Baby-Boomer. I hope that didn't slow down your plans for the new Church."

--"Well, just pull your head out."

--"But then I can't see the T.V.!"

--"The T.V.?!"

--"Get it? I'm a Military Brat!"

--"What nationality <u>are</u> you and your father?"

--"Hungarian. That's why I like the 'Lord of the Dance' song too. That's what we have in common with the Irish!"

--"What is the difference between Irish and Hungarian to you?"

--"The public speaking time before Judaism comes up for some reason. The Irish get more time than the Hungarians before the Pope starts drawing straight lines!"

--"He's infallible!"

--"He's an architect all his own—there'll be a new Church in no time!"

--"This is all good except for a loose end—the psychiatrist is Episcopalian."

--"I wouldn't worry about it if I was you—he's also part Irish. <u>He</u> says there's a loose string though. The Pope apparently likes Hungarian cookies."

--"Well, the doctor's running a practice, not a cookie jar—so he says it's not always about my problem."

--"That's not the Pope, that's those that buy into this 'Catholic psychic' business you read about at the library. I understand. If they pick your brain like that take a drive to another Catholic Church from time to time."

--"What's the difference between L.A. and New York?"

--"In New York there's actually The Apple they eat while they're putting you through all this."

--I see why you don't like Smiley Library. Their hiring practices are a reflection of Redlands history."

--"Do you see why I don't like the R.U.S.D.?"

--"Yes. Their hiring practices reflect the high school yearbooks during The Cold War with Brezhnev."

--"Oh, and I see why you like the R.P.D. They're the ones in town of the people you don't know with short hair who shower. Normally, no, the police' job is to fight crime, not to raise the bar on hygiene. So you basically donated a few dollars for soap to fight disease. That's Seymouresque. The first Redlands doctor, a woman named Dr. Seymour, came here to fight that with medicine. So if you take your medicine and support them, you should live a long life."

--Was quite a Body, which I am becoming a member of via Holy Communion, so to speak. I'm going to pray for a different arrangement on others' vibes here and about, however, esp. if the subject is The Tilma of The Virgin of Guadalupe, though I'm paying attention now. Paying attention to the art world even though these aren't signs, per se. Not like when we Hungarians made art work with signs pointing to the safe trail that would secretly escort The Shroud from those parts to Turin. Amen.

--"I know that son, I know that you have insurance—would you just play along with The President's lead to get some of these ladies some other fellowship?"

--"'Original sin'. Now that's my idea of a good conversation on the Continent!"

--"See, he had a Senior Moment smack in the middle of your prayer session and started to think suggestions out loud!"

--"Oh, is that what Mass was all about?"

--"Yes. Right after the Seventh Day Adventists had a parade through town on a normally healthy Saturday for an extra strong mug of Joe."

--"I got it. And now it's back to work on Monday!"

--"If they're getting older, they're taking you with them!"

--"Stephen...don't be so quiet, explain it to your L.A. cousins on the Tex-Mex side. Please!"

--"You guys are having 'cognitive distortions'. Modern Psychology has figured out everyone has them, but the mentally- challenged have ones really bad unless they are in treatment, like I am. So I don't need to be kissed by a magical princess to snap out of it, it's more as if a European magician cast a couple of optical illusions so my former girlfriends from Germany, Belgium and Italy could leave the good times they had in America right where it sat, before the whole Iraq War began. I only dated for a short while with Americans a year after they went home. See? They're not the Europeans that are protesting, or were, the Iraq War. They're academicians and professionals. So, first the Whites were snobs, and then the Californios were snobs, then the Blacks had the L.A. Riot, so I started going out with Western European travelers after U.C.R. It's not easy dating and getting treatment. Without simply saying some of them weren't short girls because some were, the Northern Californians thought they would try. Just then big footsteps were heard. It was Asst. Governor Bustamante. 'Bustamante going deep!(Huff-Huff-Huff) Going deep!(Huff-Huff-Huff) The throw from Senator Nunez! Oh no, looks like a bit of an overthrow, the crowd is disappointed!'"

--"So to make a long story short this is what led up to Bush /Cheney getting into office for eight years, as far as the domestic side?"

-"That's about the whole lump sum of it."

--"Well, o.k., Cousin, we understand—we'll try and move on with the cognitive distortions somehow."

--"I recommend a vacation."

--"But we always do that."

--"I know. Don't change."

-Quitting Smoking During Over-time NBA Finals Season: Dungeons & Dragons, Golf, Politics, or Gardening?

--"Up the ante, call him picky, take the job, and move the ball!"
The Sugar-Daddy of The College Employee

--Why did the pothead start the forest fire?

--Because he thought someone took his silver spoon he used to pour the sugar in his coffee.

--I saw a commercial during Bill O'Reilly on mesothelioma. I think I actually contracted

schizotheliobutkis. It happens when you don't get hired because of the California Democrats after

voting twice for Bush in a tight-knit county.

--How do The Democrats heal an old war wound?

--They change the Band-Aid 5 times until the doctor grunts, "Progress!" The lower tax bracket has to do this too.

--What are the Republicans doing in the meantime, off the subject of T.V. and radio?

--They are going over The Olde Times: they can't get enough of this stuff.

--So there really isn't a 'contemporary America', is there?

--Not that's not payin' out the arse to heal itself, no!

Cousin Steve: "So the, 'I love you *muy mucho, primos*," was the incorrect grammar translation that jinxed me from a job during Bush's second term, was it?"

Uncle Herbert: "You mean *'muchisimo'*, Stevie."

Cousin Steve: "No, no, no, don't get so close to me after all this *mierda*!"

--"Who is this new 'Seymour Morgenstern' working for?"

--"Mr. Schnivers, who caught word Stephen's writing career might get off the ground again."

--"Really? You don't say? Schnivers. How does Stephen know this?"

--"The Great Spirit of The Infidel Scout, otherwise known as Bart Simpson, who tipped Stephen off that Krusty the Clown is running a turf war against Provident Bank. Bart remains underground throughout the Continental U.S. now that elementary school is over. Rumor has it that Krusty has resurfaced in Redlands..."

Dad—this is more the problem than your investments in New York with income.

--Who does Stephen want to invite to his parents' 50th Anniversary?

--Amanda Knox. Bart says she's innocent, that it was an inside Italian media job to throw a curve ball at Angela Merkel, Sarkozy, Brown, and Obama before the Summit so the Italians could get control of the Bank Regulatory Board in the Western world.

--So, Dad, this is more the subject than a third repeat of "Street smarts", and I've been counting. Can we move on from this scene?

--"You know, we've debated and commented to High Holy Hell on the subject of diversity, but we've never really gotten to the point where we total up the points on the subject and give awards, or flip over the card, and there's an ace and a jack!"

--"And the free "Get Off The Hook—Everybody's Human" cards were distributed in the Bourgeoisies' lottery."

--"Don't criticize George, his policy on the environment rang a bell with Rip."

--"Rip? Rip who?"

-"Rip Van Winkle. This stuff was mentioned 20 years ago, but the whole family is awake now!"

--<u>From Shit to Fan: The Polemic Inserted in the Dichotomy of the American Congress During Iraq.</u>

--"Even if Redlands won't work with Soltesz, Soltesz knows how to work!"

--"My God, did you see the barrel-chest on that psychologist?!"
--"That's what we say, 'The patient got hit by the ice cream sandwich truck!'"

<u>Sailing Forward With More Wisdom</u>

The music, the food, the cigarettes, the menu of Happy Hour drinks at this '09 Rosh Hoshanah celebration, this yovel, has produced the following, despite the feeling of a headache: As far as my undergraduate work and its current interpretation, we have the father's belief this is a goose egg when silliness was the reaction to the serious questions of, "What was that about: it seems to have an infinite aperture and closure like a nurse readying for retirement who is treating, closing, and reopening a sore?" That's where the imagery of the Venetian glass sailboat came into play, the request not to drag a red herring about Christianity because they are not academically founded enough to still confusion, and the realization that other than contemporary Naple's closure with family, heroine problems and other crime, where are we now is the question. Remember that regardless of the past, all has started over and be polite and aware as you sail your own boat forward, to pull more strings so that you do not get the impression you are going back in time. Yes, these girls from the past are most likely married with children, and that's just common sense after twenty-five years.

What this is, as all are grounded in the difficulties of the Golden State's recession economics and pressures of repression-like effect, is a séance from North to South to ritualistically extend our roots forward, to make the best of our careers. As far as logs in the eye, if this is why, that is understood, but I definitely will remember how much was expected for the reception of the truth. I actually cannot stand to have to carry people at the Master's level who don't handle issues at the Bachelor's. The synopsis is not a kitty to collect when one otherwise interprets the situations accurately, takes care of thyself, and keeps in shape. So be careful;--narcissism for the middle class to bear is the shy, cunning and baffling power: One should shield thyself against it by thinking often and pulling forward, while getting the garbage off thy coattails. People want something on the subject of college every step of the way—I think it's more about protecting secrets than socializing at this point, in as far as the interpretation of past and present from previous years, where we had a generation of kings who ate off others' plates and did not know what they were doing! God hates that! So he will rebuke in an encompassing manner. I got hit by this wild boar every time I came back from traveling, not speaking of Dad. I'll not "step in it" by looking them up! There is absolutely no string to pull but the money!

Ascending Stairs Externally

So, I could, legitimately, ascend the staircase from The Pope's Rome to King Arthur's contemporary court, translating all said history between the two by the power of ritual, simply by remaining current status-wise as a writer in the English language. At this point, I point out to said higher power that as Guardian Angel, do you believe this is the divine reason the Lord placed me, an American, in this worldly position? What could clarify my doubts on said subject, now that the question has been posed? Off the subject of sin, am I still healthy enough to carry through, my age set aside, under authority of The Heavenly Court as well? All of this, the wounds on my conviction, must be embraced now. Dwell on said enigma, with passion.

On the subject of human nature, what is it Christ scrutinizes in me, myself wary of falsely prophetic and self-aggrandizing Jews and Gentiles alike pretending to have the authority to perform a scrutiny which is Christ's alone, that He would ameliorate at this point in time, in order to reveal more of the Divine wisdom on life? Is not the relationship with the Trinity personal, for such a Waldenesque person meditating on the direction The Father steers the subject's boat of faith, sun up in the window at morn by the tall, Italian pines overlooking the glass table at the second-story kitchen window?

The Longest Way Home

Psychologically, to enhance and maintain a contemporary, demon-free mental state, Judaism appears to be the building block, as Contemporary Christianity repeats its study of patterns in a historically-prefaced perspective, noting the similarity between past and present human figures male and female, from demographic to demographic.

Therefore, as has been suggested by more than one minister, looking at (a) Jewish Biblical roots, and (b) where a present, academic Jewish psychology(not necessarily political or ceremonial), meet may be the starting point for a less repetitive, forward, moving contemporary and ecumenical disciple's path to more freedom in this world. Long study. Lots can occur in the meantime, while searching. What shape are we in when we have a new platform for freedom, pray tell? When we are actually put back together we gain much, why, feeling as if another degree has been awarded. Concisely put, yes, people from the present bear distinct similarities with people of the past. You <u>should</u> be able to close the door on the past for both spiritual and physical rejuvenation and comfort. A psychologist in a small town does not really pull weeds for a living. Does the gardener, then, have an idea? Is not this in underlying manner a discrimination issue? Are you aware that religion, race, and gender on this issue seem to be the only three dots connected in an otherwise large, cultural equation? There are limits to my thinking and my achievement. Am I <u>sure</u>: Am I sure that is the scriptural study to free me from this dormancy? Because, as I mentioned—it's a long study, irregardless of my other attributes and accomplishments.

Yes, The Christians Around

I awoke after two trips into town last night wondering what on earth regarding the past had crawled into my room and my mind. I walked one mile into the center, after a bowl of cereal, bought cigarettes, took twenty dollars out of the bank, and drank an ice coffee, before another something crawled in on the subject of past roots as I got up from the table to walk the mile back. I debated the nature of this and why the checks and balances purported to manage my new life have failed. A soft-healing voice offered a swab on the headache's chemical and behavioral obstruction to say, "Don't give the Christians in the area a bad name because of the failures of a few." I hope the taste leaves my mouth soon of that diatribe's exchange of conscience.

Rolling Down The Coastline; Twenty-Five Years

Yesterday with the doctor I discussed a few things I had jotted down, as normal, before getting to the office. At this point he not only takes his paper of notes I jot down for him, I must then reiterate that (1) a former psychologist says this has nothing to do with being psychic. Yet in the neighborhood, a degree of differences economically between children and families above the lower line, are handled collectively to keep all abreast of points in progress and digression. Gossip has evolved to a pragmatic discussion in lieu of my proposed complete abstinence from the practice for reasons of repetitive behavioral characterizations amongst the haves. Hence I can sense instinctually what place the soap opera has left its plot points, and the dialogue with the Holy Spirit, in an often Waldensque apartment morning, as was described to him as unique. I apologized for my previous anger with the doctor, though stated this instinct is why, as I have moved forward and no longer feel certain issues are mine religiously, we have the proposed psychiatric dichotomy in <u>any</u> church body.

I was informed a new smoking cessation pill has been ordered: He told me Chantix is an anti-depressant the insurance may cover since Uelbutrin it does not, and we have seen where said behavioral tensions can sabotage a good beer, even in privacy. Much extra to the collective game on this subject, the Lord silently reveals, as I click on World News Tonight and Fox News consecutively, is to be processed in these two layers of the external (old neighborhood in another face of the town and the America I see, past and present, that I am connected to via the news). Amazing: It is the same law of nature collectively on tensions, secrets, and reputations amongst the America I remember meeting even years back, all reading between the lines as news of politicians from other parts of California transpires nationally. Without getting into patterns of discrimination in The Golden State from a twenty-five year historical perspective to the present, it's of humorous note how the old college soccer ball rolls all the way down the coast of the state, from Congresswoman Pelozzi to Michael Moore.

On the subject of my resilience to smoking-related problems, I believe the Neapolitan in me will see me through until I quit and beyond. I would leave it here, yet economic difficulty in the City Hall debt and the funding of The Library simply may be more affluent Christians' own rivalries amongst the city's differing polemics, hence a narrow, middle road.

"I Know What You Feel Like, Santa"

On the smoking and psychology, there seem to be ways of waking up. One of the former people in the old neighborhood may not be the sharpest person. Also, some of my classmates from here just aren't well-socialized, even to this day (Hence no idea why to keep in contact). The smoking of the first cigarette seems to go in the hole for family confusion, to fill that. Medication resolves focus to observe these things while diet, balance, and chemistry promote closures\ apertures\advancements and\or backslides. Other than talk is one thing, I feel lighter, and yes, different, though it wasn't the fluidity of chemistry I expected, that smoking had stopped assisting. And, on another note, maybe it's a different value system between psychiatry here and psychiatry there.

Wait: We've got a much more serious factor: BLOWING IT OUT YOUR ASS, BUCK PRIVATE-STYLE between patients' goals and medication's objective.

***I would swear, between Parish and Media, the University has no clues how to keep me moving up other than a new reading suggestion. As far as my <u>own</u> work—you would be surprised as to the roller coaster around the loop of past baggage and failures we do despite exercise, diet and prayer. Too negative? Missing something?

I've been told not to complain, and not to tear in to anyone all day, by polite insinuations from the media, speaking of others, though providing a mirror for my mentality in light of current issues. I actually seem to get my head put on right sometimes, and sometimes no, by programs and news, religion and University, and the Internet +phone as they prime right for the middle on these things. Told by Dad to continue quitting smoking, now that I've worried off the charts.

The Job Market And Others Called

My godfather Frank had told me about St. Francis Cabrini. A nice parish, yes, that gave me notes on Vatican II. It was here I thought twice about whether or not current world Catholicism was for me. I ventured out to a Baptist church, a small one, went on a relationship conference, worshipped, prayed and dined with them. An interesting teaching, though it was incredible how much my mind slowed in the absence of a more profound, ritualistic, and educated hierarchy. I also went to cafes for peace, conversation, and perspective, bars and restaurants for drinks and a look at the inside of my community at night, forewent donations to save my money, tried a homeopathic smoking cessation idea, and completed the rehired second phase of my Census Bureau temporary job. It was now time for traffic school, and I was wound up. I opted to go to the Morning Mass, and give all this time back to a Catholic Christ.

Watch Out For That Brick In The Courtroom!

It is clear what an entire dialogue in this journal, for publishing reasons, should include. An explanation, and trouble-shooting ideas, of the Judeo-Christian complex, based on my twenty-plus year experience as the brother-in-law of a Jewish husband. His generation: I believe the oldest section statistically of Baby-Boomer. I, the last of this time frame in American sociology. A great paradigm behaviorally I have at my disposal due to the intellectual, pragmatic, and social factors. So, this will not be a detachment, or dichotomy, such as with Generation X at times, and neither in objective fashion, we have learned, can I use empathy as a crutch. The sociological bricks thrown in contemporary, and recently historic, America, become a factor if so. Hence the point is the effectiveness of the decision-making potential, which, also, until optimum conditions are stabilized, cannot be leaned on. The seniors really did work—there is no dog crap now to step on, in the middle of the dialogue, yet costs were needed and losses? Let's not cut each other apart, let's cut the turkey!

Now, as well, regardless of economic differences in this town, put all people in the one and only spiritual realm, i.e., the Holy Spirit's interpretation, and solve the behavioral problems. If God does not challenge us beyond our abilities to resist temptation, then it stands firmly to say all the problems can be trouble-shot, without narcissistic repetition, via spiritual law, balance, and dynamics. If you believe this is Christ's power, it is. To solve this in adult Catholic maturation is to have a healthier platform for contemporary American strategies and a more cohesive worldview. Local, national, and global.

--"Hey, wasn't the deal you handle the ex-boyfriend's baggage if you tie the knot with the ol' lady? What did you think this was, a free lunch?!"

Football

--Building them up, not tearing them down
--Hustling up the plays
--Teaming up to make the tackle
--Positive plays
--Mixing it up a bit to make something happen
--Coming off the edge
--Deep drop
--Reading that
--Sniffing the thing out
--Close the gap on the hard D.
--Full of personality
--Little bit of traffic
--Make the third down achievable
--Setting to throw
--Knowing it is coming
--Line drive ball on the kick-off

Another City Calls Come Summer

A new art project idea: A standing, life-size version of The Savior in sandals and an American flag for his tunic. Call it, "Father and Son." See if you can draw Pat Gilbreath. Draw men unloading cargo by hand from an old barge.

Yes, I'm full bilingual in Spanish, for what I lack in absolute fluency. I am of fair hue save for my dark arms-- even my mother has called me a gringo on some occasion, and I believe between this and her finances I got the feeling she attempts to give others a problem, even though I have lived on my own. She's nice most of the time, and helped me learn to cook well, why for a long time helped with the laundry chore while I helped with the patio-cleaning, car-washing (Dad handles getting mechanical issues getting taken care of and insurance), and she instructs me on how to transplant plants around the property. I've transplanted a broken branch of a type of barrel cactus, on one occasion, into a new pot, watching it grow into a whole new green cactus, as well as gotten succulents to take into well-placed new beds here and there, by the sun of the palm and the shade of a juniper both. Have you ever weeded a rose garden? If it sounds like a sticky subject, send me in on the ground.

This can be the beginning of an Hispanic identity—being an intellectual is not well-facilitated, as she changes the subject from my academic background, even though she taught me to read when I was three. Taking this intuition, that it's tough because of discrimination and can lead to bitterness, leaves me with an understanding of why she keeps this edge to overcome. The cultures are not always easy studies. I do not like being failed on the subject of her vows, when college produced research on my part. It is a complex issue, and I believe with Catholicism, I have an opportunity to recover from both Anglo and Hispanic discrimination as a bilingual, bicultural individual by experience. Yes, all this, before the subject of the third language enters the dialogue, as well as building blocks in a fourth. So, Hispanismo('Hispanic thought'-- more influenced by the extension of tradition in its potential to accentuate the straight line in assimilation without a drying out-- as opposed to the more renowned, contemporary discussions of Latino modern identity influenced by Liberals) must be recollected and juggled, along with Europe, in a worldview.

Well, an hour or so later in this same day, I am turning on Channel 28 and being passed on the matriarchal sufferings of Chumash women who lived in a time-frame when much California history was lost. In so many words, they lived at the Catholic missions, and some converted, though others experienced the adventures and tribulations of the suffering that came along with the changing, in a historical context, of the path of The Golden State. The lands sold that were for the Chumash at the timing of the Independence of Mexico. The theft and rape of Santa Barbara paisanos originally invited to stay at Chumash domiciles. The retroactive crimes, and imprisonment for such, of Chumash men who passed away behind bars. The discrimination from people of wealth toward the Chumash lady on a quest for the balance of nature in Santa Barbara proper for her own life. The relationships with men these females of said descent went through, and the children they had, via their often brief marriages.

The anthropologists from the University, who documented these things of their culture and language, took a Chumash woman to some of the larger

American cities(Philadelphia, Staten Island in New York, and the Smithsonian in Washington D.C. to complete the research), a world of excitement for her to be out of California. As a Californian I have travelled to Texas. After my cousin originally from those parts, in the U.S. Marines, invited me to his retirement ceremony in San Diego, I went to see a San Antonio and an Alamo with a different facet, the history and tradition, than the border city of Del Rio in my youth. My family perseveres on a wide American scale, yet it perhaps is not talked about amongst us, the spirits that the Chumash women experienced. Which very well may be the same spirits that visit me in my Redlands abode, those of the Chumash that had me experience many relationships with women, all be they foreign, in my thirties working in Santa Barbara. I held back the Texas thinking of what I lost from my California life to gain the knowledge o f love and healing of the soul in that city. In fact often it is as if neither family nor, recently, University understand. As Dante would put it, or analysis of him would, I am a Romanticist by historical connotations, Materialist in tie to such with a keen faith in future understanding.

Many of these women took quickly to books, not to conflict, yet preserving the land before education is the Indian way. I asked for Santa Barbara's opinion, and my Hispanic mother has honored me on said condition for returning safely home, believer in the Procession of Saints past and present she will always be, born a native Texan of Mexican descent. A compromise with the spirits of my great-grandmother born in the parts of Coahuila, Mexico, the state who suffered not unlike the Chumash. I see Judge Sotomayor in said light off the didactic subject of the issues and good vs. evil, knowing economically we have started to earn Christ our Savior's forgiveness as a nation. I became a member of the trail of Reform in America in the year 2000(yes, before September 11th, 2001). I confessed near thirty relationships of a short month or two, over the course of twenty years, that had not brought the Sacrament of Marriage to fruition, in the Episcopal Church I was absolved by. I was absolved as I confessed the sin in order to convert to Roman Catholicism in 2008.

I tried to maintain serious relationships on a pattern striving for the opportunity to gain symbiosis, for monetary reasons of my own not proposing outright (for in my family, to marry, you must be independent, though to be an adult, no), yet the spirits bothered my relationships with twists and complaints, small arguments crescendoing into a polite end of intimacy with pretty females as well from California and other states, who enjoyed not only my intellect and work ethic, yet my heart and my festiveness. I am now with only one date, albeit arranged, since the U.S Armed Forces, whom I support, invaded Iraq. I pray for them.

Despite our family issues and the spirits I am true to the U.S. flag of Ross, not alone to that of the Golden State, and if I suffered or still do, my leadership in those other American cities is brought to me often via technology. I miss them, the Santa Barbarans and the Americans plus cities from my American youth as an Air Force Major's son, constantly relocating as our family would transfer from state to state. I am alone in the City of Redlands now, the home of the Smiley Brothers' treaty between the Federal government and the Serrano, another tribe who suffered, yet was finally granted land through the Smiley's efforts. I earn my meals and I walk past their historical A.K. Smiley Public Library to fight the common cold, dreaming of a more romantic California life married to someone if I can earn what it takes to settle it right this time.

I have given birth to twenty-nine artworks now, since making the transition from Episcopal to Catholic. Christ brought these works out of me, of mainly conservative yet articulate hue by design of Christ, absolved an artist who had his first relationships with fellow travelers in Morelia and Mexico City, while seeing the murals of the great Diego Rivera. As well, I was at the Lake of Patzquaro, viewing the Indian artisanship of the State of Michoacan', while having saved my money to eat Enchiladas Suissas from the skillet laid on the table in a restaurant over- looking that lake. It was more important a meal than with my travelling companions from the University at the five-star, while viewing the cliff-divers, before purchasing a hammock I no longer have from a European in Puerto Escondido. I now have a fold-out couch I sleep on just like in Naples. Psychologically, I remain three hours south of Roma.

Does Bart Have Them?

This Nicotine addiction is about people who know things about me that don't reciprocate. In fact most have been on the wrong side of the law at one point of their life or another. They've all got smoking as one of their past or present issues, and both night and day on sex. At least five have a weight problem, and some with weight and an anger issue, even one, who wants to quit, who is a virgin. In defense of my soul, I've witnessed to all these people. Most recently to a heavy-set Jewish grandpa with a mental disorder like mine, but successful though divorced father of two working children. I heard a voice: "That's what that's like, you playing Dad," and a pull of smoke from behind me made the comment. "That's *le streetsmarts*, Stephen."

It took one to give one at Brand New Day. "Terry" brought up his involvement in Cocaine Anonymous but gave me advice on how to get a job. But, while I was trying to quit smoking, he brought up closed systems, with my Christian background on these people in this town. Hence putting the foot down can make me restless. It hasn't been a highbrow discussion in Redlands from the beginning. You would need to know the inner critic's narcissism and jealousy as lived by some, argumentative for the sake of their own profit. I looked up suddenly at the silhouette of the angels holding hands to make a clay candleholder of four little lower orders in a prayer circle, and behold, a shadow of Bart Simpson had joined their circle! Going where no God had gone before. We would keep praying to quit. We have that many pieces to the puzzle of addiction as it sits.

"It Takes A Support Group"—There Are None Now

Why don't people reciprocate? Because the party does not work for them. We all do that, or have done that, in this process of moving forward to better moments. I actually had a tail-end of a party, where the leadership I provided by the availability of a place to come was utilized, and then I moved, and it worked for a time with the new neighborhood, but there was a competition that led to a crash(other than vehicular) when new people were grouped together. We-had a party-crasher that wanted to hold court for bringing a case of beer, when beer supply was not the problem. Guess who? The leak in the parents' neighborhood!

To start over from here, I do not see other citizens getting inside pressures on their socializing efforts. No apology has <u>ever</u> been issued. I'm absolutely sure to "slip" means to include rehab reflections in the dialogue of an otherwise progressive Christian life. The stark dominoes on narcissism keep getting me for what the fear is—parental flip-flop, which I <u>know</u> now due to newly admitted evidence as a conservative compromise to avoid (not a fun movement). I mean, the word is peace, not hernia. To move from here on smoking cessation, now that alcohol and sex are <u>characterized</u>, not demonized, initiates new progressions.

Christ And Cigars

When the environment is being manipulated, you can keep moving, and take time to slow down and get your center. You have some suspects, or the spirits of such acting as impressions made on your memory, as to what the resilience in the environment is. Suspicion is currently natural in the Redlands area due to recent robberies. Although the two issues, the personal resilience you face and that of the robberies' tension slimly would be connected, you still may be aware of your own tension. As you know, Los Angeles can be a great place, though perhaps for your life issues God has been elusive about explaining why there's no place like Redlands!

Is heavy smoking a domino effect of issues and experiences, psychologically? Is <u>this</u> it more than physical addiction? That, I recall, was pointed out by one company offering a product, in so many words.

As far as how Jesus enters into this subject, He is a universalist. He made the whole thing!

The Perp And The Narc—"Sogni D'Oro"(Golden Dreams)

I dusted off my sandals, as I mentioned, and listened to the Baptist from the South for awhile. One of the ladies with a very strong and hearty singing voice as well had a blessing for spiritual fortitude, which, in retrospect, came from absorbing more Scripture than I do on a regular basis. A young father in the parish attempted the same image of fortitude with Scripture, yet he lacked courtesy, which she actually, all finances apart, had become rich within this part of The Body.

Upon returning to regular Morning Mass at the Catholic Church located in The Historic District, I reinitiated talks with my Godfather. We had breakfast at a family restaurant one day, and he talked my ear off, in stubborn fashion. I actually caught a good dose of the flu that day, between him and the coffee shop I went to in the evening, especially when, late at night on Halloween, I heard an aggressive man make a young lady entering my apartment building yelp. I snuck in the bathroom and called the Redlands Police Dept. on a potential date rape. They responded quickly, and now, come mid-November, the building has since been tranquil. I, however, got run down, despite my two-mile walks for fitness. I finally got my flu shot yesterday.

Getting back to the observation to be made, there are two paths in Christianity at this time. Against the nice guy even though he attempts to follow well and be supportive alike. Hence, what is correct? And, after all, do not both paths have their successes and failures in Christian life in the homestead? Is this the true root to division in America right where we step onto the carpet? Both theological and academic believers can be snobby and lack refinement. True believers with strong hands in both categories who surpass this to a state of more humility only increase their strength and testimony, however subtle, and should not get confused by the hairs others split to decide which path to take, or what to pack for the trip. As if I did not need a huge art collection for my testimony. Of course, the catch being, I created all the collection, and now may fund my future eventually by selling replicas of my Christian and patriotic works. First, though, I gave my parents a computer's C.D. with all the drawings stored on it, in case they want to send a gift copy of a nice drawing to my aunts and uncles in Oklahoma, Texas, and Louisiana, Virginia as well, why Washington state and Arizona too. Grandma and Grandpa from Texas sent her little disciples all these places while on the Hungarian side, we are in Connecticut, Florida, California, New York state, and Australia. I sent a couple Hungarian works to my Dad's cousin, Uncle Tom and his wife in Connecticut.

Preferably Prior To Conversion!

Idea: The Roman Catholic Rorschac Exam

Hungarian Picnic!

Sometimes, though I per se' am not versed in actual astronomy or physics, I have noticed that as far as collective consciousness, or quite simply collectivity, perhaps leading to a happy hometown, the following is of interest: The morning sun, peeling off the layer of overcast weather, actually is a point in the sky, created by the Almighty, used to focus (or refocus) a Christian mind with God's opinion, though I won't speculate it to be other than the sun centered amidst Earth's orbit and our heads spinning accordingly. To look into the sun is harmful, though perhaps to observe the pattern of it falling into an overcast layer of nubuli from previous low appearance on the horizon, and then reappear amidst the crests of the tall Italian pines, is simply an observation we can make as we write by this morning light. It seems the layer of stress created by obstacles in our path is used by the Creator of Nature in a certain symbolic state as the overcast, damp layer on the landscape. It actually can be an atmospheric pressure, the layer, or is, in assimilation to the changing season, all this simply the obstacles on the plate of pragmatism in the morn? Who can break fast as such?! What a stomach one must have. Changing the subject, I believe we do not speculate God's eyes peeling through the circle we cannot grasp with our own as if the sun were His inter-dimentional peephole (assuredly he only watched at night through the lens of the moon!). Neither does he psychologically shock us with his power if we stare straight into the sun to peek at the Heavens! In case you, as reader, wonder what in specifics it is about the right atmosphere in which to write that is important.

As well, after Madame Sarkozy's denunciation of certain Papal policies and the Brits' insistence that they get invited to the international remembrance of V-Day, I began to wonder where Tony Blair left off with France, after conversion: "No more private parties?! Why, assuredly. A little privacy. A little privacy. A little more privacy, baby-y-y!" I think this is the power in time where I needed a Hungarian Picnic, followed by the Anniversary of The Wall falling, Pink Floyd's or Berlin's I care not to speculate, though I like Madame Carla's music. I hope it stays acoustic. What an afternoon tension-breaker or early morning gentle movement. Now, in one small flow, we may hold all these celebrations of remembrance and Masses of Europe, without them getting negative about "the Americans" or dwelling on our own issues, without reaching for the phone or the beer, without going through the fridge like a bear, and drink that particular cup of imported Medaglia D'Oro(Gold Medal coffee, for those who do not speak Italian), that cup right there as the sun pops in and out of view in the Redlands morn.

The Nation-Wide Climb On The Subject

The premise with which I inaugurated Tales of a Schizophrenic Savant, once I had needed to pinpoint the stagnancy above and beyond my own psychiatry and discrimination: Sociologically dysfunctional, leading, again, to intellectual stagnancy, and in turn economically challenged. Who? One and all. This, apparently, has been a two-way argument between intellectual and majority, leading to criticisms on both ends of discrimination. Hence a wrestling match academically with not only the moment, yet forward progress, as well as passing, with conclusion, to Catholic standards. Quite obviously Catholic would opt out of the challenge for a Catholic educational view, for self and family of the future. The collegiate system as seen in their own dysfunctionality between social and academic ties, has asked for a Brain Default Mode Network (Where friends of the past, in a cortex network reinforcement context, actually are support as memories, whereas before the challenge was too severe). As well, a Crystal Ball Theory exists (actually a sign of depression or arrested development, but where we can all get the jist of where we are at the moment, in other words, a domestic version for the educated mental health patient less complementary than the Catholic global and universal explanation). It is a set of explanations for the secular hypocrite, strong the stench of the sin popped from him, which symbolizes their failure to intellectually and socially, prior to medically and physically, adapt to the exigency of the corporate world's efforts to turn over a new life.

To rise from such morbid Western circumstance and stigma, created by SSI\ Social Security for the American worker, regardless of otherwise fine repute and amicable competitive esprit, is not unlike scaling (yes, taller than El Capitan in Yosemite) a virtual and invisible new Everest in the arena of other socio-political and economic events. Do you troubleshoot the village at the base, or begin to adapt to the ledge upon which the climb progresses, constantly looking from the weather's signs? Snowed in, in effect, is the sensation, as all negatives have been laid flat out.

The sensation augments when, taken into consideration, the request and procedure of logically troubleshooting the inefficacy of the intellectual-majority relation can raise question and lose more economic time. Hence cultural sacrifices to both tobacco and liquor quality are made, in fact quality establishments on a regular basis almost remove themselves from the dialogue on their own. Can we actually progress forward with said questions raised, or will this further chasm the dichotomy?

As we look at all the options in disabling the dysfunction's hooks, the Church in turn has continuously more to say and do—Pity one cannot work there. In effect, to laterally move from a sociological to a theological framework is a transition all its own, and the Church's transitions juggle further non-academic questions as well. So, <u>where is</u> home for the intellectual in pursuit of a Dantesque Romanticism? Good question. Especially since Dante and tobacco go hand-in-hand, the change stemming from a Renaissance root in dialogue with the Almighty, and the intellectual is, again, sacrificing. Why, for said reasons peace and discovery groups in similarity induction-wise to The Priory of Sion and The Masons formed. To be a poor one of those without mate is not a subtle stagnation either. Hence the village, once some dysfunctional side is waved away, may manage to slide said persona back to privacy, with consistent yet spaced-apart contact with the Church. Similar to a priest. As the Church pointed out, all believers are not just brother and sister, yet priest under one High Priest, Jesus Christ. So, to support the priesthood one must seek services from, yet avoid, the dysfunction side associated with the secular majority. Why, this is getting to be mental architecture! Quite difficult, once some sense of inspiration from Christ has attained a foothold, to feel at home with an un-wanting yet congenial world. And work, Scripture itself never seems to preface an opinion, only deduce one soul's behavior.

All this time, while non-fiction climbs Everest, where in the world did fiction get its respect? It's as if this is a bake sale, not a market place. Oh, for the childhood days when fiction came easy and spread love, not just lesson. In said sense, while questions arise bridging majority and a Church with intellectual, where is said love? Have we not anesthetized it, through the ritual of the seeking of an answer? Stands to venture a veritable Madame would have to offer the reward, to reaffix love in balance after the climb. So was it love that got us this far, and approximately here? Very solitary this walk with Christ, in fact, there is the hook for nicotine again.

Now, you look at <u>where</u> in the Continental U.S. this societal dysfunctional side is noted, and you watch the weather segment on Channel 2 news with Katie Couric, with the sweeping cold facing a significant section of the nation's constituency, and you wonder. Have connections between warmer American areas such as Redlands, CA, and colder areas suffered so much, given the two poles' other influences, that this is not a picky, rich female's argument, but an environmental oxymoron that has some Americans seeking Santa and their "child within" of youth, their youth memory amidst a wave of an information assault, while absorbing a well-established but facetious protocol with documented world news? No, that's not it, as 1 out of <u>all</u> 10 Americans are out of work. The future is bright to grandchild and grandparent—it is the middle phase of adult, 27-40, in this country that is efficaciously calling for an appeasement of the pain of the struggle, while the other two age groups say, "What pain? What struggle? The future of America is bright." Which, does it not, bring us back to strong discrimination for 27-40 year-olds and a smoking issue to appease it? Since we all wake up with intent of achieving.

Don Stefano At The Wedding Of The Lamb

This Redlands life had been going on and on, amended, abridged like an ethnocentric Italian author's version of me, the American female, and a bastard of an American nation, until finally I realized: This is not a bad book I am a character in, this is a wedding album, of a marriage between Christ and the Church, and Jesus is looking over how much disciples can take as far as their role in the wedding. The entire dynamics of contemporary Italy lining up with the bilingual American male require a certain flexibility on both cultures' perceptions of the disciple to maintain a conversancy that gleams the Lord's love during tribulation, not simply a rote and long-term memory and a well-kept "temple" by scientific scrutiny. So I'm not a character in a book—I'm a face in a wedding album, other than mine, of a sacred and powerful sort.

Humility

I figured out, besides realistic hometown issues, where the tension is coming from, and, the energy goes.

What The Bar Should Be About To Me—A Heart-To-Heart

The discrimination continued in The Roman Catholic Church, and it was not the devil. Christianity, outside of nature's way, was beginning to have its problems across the board, in every sanctuary, across the globe. The world showed signs of repentance, reconciliation, and reconstruction, yet a straight arrow into freedom and peace became harder to shoot on Sunday without a turning over of the tables for selfish reasons somewhere in America. The procedure was downright carnivorous, the appetite for winning the competition, and secrets were juggled more than rules, as if the power of such finally brought an accord with the modern-day knowledgeable and a Merlinesque code of wisdom. Eggshells to walk on had been left on the ground from one end of Paradise to the next, after an Easter Sunday egg hunt and feast too rich for the palate of a healthy cook. It kept bringing to mind questions such as which horoscope writer is having an affair now, what on Earth calendar year could it be in China, or Israel for that matter, before we began to split hairs on other ends about the particulars of the new church season and its customs. It must have taken years of private theological research all over the world to arrive at a current Christian psychology so complex, money for a spicy and mature relationship became a soldier's surgery on the time bomb the issues metaphorically had become to the straight White male in the economy on the pathway of Christ. Better off hiking barefoot over the broken beer bottle glass in a hidden Californian Indian canyon.

To say no incongruencies between Catholicism and fair Christian play in life had not been noticed would be stretching the truth as far as the $200 a month I had left over after the rent had been paid. To actually gain aperture as an artist to the inspiration Italians of antiquity had, in the current American Catholicism (without developing an aspirin addiction) become the least critical danger I saw to avoid by leaving this faith for better times. This simply had not been the finest standard for a Democratic good neighbor policy I had seen: Without regretting the duty politics one more time, I remembered what had gone wrong with a supposedly promised productivity in Italy (when things were not focused on the bright side), and thanked God for keeping me here to see things more clearly from start to finish. They would need to substantiate the role I had been asked to play while speaking this language with considerably more compromise on quality-of-life maintenance themselves. The Hispanic Catholic as well in America became redundant on the deductions of discrimination, economically, as well as evil spirits and\or demons to explain religious context of the soul's challenges. Readings from time to time of Dante were almost a promise that took an eternity of sacrifice, as if the Christian journey was the reader's time borrowed by an author now in Heaven who once there, could not honor an Earthly timetable. The priesthood fell short on the logic to explain their own misconceptions of what human religious direction deserved, if Christ loved and rewarded in the manner many now understood through maintaining a belief in connection with Christ in their abodes. My new decision was to use the thought that came with faith to continue higher education and its' mission without getting into the conspicuous survival struggle of overly-direct Gospel proclamation, until I had more disciples than myself agreeing we were worth more intellectually, and a team had an understanding of how to let the individual cultivate his dream efficiently.

All of this said and done, of course, after the damage the secular campus had done to said Christian scholarship in a modern, technological desert, as a path was walked with thought of God in isolated manner, it tucked in the very look of the notebook with an artistic glow, to the response of the resilience. And whether it was Irish or otherwise, the double-standard in pubs on how to cultivate amiability packed the downtown business community with an economic problem best avoided for a higher-costing plate's less insulting talk. To think military men were boxing foreigners in earlier wars while we boxed our own after ordering a chaser. Scars of irritability seemed to follow the Western World drinker like a terrorist's demeanor. My loyalty to flag and cross turned a tide of said nature days on forward to a keener look at the power of Christ's chemistry, which for those who had filtered water for coffee and tea could still bank on love for life until the next glass was raised. When I could stay in the present without evaluating my past this Iraq War was then an out-and-out victory, for that was the time it had won me to do so as sits.

President's Day

I feel I became pigeon-holed in order to be carried for in academic circles, more often than not…similar to how I feel about family? So, then, with genuine belief in my ideas coming from my mind in order to proceed, could the diagnosis have been avoided from its stigmatizing under-current taking center-stage to produce schizophrenia? I could never know the truth, now far-removed from the actual time period's reflections, actions, and events. However, I could continue to stay on the same page and build up, just as this phenomena is seen in assassination attempts or other political and social occurrences in America. Hence I am back at the beginning, an interesting mind set aside, being driven cordially by this nation's leadership in various levels that filter down to the present space and time. To take initiative is a forte'. To say I'm dwelling on it when it doesn't feel right—if this is Christ protecting me from disaster as potential actions from bitterness of others is imposed on me, more respect for Him shown and the surface of somewhat of an awe-inspiring fortitude. Lots of theories on these feelings' origins, to mention it, exist while I'm politely acclimating to my age and situation, religion and road. Am I reading Einstein for an academic reason tying in to my own situation? The original idea was to open new windows in my mind, air out my thinking tools, and proceed forth less tenuously in a literary sense.

--"Now we know Hungarian interpreters are pretty hard to beat, but did you hear the one about the Hungarian Bobby?"

--"He crawled into the pub to address a complaint of 'stolen fish and chips,' and (1) wound up diving in the water to retrieve the plate, after the perpetrator escaped in a stagecoach with a screaming cocktail waitress; (2) Found them(the fish and chips), and swore by it in court even though the establishment complained. They were behind the men's lu!; or (3) Caught the perpetrator nibbling on them in the alley, knocked him out, finished the meal, tossed the plate to the brick wall, walked back in with the perp's wallet and put a pitcher of martini's on the stolen credit card, a round for the House!"

Enemies of Redlands Past And The Latino Precursor Now

A huge, bacterious air and tension pervaded the kitchen window along with the piercing shine of a freshly risen sun, right at the table with my glass and metallic candle-holder carrying the two 50-star American flags. As I settled in the adequately-shaded sofa chair, an external, course pressure in the ambience let me know that somewhere from beyond the windows, a person was thinking right now, "I don't want America in control." I had heard of Marxist attempts at deconstructing American strength, yet had never known this to play part in the Redlands area. Back in the mid-1990's a retired USC professor had mentioned Kirkengaard, the Dutch author of works on Christian existentiality, available at the public library. In addition he lamented about our system and appeased Communist thought, all in the time span of three cups of coffee while mentioning the difficulties and pragmatics of the business side of screenplay writing, The Writers' Guild America West, etc. He was elderly, used to walk the Historic District alone with a burdened brow and had not been seen since that time. Neither to mention it, had any like intellectual mind ever conversed on said topic again before or after my former President's talk with the Russian Putin, prior to The Surge in Iraq, in Maine on C-Span.

Why would morning prayer in the studio, not hung over at that, initiate said memory on Ash Wednesday, 2010? Certainly nothing in the news preceded this, and my morning call to respond to a postcard in the mail from Boy Scouts of America was quite the opposite pole represented in the dialogue internationally, as an Eagle Scout myself. As abstract as the aforementioned environment for dwelling has been, a needle-in-the-haystack anecdote as such was shed off my new Catholic mind, as well as pondering if an external existence of Malinchenesque(Malinche, Moctezuma's queen) nature could have had the wondrous air bringing in this realization that indeed, despite loyal and comfortable Redlands ways, an evening lay nay.

In the latter respect, despite what surface argument there was, in other words, a stringent grounds and understanding in encounters (passing by with at the most a nod if not looking down or feigning no notice) of male White-female Hispanic version, I could hardly reiterate the said malicious undercurrent the particular queen was characterized by, to be known. To say one noted that currently an apparent belief in the need for simple and respectful distinction between two cultures is prevalent sums it up. Assimilation is a guise of a tool of power which can be used to abuse, not meld into a mentality that delineates the course of a mutual esprit. Yet Catholic is Catholic, Scripture is Scripture, right alongside America being America, as both history and present know Holy Spirit interprets the moment, often with humor, so that I still smoke, and other than Hispanic culture at its other center most here do not. Black and white, no cigarette, no

173

Malinchenesque air first thing in the morning. ?<u>Cafecito, alguien?</u> ("Coffee, anyone?"), as the grandmother hosting in Morelia used to say.

So There Are Critcisms From Love To Life

Do you ever say something to yourself in the mirror, as if putting on the scene to a play, and, after delivering your "lines," so to speak, there is either a forward progression or quite a punitive tension? I wonder if we are on stage for The Virgin Mary, and her court, talking our soliloquies through the hallway into the bathroom mirror on improvisational grounds. After Mass in the afternoon, I prayed my highest priorities until the tension went away, seeing an image of Mary's shadow in the backdrop of Charles Wilson Peale's painting, The Staircase Group (Peale, 1741-1827). At this subtlest of still moments the birth of air was unwrapped from me in an infant's wheeze, as I realized I should begin preparing <u>anybody</u>, soul and mind, for the chance to give life to a healthy child, and stay on this path, taking the same abridges as others despite the thoughts of autism, in preparation for The Lord's mercy and divinity in the future, if doctors may and will continue their resourceful quests.

Quietly, as if whispered from my very nest by The Virgin herself, I thought, "Keep your Dantesque new beginnings on the ground, not in the clouds", and thus was explained the answer to my prior criticism just days before. I, upon return to Catholic pews after a two-week sinus problem, believe in due time Christ will explain, whether through the Holy Spirit or Mary, the not so contrived responses to my other criticisms of the Roman Catholic Church, provided I stay a disciplined disciple of a busy nature.

The Nitty-Gritty Of Martyresque Procedure

We have a simple example where when we want people to face the truth of their sin, we moderate the pain with tobacco and its ingredients, though have to quit smoking, rather than over-explain through reading <u>The Culture of Narcissism</u> (which has been calling us from the unsolved, silent nature of our San Francisco past from time to time). God help us for fear of being painted into a corner. Isn't that where I, as a young intellectual, was painted in the first place, here and there? As an intellectual I seek a broader look, and fine, from a Catholic perspective, but it's not flames, it's progressive, apparently, the burden of nicotine (not the movement), and the pain is Hell: I'm actually there, Purgatory, not Earth, Earth is the paper, serving me time again by using that perspective to make penances to end the ferocious enigma of manipulation, and realize it: <u>It's for sins in a</u>

previous life (point of procedure for the theory to weigh it, in connection with reincarnation's thoughts, equating to Dantesque Catholic levels of comfort).

To realize this is to broaden the look where twofold, you encompass the religious explanation into the whole concept of why you are here. Schizophrenia, to describe an intellectual's look at the area of California's main destructive influence, necessitates a look at the Greeks, not the Jews alone; maybe there can emerge a better equivalent of democracy and comfort monetarily-- I was obviously an oppressor previously, looking at my pocket book now, though I would not say a prison guard.

Am I in denial of St. John the Divine's time-table as it relates to contemporary end times? Am I the equivalent of a futuristic prisoner in an end-days Rome? I did start this discussion of enemies off with a dispute over Jesus both with non-believing Gentiles and potentially-fallen, wicked, two-faced, end-days people masquerading as knowers of Christ's love. Since I put a halt to my love life to accumulate an equitable amount of money for a proper marriage, and have been stopped (for whatever reasons macroeconomic, microeconomic, or personal ones from discrimination to failure), I now live in a glass menagerie of what sex is about in a healthy relationship. The menagerie becomes shards of broken glass tied together from the loose strings of the manipulation I am experiencing in Hell, and I have such a crown of thorns replicated with said artistic medium on my head, with the reputation the mentally ill have on the subject of healthy, happy children of said ideal Matrimonial Sacrament as the cross I hang on. Contemporary Greek and Jewish-influenced California, complete with Angels and demons alike, warring against each other while observing the imprisoned Christians seeking an answer to their punishment, for it was term in Purgatory, not Hell, with chance of freedom: maybe this is the pain of Purgatory, and Hell is eternal fire. Maybe? Am I in Hell, Lord, or Purgatory? The end days St. John the Divine spoke of?

This is equivalent to the pain of what I'm calling a Purgatorial crucifixion in celibacy for aforementioned reasons. If I'm imprisoned now, non-believer and masquerader alike must remember my earlier evangelizations, to say the least. Which brings us, I believe, to the evil we prayed about in Mass, the others who were worshipping Saturday morning with the exact same impression I had of last week, for whom else this sermon on praying for our enemies, by Father Mario of Argentina, was given. How the West Was Won…If You Have to See a Psychiatrist!

Coming Forth In The Heat—To Keep Thinking

To stay in the present is the desire. Therefore, to psychologically go back to the roots of nicotine addiction, for extrication of such, is this the strategy?

To change the subject, I now have attended another Saturday Mass led by Father Mario from Argentina. What a week leading up to it! I see more clearly this opaque plate of obstacles forcing a pre-mature deduction of discrimination in The Church. And in the subject of sex, child-bearing, and Purgatory on a contemporary basis, <u>look what I read in Holy Scripture, quoting the location of the verse by name and numbers:</u> (Matthew 10: 34) *"Do not think that I have come to bring peace upon the earth. I have come to bring not peace but the sword."* The quote goes on to divide man and his father, and put a daughter against her mother. But what of, in contemporary California, young male and female of different families, looking to aspire, marry, and bear offspring? To say said sword smitten here did not influence early schizophrenia is blindness.

And now? Schizophrenic straight men, <u>and</u> women my age(over 40) will bear an increase in risk of autistic offspring! The Almighty's wrath has fallen upon our high school and collegiate disagreements of the topic, and if you were there, yes, underlying motives are the head of the serpent we see St. Patrick pointing to, so to speak, as we approach his patron saint day in 2010. Nicotine, alcohol, drugs, suicides, poorly respecting and arrogant, over-competitive youth to say not the least this, the sociological trail to the hard crime world, why, a repercussion of an age of Californian now under scrutiny from The Divine. Second chances? First, before I raise my head to assume it, I have taken the time to understand the wrath of God amidst our denials of Scripture's worth to avail us of the real problem.

I had an argument with my parents who claim they've now been through certain problems for years, that I should live within my means, and that despite our words and need to heal over time, they will not withdraw their support. That's what Father Mario's sermon was about: Not focusing on our <u>(California)</u> pride if we intend to receive the mercy Christ said He died for so <u>that</u> we could receive it from God. As well, the Holy Spirit through the course of The Sacrament of Communion exemplified why, what, and how we constituents were receiving grace to handle our life's most ardent obstacles. After that, I struggled forth to again apologize to my parents, since a humble spirit, not my pride creating the division in an attempt to champion the verse of the sword and not peace, was His counsel. So through private reading we understand the problem, maybe not the strategy, and through the Priesthood's tradition we understand Apostolic secession and the representation of Christ in our day, to heal schizophrenic and elderly family from the wall they have offered to buffer some in Purgatory from the demons that would rubble us societally down to the last earthquake's pressure, by absorbing their impact in daylight and spirit world, by continuing treatment and correcting our own wrongs amidst our enemies of other persuasions.

Will Christ simply preface further division by sword? Who will my fellow Catholic be then, and what of my family, my future, my livelihood, my addiction to tobacco smoked with too many drinks in hand? (Matthew 17:24*) "Then Jesus* *said to his disciples, 'Whoever wishes to come after me must deny himself, take up his cross, and follow me. For whoever wishes to save his life will lose it, but whoever loses his life will save it."* I guess that's the story, not talent, money, marriage, and children, losing my life. I think in that light it's been a long 22 years of cigarettes, and I know angels' opinions on God's reasons through nuances on how they mold my feelings to comfort and serenity at home. So from humor to hierarchy, Seraphim to saint, I stumbled that many times to get to 47 years old with this cross, for this State, from top to bottom, to re-begin with The Gospels I read over three times while staying south of Rome for ten months.

Father Mario also spoke of a Pharisee set aside, who debated with another Christian, a tax collector of humble spirit. It seems working for The Census Bureau has been a little of both. At this point who knows more of Purgatory than those not leaving? It appears as we noted to check Iraq, a group claiming a different strategy on California, through the self-fulfilling prophesy of the medical field's current research, has gotten us checked in return, while providing little more than chance romantic encounters and $845 a month with which to pull up to a pre-established statewide bar on work ethic. Those currently in treatment are seeing shots taken from Social Security employees themselves that this is more known as Welfare than Disability, and God's wrath nationwide economically for the lack of respect to go after the reputation of what is otherwise a military officer's son at the wallet. California, Dantesque, or psychological scandal? Culture, mentality, and social norms? I knew these in three cultures before diagnosis so, en fin, whose wearing the mask? My own Catholic in the process of bringing me into the Church, had the manners left to dump that one on me. The psychiatrist from Episcopal side dumped on me (1) you know too much; and (2) "don't go over my head." So now you know what reading Dante, speaking Italian and Spanish, and being a blue-eyed mixed-ethnicity birth are about in California.

Would searching Purgatorio for an analogy or metaphor actually keep the very earth from shifting long enough under foot to do Twelve Steps and abide Twelve traditions? My father's word's: "You cannot have the stone back once you throw it." From the perspective of the Conservative votes I placed, I don't need two separate interpretations of the old-timer I am now in this state. So, that's why, despite words, Catholicism can separate opinion from truth through respect given by attending Mass to learn more about the week-in, week-out feelings Christ actually has for the people he died for—before the Resurrection on the Third Day. Schizophrenia on childbirth not unlike Paul being struck blind while on the road, for killing Christians, and the former for extensive revelry in the University? Statistically no, not a patriarch or matriarch on the subject of college revelry despite similar ideas. Truth be told one political party likes their theory: mental illness is caused by drugs. Another party likes science, i.e., it's genetic: for the raison d'etre of how their parents' family trees line up (however, not always a chromosome 22 Delesion—a complete book missing within a person of this aspect of hereditary mapping).

To finish this aspect of the discussion, 5% to 16% of children with a parent diagnosed will get a mental illness(84 to 95 out of 100 children won't, provided a healthy environmental factor situation), and one says 1 in 50, another says 1 in 100 for an autistic birth. UCLA currently, after tough economic times, has relied on Federal Aid for much support in psychiatry, and is investigating DNA mapping, the call by both doctors that I talked to on how to avoid natal problems offering statistics. These are the beliefs, concepts, and issues I work with. Nicotine, however educated you are, isn't academic: you need help to quit, from the Holy Spirit and his friends. I'm giving it my sixth try in 22 years. I've tried five times in the last two years. I've managed my record to be nine months. I write and pray, walk and eat well, to make it with my valued stride that was a Federal employee rehired twice. I chose to quit now, not take the third offer (a commute to the mountains to train and work there, 20 hours a week).

In any event, it's not completely sad, as far as winning the West on this topic. God told George Bush, Jr. that I'm currently with John Wayne on it, and Bush said, "Then Steve is good support." Apparently Jews weren't a big part in winning the West, but they've been here a long time. True, they've been in Italy, where Purgatory was written, and just about everywhere a long time. As far as the West, the Serrano have been here a long time too. It's between the Serrano and the Jews whose been in this the longest. Yes, right, and I'm a gringo. So, uh, team, without overstating the obvious on cultural differences, the Mexicans and the Gringos will be a part of the same Roman Catholic Church now. I do not personally think Sister Mary Garascia, who is Italian, is talking about this division right here, and quick team, LOOK, kill the joke about Gringos before all of Jews For Jesus in San Francisco gets San Bernardino's extra cash while you're buttering up the Serrano!

So, I had to do Wayne's dirty work(It's okay—we had that worked out from the beginning on these things—we're still a team for the time being), when San Bernardino changed the subject from cash. He said, "Are you sure Garascia's interpretation of Dante is not slightly different than yours? She's the Italian here." And I said, "San Bernardino, I speak it!" He said, "then you're a sanctimonious, over competitive gringo like the rest!" and I said, "No, the Lord's not going to send John Wayne to do your dirty work, San Bernardino, and welcome to Purgatory for the discussion while you're commenting!" So, that's whose here at the Conference I went to for Wayne, San Bernardino, myself, and my guardian angel, to make sure we both have an agreement while I'm helping win the West as one of Wayne's men. So, smoking, the tip of the fuse I'm handling for the time being. And none of these groups is actually alone on the subject.

I put the fuse out, in the dark (maybe a sign of the future, although I do not wear shades inside, these new environmentally-friendly lights are so bright I smoke tobacco with them turned off), and there it was, a more gentle feeling than the last time I thought about it, the thought of the Tilma, La Virgen de Guadalupe, The Patron Saint of the Americas, saying, "Most girls end things quietly—these leave messages around town." I actually had no break-up in the recent course of the Iraq War, though won't say she's lying. Besides the bar to pull up to we have the guard to keep up to, though 47 is not too old to have children. So yes, as I explained to San Bernardino and Dante both, since my angel already knew, this is when we miss our vacations of youth in Texas. With family and old friends of the family we would meet in Del Rio, San Antonio, and Austin, why, if I had a candle, I would turn the lights back off after lighting it for Ciudad Acuna when the border had no threat, that they could return to those times, regardless of what's occurred in the meanwhile. There are differences of opinion on what's going on in California, and that's what I like about San Bernardino—he's from the Las Vegas area and had experience in the desert before his life work here.

I Started When I Was Twenty, In Mexico

Since you feel you received too much pressure in growth stages at this particular station of your family, and it's not supposed to be on the plate but over, (1) Read the assignment by Dr. Aylmer, and (2) on your own part, see if you can write out the elements of this, in effect, counter-intelligence and what you see as left now. Are there other factors in the region slowing down military families, and this is both your exercise and experience as what its affects are like? Explain the reasons why and repercussions of your discrimination received. Put it in the here and now. If you think it is soluble for your goals. Secondly, if it's really not on your plate, is it the repercussions of smoking's severe affect, acetone and paint thinner treating the tobacco at these companies included, doing this to you? Screwing you again?

(Doctor's voice) You lived at Harbinger Home at election time for Arnold Schwarzenneggar's first term, and took your roommate Greg to vote for him along with you. During this time frame, you claim the political pressure in the Southern California region, particularly for being a supporter and actively praying for George Bush, Jr., slowed you down from quitting tobacco. As you say, factors everywhere. Put the counterintelligence solution and the pressure's\factors' resilience solutions together to get an "A."

Well, to beat that nicotine craving, after my walk downtown and the barber was closed, I used the music to read between the lines on what I had seen on my walk: It kept my spirits up despite an initial tug-o-war.

Let's Have A Serious Talk

What was the Devil's strategy in the counterintelligence sequence leading up to? Less than innocuous terms for the P.M.S. of the peri-menopause facing the women ahead of the large population of California forty-year-olds, who faced problems with a one in 150 chance of an autistic(biologically) childbirth. He, Satan, is stacking this up in long-term fashion to destabilize the future of California generations. Let's look at this counterintelligence: Is it a simple lack of bipartisan stride in politics, and if so, how does California avoid being strewn apart? Apparently there are many more joining the ranks, in theories of evil, of the early persecutions, at least in my life. Why, it puts every male from the past in the camp of the Devil's strategy to keep California socialization on high school, not post-collegiate, terms, plus economic factors statewide. Yes, this is a trend in our thinking, and when our roots have this nature, the struggle to stay in the present is the tug-o-war, to put it in one comedic author's view from this time frame, of Polish-Italian proportions! Maybe this axis in the dialogue is where one politician had to apologize to Sarah Palin, for his tort about retardation!

Psychologically, on an individualist, not collectivist basis, this is more than a bipartisan problem. The redundancy of the current collectivist psychology takes prayer, being based in the present, continued quality therapy and study(childbirth is a holistic enterprise psychosomatically—it does not just depend on the schizophrenic male seeking aperture, it depends on the wife's genes as well, plus other factors, though studies of autism are initiating now: not much is known. It is not the case with environmental studies, where problems were brought up twenty years ago, businesses handled their version of it in confidential fashion, and then George Bush, Jr. claimed to enlighten the public by being the first President publically to grab the bull by the horns on the education of the past).

Yes, fact, 20 years ago San Franciscan collegiate educators had introduced their students to the exact same population\corporation dichotomy with a mention of Global Reach, to have students face corporate exec's children as if the students were God for the buck they nonchalantly trumped campus life with in their hand. For a son of a man of rank, this is extension transference in other than a cultural context, perhaps an academic segue with an introduction to the true obstacles in the whole Sierra Trail now coming to light. Without so much as saying childrearing is a fork there, for Purgatorial terms are not, we see the sun rising as the squirrel climbs the tree with the nut in his mouth. The squirrel will be the psychiatrist, mine in particular part Lebanese, part Irish, without being so anti-Semitic as to further examine the stereotypical nature of the Jewish psychiatric side in the Jewish-Gentile complex, for he, that one at the base, works for a living, has been busy, and has years of experience in his practice. Makes the Episcopalians' part in the dialogue (my psychiatrist's religion, not his <u>practice</u>) pale like the overcast nubuli dissipating with the sun's rising rays. He, the Jewish psychiatrist, will be quick to point out to you that the Van Winkle family is not of Jewish background, no more than Einstein was Dutch(although apparently his bohemian lifestyle was a milestone for Dutch-dating, as his biography by Walter Isaacson describes the monetary problems facing his post-collegiate introduction into research and depression-like situation in employment). I will reiterate here, Catholicism is what I'm smoking with my morning cup of coffee, leaving behind the Episcopal favorite of Earl Grey Tea! Yes, Earl Grey, apparently a distant third to C.S. Lewis and Dr. Elyn Saks in suffering during youth in Great Britain.

In all due respect I believe, as Catholic, that since it was brought to light years ago through Dr. Armand Nicoli, Jr.'s <u>The Question of God: C.S. Lewis and Sigmund Freud Debate God, Love, Sex and The Meaning of Life</u>, and Elyn R. Saks' <u>The Center Cannot Hold: My Journey Through Madness</u>, that Britain has improved on these issues of treatment of bright children in schools and treatment of the mentally ill in the system. British comedians in my parts have brought to light in healthy fashion their own pressures and obstacles romantically, and at this point it became a thought as to what Continental statistics were on autism, the forties, and schizophrenia in current fashion. Certainly on the subject of corporate issues the Exxon Valdez was the heavier of the Lord's wrath, than the "Earl Grey Arguments," in which Christ brought a sword to our very own parishes in America, plus unmasked over and over Catholic problems of the past(for which they paid) with treatment of children to courtroom proportions, prior to my conversion.

All this over and done with now we are reminded that God chose America, Great Britain, and many allies to wage a Holy War against terrorism, and for that matter, all our enemies in not preparing for another generation, regardless of our particular country of residence's environment, and as we get back to California, we perhaps see Bush, Jr. as noble leadership waking us as Americans up to the unconscious sin which caused the tension our enemies struck us for in the first place, from The Alamo on childbearing biologically and economically in California to The Sierra Trail of a medical community whose wing we are under in The Golden State, ever mindful that like the Texans at San Jacinto, Sam Houston of modern medicine assuredly will rise, and that yes, the Alamo, researched to the truth, was not Texans alone, though an international soldiers' outpost.

As far as a counter-intelligence, it could be said this is the suspicious facet of schizophrenia. Many have not contacted me for years, it seems to be a spell (as Father David Caffrey put it) that a prince is under until he is kissed by a princess, and he is referring not directly to anyone in particular other than a Scriptural reference point he was making in a sermon on that day's Gospel, back in circa 2007. It smacks of male and female both who do not carry their own baggage while complaining on the Trail, the spell, as would be described by a mental health practitioner's words, one I've met that served his country.

So the inner critic becomes the thoughts and words of the tobacco addiction. Will cause its damage over time, and in substituting alcohol with tobacco in words at meetings, it is sly, cunning baffling and powerful. Damage to the nation puts it in the category of weapon of the dark-side, a "silence of the lambs" approach to the real life we should have as citizens. If in a strict medical sense, painful to quit. Gets habitual for the mood to think clearly, yet it is not the center of this, it simply shifts the power knowledge, love, and truth with a shove from its position at the top of the emotional charts. In following Christ, yes, it will follow too, and closely, yet you are cut off from the part of The Body you most comfortably fit. If health improves sex improves. Does sex improve odds of beating addiction? Apparently not, not at all, from what I have observed. The hard-boiled egg on mental health, children, and sex means simply that maybe, maybe not, God could have another route than carrying the bloodline, all said and done before other worries. Yes, some women can't have children either or don't want them for certain reasons. Large state, California. No, smoking hasn't prevented that, though science's cause should be listened to and is growing. Some find certain of my qualities interesting and will engage, despite finances, though many reject tobacco, as much as excessive drinking. Not necessarily impatient or less flexible the American than the European. Loyalty could arise in an American, more so. Lots more on the plate than a foreigner who smokes this time. As much esprit to rapport and circling. Young women along the way. Who knows? Maybe would date seriously even though had hoped to grow into a mother's shoes. That mental block took awhile, most definitely. So, it's more first the addiction must handle the high waves in cessation. I'm only 47 and three months.

If Dante Saw Us Now

Dear Lord Jesus, thank you for being my Savior and friend and thank you for the beer money. I want to cooperate but I cannot think clearly, and that's before the beer is drank. I thought smoking was helping me think, though I realized, with the help of others, that my problems with thinking clearly and being a loving Christian are the same past and present! It is American history now in the Virgin Mary's eyes what I went through growing up in this country, how the problem was solved with Allies' girls, what the impressions were on my fellow American after, the war we fought, and how certain American families paid for the wrongs while we supported the troops by praying, "Hail Mary, full of grace, the Lord is with thee. Holy Mary, Mother of God, pray for us sinners, now and at the hour of our death."

I see what is being politely insinuated with some sort of saintly interjection: Different drinking friends are like food groups. Furthermore, they let some people have it in this County on the call, though others are kept separate because they are intellectual and we need to increase and refine said output to meet, as the Dean recommended, "other side of the freeway" standards. For example, what are they reading while waiting to get a haircut? It's a mystery. Well if it's a mystery and you don't know the end don't comment on my drinking and writing—my personal guess is I have written dramatically less material on marijuana legalization than what is required to be a bell-curve beating published Progressive. When break is over the discussion on Purgatory will proceed. Pray, now, for the volcanic ash victims and all in Europe who want a different take on it than Infernesque, meaning the first book of The Divine Comedy, <u>trililogically</u> speaking.*

*In the sense of a trilogy, as defined by Budweiser's Collegiate Pocket Dictionary.

The truth is I do not look a day over thirty. The past is not that long. Lots of jobs. Not stupid—hustling on. Could cut down now and get a fresh start after rent is paid and the month begins. Quitting is not advanced academic psychology— slower than I have done it.

<u>Allen Mandelbaum's Translation of *The Divine*
Comedy (Copyright 1995) *Purgatory*</u>

As I read *Purgatorio*, it occurred to me that the situation facing the Golden State is provided a rather literal and religious analogy of the rise of autism. Mandelbaum's translation is known as "The California Dante" (translation of 14th century poet Alighieri—there have been other translations). I will attempt my best for now to describe my interpretation of this divine analogy. I will write out what I can see in the Cantos of *The Divine Comedy,* or you can follow with your own copy, and my material, in cross-referencing the two.

(1) The Proem, or prefacing pages, by definition:

(2) The Prose describing pre-dawn, therefore sky is a purplish-blue with light incandescing, from the other side, a shimmering center to the color scheme, plus four stars. *Pur* –*gatorio* Canto 1: <u>13, 1:22</u> from the Southern pole, and perhaps Mars is east. That should be the right already though. Sun rises in West, where oriental sapphire horizon is, planet spoken of may be opposite though not mentioned as such, to go without saying. Four stars, southern pole, symbolism requiring astrological analogy. <u>1:23</u> other pole and <u>1:29</u> repeating this… <u>1:37</u> Four holy stars, possibly four Gospels. <u>Canto 1: 73-75</u> death for freedom, or faith, for reward on The Judgment Day when all are raised from the dead.

Seven realms of *Purgatory* 1:82. - 1:84, permitting transcendence where Cato, who perhaps is a lone symbol of the Communion of Saints, talks to a Virgil who represents Dante. They would like permission to mention his name, or for this day and age saints who come before us, in Hell, or the ordeal we have traversed. On another note, it could be conjectured that Purgatory is not between Earth and Hell, but between Earth and Heaven. Now Purgatory is above Earth. It is still a transcendence, despite sin, from Earthly life, prior to Heaven.

Purgatorio Canto II verse 8-9: Fair white Aurora aging. Day to night. Helmsman angel, new souls arriving.

Verse 12: I pocchi rimasti (the few remaining): Those who go in heart yet stay in body.

Verse 17: A light crosses the sea swifter than a bird.

Verse 21: Not a shooting star: It gets more bright.

Verse 22: To its ides and below, whiteness (purity).

Verse 25: Virgil says not a word.

Verse 26: Like wings at first, it is the helmsman angel.

Verse 31: With scorn for human means.

Verse 33: Needs not an oar for these distant shores' crossing.

Verse 37: Described as a bird "divine."

Verse 44: Blessedness inscribed upon him and a hundred plus souls within his steering. ("The late bloomers?").

Verse 46: The spirits sang symbolic of the sacrifice of life of Egyptian proportions.

Verse 49: He blesses them.

Verse 52: They seem not to know the place. Trying out new things with the eyes. (Modern Medicine's new arena of knowledge).

Verse 59: They went to ascend the slope (of autism's opaque field—little is known).

Verse 61-62: To analogize, the leadership is not familiar with the issue as well, though as always perceptive.

Verse 63-64: The first path, according to Virgil, (who is symbolic of modern medicine's men) came with much more difficulty: This ascent, discoveries of autism's cause, is like sport compared to other topics in psychiatry and, for women over forty, childbirth.

Verse 67-69: The souls turn pale out of astonishment to notice Dante breathing.

Verse 79-75: They stare hard at Dante's face, having forgotten to proceed to their perfection.

Verse 77: A spirit moves for ward to embrace Dante.

Verse 79: "O shades (of the spirit), in all except appearance—empty!" Symbolic of the empty feeding of the romantic nature in our said current time.

Verse 80: Describes a gesture of Dante to an approaching spirit, that of Casella. Could it be symbolic of Agape after war again in America, like a long lost comrade or in respect to the fallen? Being greeted is a blessing from the Trinity.

Verse 94-97: Matters not this crossing, but the justness of will to Casella. (Matters not the Purgatorial subject of children to him, but the will of the believer).

Verse 98-99: Helmsman Angel accepts all who would embark; most tranquilly, therefore in control of topic.

Verse 101: Waters mix with salt: The discourse can have a random side. Or too literal? More figurative in speech for poetry, or a metaphor? Fresh water with salt water, therefore Native Californian blood lines with newcomers from far-off places.

Verse 111: A tiring field of study, autism, without a doubt. Casella could also symbolize the private doctor's perspective, not a plain comrade.

Verse 115-117 Song comes in to change the subject from fate to sweetness, before they are motioned on with that in mind not to be lazy...a laziness that, in Verse 121, can prevent all involved from perceiving God's revelation (or said topic). Verse 111 talks of just love channeling through the mind.

Verse 133: The pace is hasty.

CANTO III

Verse 2-3: Racked at the mountain by rightful punishments each petty fault becomes a harsh rebuke: The patient's personalization.

Verse 12-13: Too forward and needs then to widen its attention. Patient needs to see problem in larger light and commonly narrows bridge to focus solution.

Verse 18: As the light (God's revelatory answers) shines on Dante, a symbol of our time, the connection shatters the rays due to the darkness in which he has been forced to dwell. Other than the autism issue otherwise a healthy category of 40 year-old. Almost timeless, appearing younger.

Verse 24: Will be guided regardless of outcome, hence mistrust not necessary.

Verse 25-27: The shadows cast, the body from within (symbolizing the actualized adult patient) belongs to Naples, buried in passages of evening. But during day a commonplace citizen.

Verse 34-36: Foolish is he who feels intellect can reach what the substance (love) of the Trinity does.

Verse 37: Quia? Confine ourselves to such, for if we knew too much Christ would not have been born of Mary.

Verse 52: Looking for said ledge where he, commoner, no wings, could climb.

Verse 58: Band of souls, unhurried.

Verse 73: By ending well, it is talking of the generation's guts to lay down the law, in passed word, on not only this topic and more. Whether or not God's revelation is their own at their age, they have an ounce more wisdom on the whole subject to both buffer and safeguard.

Verse 77: It's true: Concern is heightened when one watches failed ideas continue coursing through the slope's sediment until such is too weak to climb.

Verse 112-145: A request by Manfred to tell his Constance of his body at this passage.

Verse 133: Despite Church's curse, there is no one so lost that the eternal love cannot return, as long as hope shows something green.

Verse 137: Only fitting prayers may relieve a stay here for dying in contumacy of the Holy Church.

Verse 145: The band of souls moving at commoner's pace advance more quickly possibly if Constance is told of Manfred. Who could Manfred\ Constance be?

4 of 33 Cantos: Manfred, fallen by two blows, his daughter, Constance, a queen of two provinces.

Does not Manfred symbolize God's symbolic champion of medical research himself? Firstly, a psychiatric blow, secondly, the chest, the heart of the matter as the band of souls, Dante and his shadow alone, stand wondering how to climb the Purgatorial Mountain of autistic birth?! Constance is the champion's daughter, whose trust is put in the care of Dante's loyalty in the quest of medical answer. For her to know why both here and there, two provinces are under the same medical scrutiny is the champion's wish, as he warns of contumacy. Thank God for confessions and service to Mary, conversion, and Communion.

If You Missed Something In Psychiatry, Here It Is

I began to seek a serene, meditative tensionlessness in my mind as it connected to the eyes. As they looked out the window and about the room while I sipped a lukewarm coffee made yesterday, voices began to have a conversation, only the conversation made perfect sense as I sat in silence, breathing in and out not so much in an exercise of such. So the first voice said, "It's a higher level of consciousness when you get used to it," speaking of non-smoking. The eyes then looked at the current research I had helped the two young female psychologists complete for one of my Alma Maters, U.C.L.A. (This had really gotten some air in the pipes by taking a trip to Westwood Village—a voice there said "Continue to recover from the initial onslaught of a cancer-causing habit—look at your Dad, seventy-six, quit when a young fifty-three and a master of L.A. traffic stress. You're last drive on your own required over half a pack, back in 1998).

The research was using my particular case study to derive a much shorter battery of tests to determine a diagnosis of schizo-affective disorder in patient Omega. However, it required I reside three days overnight at a Westwood hotel, in order to be on campus, by foot, by 8:00 A. M., after continental breakfast, and remain until 3-4:30 in the afternoon for three straight days. Under Dr.--patient speaking terms, that is, in an exhaustively thorough milieu of psychological batteries so that all bases were covered before the later shortening of future patient Alpha and Omega's exam, to determine the occurrence of a mental illness. I did not get real sad when it was over, yet thought, "It may be awhile before I'm active in something on the campus again."

I saw a young mother with bright celeste eyes like mine in the waiting room on the second floor of 300 Medical Plaza Drive, holding her baby. At first I thought, "This is how far behind you've gotten because of what's happened to the mentally ill in this country, which, you must remember, many feel started with your own errors in the not-so-difficult environment." And then, looking at the shape of the mother's face again and her street-smart jeans, the Christian existentiality hit me with a powerful, healing impact to the tension and lack of focus\meaning to my mind. She symbolized the large angel of maturity I had hanging on my wall, a framed 36"x28" drawing of originally Madame Sarkozy with her guitar (yet the face came out an angel when I drew). Her celeste eyes and light shone in the Madame's studio window, and, as in the photo, it shone off her shirt a breast. I could not capture light on the shirt with a pen: I drew an exposed breast and a wing for the arm tucked behind her hip on the cover of Quel Qu'un M'a Dit (Somebody Told Me, her album). My Census Bureau supervisor said my drawing, "Some Angels Hide Their Wings,"(the drawing is very light, as if she is on another plane), is like fine art.

I made fine art without a lesson! So, as this maternity angel flirts with my imagination in my studio apartment, telling me her interpretation of these so-called environmental errors of mine, the lady in the waiting room is her, holding me in a brief look at an angel's perspective of schizophrenia in the concept of Biblical eternal time: not earthly hours. The celestial eyes of the mother and her hair color are exactly as that I've always envisioned my angel having, as she reminds me of a young woman who was hosted by a family in Italy the year I was: light brown hair and bright blue eyes. In this sense, earthly time symbolically is connected from before schizophrenia to the present: twenty-nine years I've spoken Italian as my second language, and UCLA the school that taught me proper grammar over and above Neapolitan-influenced language. It is now said I have a Northern Italian accent of sorts, my speech has changed so much and so has my thinking. To me I still feel Napuletan' in America but am not—I have not seen Italy since 1990 and the changes before and after Iraq, when I lost contact under the Bush Administration with all European-residing old friends, as well as a squeezing of my American friendships through a whirlwind of vigorous treatments, from Loma Linda BMC to Cedar –Sanai to San Gabriel Medical Center, before being fingerprinted by the F.B.I. and told to stay under Federal Census Bureau employee oath for the remainder of my life(or face federal imprisonment). This preceeded my four-mile walks, up steep driveways and streets, past loose dogs, door-to-door to prepare people to declare the terms of their American citizenship when others came by to finalize the information for the 2010 Census, under President and Commander in Chief Barack Obama.

Yes, Census Bureau money doesn't survive when Detroit auto companies push the country through over-haul, though I'm an Eagle Scout and tiger too, whether or not "tranquilized" by the sheriff (analogically-speaking of medication) "runnin' wild through the backyard of the San Bernadino party scene before runnin low on auto-repair and rent money in a now-past day of Venice Beach"… The mere mention of the word "Venice" having me communicating in English and shaking and moving like Napuletan' on vacation during Carnivale, after a Peroni on a slow-rolling Italian ferrovia (ferro being Italian for "iron" and via meaning "away", the word translating as "railway") in the same language. And then there's "Couchette," Italian for "sleeping cabinet", to let you know they can and will both welcome all of France and cross onto their side of the Riviera too! Tough Love Tourist Agency, si parla? The pan: Southern Italy, and the fire, readjustment to being somewhat like a Native Californian again after that cultural immersion, having been reared here in California since the age of eight.

So, what are the voices back in my apartment saying after this look into Cinco de Mayo 2010 in Westwood? I had told one of my doctors it seemed as if luck for schizophrenics had gotten bad in the Bush, Jr. years in The Oval Office, though it is not as if he had any set policy or ignorance on the subject as a Texan! It seemed that during this frame of time, a voice explained, many problems with American writers in my neck of the woods arose, from a small-time agnostic rebel stealing a book from me to further his own career, to the writers' strike in Los Angeles. I had gotten turned down by USC in '98 for a Master's in Screenwriting, failed to get a good domestic footing in relocation attempts to Los Angeles to write one there, and gotten accepted to Johnston College for their Bachelors of Arts in Writing For T.V. and Movies, yet now during Bush, Jr.'s first term (whether the law came into affect while his Presidency presided or not), funding was no longer allowed Federally for students going back to school for a second B.A. or B.S. I remember that's where the seed of suspicion had come from, for I had a high school classmate who did get a second Bachelor's in California, whether it is common or not, to change careers on said terms (from, I believe, Social Sciences of some sort to Nursing).

The voice then claimed, "This string of bad luck is not only one you, but one people around America began having, in looking at auto and bank bail-outs plus insurance problems. Economic repercussions of the caveat we received: you were issued a caveat." I keep thinking the caveat is from George Bush, Jr. for my first book, which won an Honorarium and speaking engagement from the Chair of Psychiatry at UCLA—but not a genre supportive of more Republican works needed to maintain proper constitution of our Democracy…hence not published by decree of the likes of Crown Books on the open market. This is when I am reminded of the voice I heard at UCLA yesterday, by the cafeteria, as if a doctor having lunch at another table was commenting on my deepest worry: "You'd be surprised about what's really behind the subject of the mentally ill and money, say, to raise a family." Jesus came to mind over the course of the past two days to tell me that, "as far as how it most likely works for tutrici (tutors), the less you pinpoint discrimination the more it goes away," which reiterates my parents' words, "Don't get them believing you know they think like that." Or, putting more concisely I say, "Don't get them believing they're like that when you are upset."

In any event, no, I really have no idea what this caveat is for, or who issued it, but it's here like a chunk of dirty water ice from the lake after the snowfall—That is, the snowfall at the South Pole, normally fairly clean waters, yes—the dirt color in the ice must be Washington Elite Penguin Poop.

Or Is It Not Her Move, And Command?

This was Alejo's (author of *The Lost Steps*) interpretation from the skies, in the procession of saints before the throne of The Virgin, angels bowing and rising in songful worship nye: In Hispanic and Italian Studies, part of the mystery of the schizophrenia is the two different interpretations of sexual tradition. Plus, it's not just that they are family. In the Italian, the man is seen as gallant, and in the Hispanic, when elders' conservative wishes are observed, he is seen more complacently, except when with his own. On all ends it depends who you talk to, though the discussion really ends there, and this is a happy medium of a concept in an, above all, Catholic movement of love beset by its own rationale. Or, as Einstein might put it, 'with a modicum of irrationalities that actually have an advanced connection in The Body, within said framework for the development of theory in both scientific and social scientific faculties'.

When She Had Wisdom, We Had A Discourse, But Now, Post-War, Zilch

While multitasking in an attempt to both clean and organize my quarters, it occurred to me hygiene is a factor in eco-psychology and academics. I will have to inquire about details from my psychologist, Dr. Aylmer of Harvard University, on location currently in Redlands.

Speaking of hygiene and summer, though not normally tied as topics, my current staying quarters in its present condition and what I do to maintain it brought back reflections of Santa Barbara. My occupation as janitor for the Banana Bungalow Youth Hostel prior inspired thought and humor, what with its combination of cultures and personalities in the clientele, plus in the staff (ambiguously said, this ending predicate with its prepositions and subjects!). Yes, to concur my former boss, another World Cup fan, may not have left the area, though may be the same part Brit-\ part-Aussie that I overheard on the phone. I had been transferred from Reception to a certain department of Brooks Photography Institute, while searching for a former A.F.S. exchange 'colleague' from Pope John Paul II's Italy exchange time-frame, when I could've sworn the voice was the very same Brit. As to why my boss handled moving on this way, <u>in cognito</u>, I would have to say brings to mind the other definition of Australian Butt-pilot, i. e. one that flies the plane with his head up his ass, as well farts backward, in contagious form so that the whole hostel, in turn, begins to fart backward themselves.

As I began to make a mental note of what is necessary for my apartment's further sanitation, still in the Dickie jumpsuit (?correct) my parents gave me, the voice came to me, "Admit that. Admit the Southern California female is a burn." That would simplify the reason for the occurrence of several intangibles within past and present events. I'll draw a straight line to news-breakers on T.V. within the demographic: The cause and effect item of point denotes she is behind the scenes in the vicinity! She is the type of lady who has Bonnie and Clyde memorabilia, to paint further. A relationship with any foreign woman, other than one of only Hispanic origin, is seen as a threat to the territory she had acquired with her social capital. Her reputation has preceded her in order to warn other members of the clan that this must be a movement with new rules, inventory, and strict observation on a concurrent basis.

"We will force him out of the pocket of self-control in a marriage by insinuating he is repetitive: not a normal creature of habit, simply a creature period," she proclaims. The overseeing strategy she has implemented is to disconnect street-life in various communities from comprehension of urban savoir-faire, while at the same time now being positioned to remind him this is the exploration in direct fashion of black and white truisms. Hence, this is the balance in the working world that has both social and economic repercussions. I would like you to observe the other culprit strangling small business besides government regulation, a scenario complete with insiders. This is a bum, not a <u>femme-fatale</u>. She has completed the deconstruction of the contemporary Christian male as her volunteer activism in a more serious America. It is her water. I now have an idea of how to carry it for her (as my parents age with comparatively brilliant new responsibilities for me) in two large buckets hanging from a bamboo pole held over the back of my neck. Before aperture on an environment that is anxiety-proof, the Priesthood innocuously prepares a sermon on "getting back to our Jewish roots," to cover up the trail marker left as a sign for her escape route, in order to continue her discussions with "Clyde."

On a nice day in town she will leave you with, "Do you need more pork, or may I prepare you some more?" Men at public gatherings from the theatre to the university banquet hall have been informed, "I am not going through the change: This is P.M.S." I was hoping she could keep this in a private conversation with her medical professional and family—I am neither married, nor divorced, from her, but still on speaking terms. This is what she will change the topic to from schizophrenia when it surfaces. As co-pilot she will now assist the Australian in landing the plane down at Santa Barbara in Goleta.

No Moodswing Here

Healing on mental illness is spiritual when the behavioral problems are handled, and scientific when the talk is of healing your body and concentration. You cannot simply clean up a reputation where people talk behind your back, and word gets around, with an investigation, a lawsuit, or revenge. One has to focus on what Christ wants, for maybe the whole town would have been in a worse position, and all Christians were persecuted to some degree, though you stepped forward, carried the cross He gave you and agreed to psychiatric treatment.

As well, how family trees line up to produce their heir's genes is something that is just now being studied more closely, and if you have a gene as such that produces an excess flow of Dopamine, medication can control this natural chemical on the brain, so it doesn't make your synapses fire too rapidly, at first producing genius, yet later hallucinations or mood swings.

America! Hail Mary, Full Of Grace

My father a couple of weeks ago stated, "That 50 cents will get you a cup of a coffee," speaking of the high school alumni followings as we approach our fifties. The days themselves are full and weeks arrive to Thursday morning quickly. In this context, I look at Brain Default Mode Network as a theory again, as Dr. Aylmer and I have discussed its ramifications in a small town where parents of old friends have "senior moments" in retirement, including my mom. In fact the seniors do this as a team together on the subject of the good ol' days, to music with dance and wine. God in his omniscience must be discussing the tragedies and paradoxes, as magnified by Sandra Day O'Conner's husband dating of an old high school friend amidst his Alzheimer's disease around the time of her retirement, and Sarah Palin receiving a V.P. running mate nod as her daughter grew in popularity from a see-saw high school rapport. One can imagine The Father's celestial discourse with the Court of Heaven (Mary on Her Throne and Christ on His at the right side of the Father, along with all the angels and The Communion of Saints) has some pragmatism for those of us who do great with a modern view yet took Freud's rap on childhood issues as well. Hence thoughts recur of the past, as a cortex network system to alleviate any stress in the present is actually the opposite(a sign of a medication side effect, moreover a heck of an environmental factor in symptoms management, what with the Seniors and their modified Raves, before we're all back to their Bible Study!)

So God in His Divine Comedy has produced a likeable female millionaire running for the California gubernatorial role who feels we need a business perspective, and warns of racial profiling, at the same time that San Bernardino County has lost several funds for Sheriff activities and programs to curb the gang violence which is reported in research to have international roots as well in some instances. Her thoughts are we need a business perspective. The Alum are not only raising little angels and demons, yet running in and out of state businesses with roots here dating back to the momentous time frame. To see the problems certain discussions can present to a business, not just a family after the business convention is held in lieu of a reunion, is well, business savvy.

Alluding to the fact that this could surface in a much different context within a discussion between caddy and Senior Pro on what prehistoric club to use to drive this one to the green, I would have to say that ought to shed some light on why the mentally ill actually cannot discuss golf, if this paradox on moving on with America's recession and real estate slump is the cruxt of the problem. Thrown out are collegiate and post-collegiate discourse from society to hail the high school greats, while dragged is a red herring about what discoveries were concocted via a hidden grapevine on the environmental factors from early adulthood growth in the twenties and thirties. These factors are reinterpreted as Catholic sin, to stimulate the economy, all the while those that are actually doing DNA mapping with bonus business practices assure any potential parent from this 70's and 80's hey-day that their parents' grandchild will not have autism (You've got a now-larger straw to break the camel's back than the one used to draw coke into Al Pacino's nose in Scarface!).

In defense of my vote for Meg Whitman to pick up the apples, put them back in the cart, and make sure we do not go through this in the future by turning it all the way right-side-up and pushing, the medical perspective is not only separating the prophets from the parodies, yet does have Bush-like quirks too: a business perspective gives us room to breathe one and all without running into each other when shopping until the Shit-Eating –Grin Show is over, and we can seriously discuss discrimination against the mentally ill, on its oh-so-succinct level, as a detriment to the future of the Social Security system of the United States of America, since this is an allegorical blood-drive that they volunteer for, to get work which will give them a serious shot at leaving their receiving of SSI benefits behind for a good reputation in the working world once more (at competitive salaries themselves for their own homes and families' college plans). This is what angels seem like they're discussing up there when the Virgin on her throne throws out this disgusting version of modern dialogue on romance.

Let's get back to the truth: If you thought the beach was a great version of life, you would love the garden. On that note I can picture a drawing of Sancho Panza and Don Quixote being knighted by King Arthur and Sir Lancelot on the plane of myth after an explanation to beat all time of the Spanish Armada, in a space of The Lord which has no past, present or future, only the prayers of a battle and God's creation of Earth in rotation. Are Spain and Britain going to be okay after a World Cup in Africa, the discovery of an untried Nazi, and the secretive baby burials in France and Germany by women with "pregnancy denial" as their psychiatric issue? God Bless America, in quicksand or not. As far as the Pope, a little bird told me what he said when Arthur and his Roundtable arrived in Rome to report they had succeeded in conquest in a long, drawn-out Crusade: "I'll have another!" And on the serious end, Isaiah 10: 15 states, "Will the axe boast against him who hews with it?" Wherever America's enemies are now, champion Mary well, and "Thy will be done." Poor or rich, conservative or liberal, we've been blessed by Her with another day in this country, and I am going to live it.

Would Have To Be In Israel Now To Understand

"Schizophrenia occurs when ashtrays mate."

The rest of it, professor, on the subject of Jung and Dream Interpretation. You keep on spiritually guiding us into O.J. on the field.

The Culture of Narcissism: I think the ethnocentric make America a cool place to live. I've helped a Jewish family as a military brat-in-law. On the topic: don't make comparisons. Use Hebrew for a bar.

What Pissed Off The Kobash Kid To Entertain Life Underground

For Dr. Murad:

There is an idea that the development of schizophrenia has environmental factors, and is connected to the birth of autistic children in the schizophrenic male. Pot, to me, is an environmental factor, health-wise, socially, and academically. Is this the possible cause of rise in autistic births overall, because of a deterioration of the environment due to the 30-year existence in California of the layers of the environmental factor on a cultural and psychological level? Also, be aware one pot-smoker academically rewarded for his ideas in a paper, and another pot-smoker penalized, not because of pot, though because of the professor's viewpoint. UCLA states they do not know if schizophrenia is environmental or genetic. The Genetics Center at Loma Linda University states environmental factors will trigger schizophrenia, depending how, genetically, the family trees line up, though in specific cases, such as mine, this is researched genetically (especially on the subject of future offspring and their possibilities of having mental illness or autism).

If psychological support "flip-flops" and academic challenges and/or jealousy factors exit in a home environment, coupled with marijuana as an external and/or internal environmental factor, and, last but not least, the "money is the root of all evil" factor in the community (as a social fabric) all exist simultaneously, then you have a predisposition to the onset of mental illness. I am talking science, not religion—do not slander the very concept of dharma by mentioning it in a scientific discussion. At first I thought Loma Linda was correct in its giving of causal effect to both, though perhaps in the end if this were based on an attitude, marijuana is the sole cause and a succinct slander has occurred in the social fabric of a closed environment, for political and/or personal gain, and therefore leaving it an open book is good, as the UCLA thought clarifies, and maybe we do not know the exact cause, hence, have the Holy Mystery. (Nothing like training for an Acapulco-style cliff dive during my forties on the subject of childbirth, to the entertainment of psychiatry's finest from Lo-La to L.A.!). Curing the mental illness at a greater level of functionality in solving said chapter of the mystery by producing an economically free patient, is it possible we would be again at the starting point private practice had before either UCLA or Loma Linda stepped in, i.e., a patient who could start fresh in the old environment with a new psychiatrist, slowly having him try cessation from medication?

"I will tear down the Mitten Building and rebuild it in three days…by putting big-screens on all corners of downtown Redlands. So, another one of my jokes you did not understand?! "Over my head" as the Redlands expression goes that has sent the City tumbling into debt? Well, you're a talented bunch one and all. The moral is, if the media is throwing bricks from all sides on the subject of talent, get inside the bar!" Love, The Kobash Kid

--"Batman, the Kobash Kid has struck again!"

--"This is no job for a Polish-Italian joke-book, Robin! Off we go!"

* * *

The subject at the Staff meeting came up of changing the slogan for the cable commercial to, "The Magic of AMC." I'm the interpreter who had to be present for Milan T. Vu's corporate executive…Since I'm American and White to AMC, I didn't bother translating the whole meeting to him. Since they think I'm "Elrond" on this subject.

Frenchie and The Mormon: "Why do your legs shake when you sit? Are you scared of something? OR have you been on drugs?"

The Interpreter: "No, I had to translate for my Dad, The Major, the expression 'Rompere le palle ('Breaking the balls'),' and one thing led to another."

--"OKAY, gettin' a little warm in here. And a little stinky."

--"O.K.,.it's always been stinky in this town."

--"So what does that mean? They're gettin' a little warmer—and there ain't no global warmin'!"

--"Wow, that Katie's a real existential one! Didn't know she had actually polished off the whole 350 pages of Kirkengaard before changing the career to newscasting!"

***

--"O.K…So we had to invite the whole gang for the cross-cultural worldview discussion!"

***

--"Steve, you can't be a prophet in your own hometown, but you can be a comedian!"

--"Ha Ha Ha! I'll take it! Any job you can give me, good ol' buddy ol' 'Home Sweet Home'!"

--"Oh, is that where the bull's horn went in? In the "work hard" artery?"

--"If you cannot get out of the box of bein' sexist, you must have gotten the rap from a clean old man this time!"

--"I got it. The young men who didn't make the rank of Eagle are the Irish, still holding, and will be advancing through the economic field, after this last version of me looking at my account balance, in which I have been dismissed back to dawn. Must have been the "trustworthy" part the country did not grab on to! That's when the Census Bureau called and said, "Since you are squeaking so cleanly from here to the earshot of The Washington Elite, you can be fingerprinted, sworn in, with the passing of the F.B.I.'s graces, and have the ball back. The

'Certificate of Appreciation,' despite the misspelling of your last name on it, <u>is</u> symbolic of the game ball! The second time they hire you no, no game ball, but way to beat those guard dogs on it, 'shy' being a factor of staying a free American in their book!"

* * *

--"That was a <u>war</u>. Who's been playing the role of the ghost that plagues the apartment in this mysticism marriage to the Theatre, in which we get the house blessed?"

--"It was the cigarette's preservative, Father."

***

--Next on <u>Trinity</u>…"A repulsive female pressures the men to 'come out of the closet,' only there's no closet to come out of!"

***

--<u>SLOW TO RISE</u>, a class team-taught by Newt Gingrich and the Senator from South Carolina.

***

--<u>The Out-of-Staters' Interpretation of The California Collegiate Scene</u>, taught by visiting professor Horace Nunn. Dr. Nunn is a visiting Social Sciences Scholar from Notre Dame, touring the U.C. system of schools and universities.

--"What was the letter I sent you while on Crusade, Pope? Help me figure it out. I can ask the monk to resend it from Jerusalem. Is 'resend' a legitimate word, Pope? There's 'retain,' 'rescind,' 'resent'…Never mind, I looked it up. Sorry, Pope, I, Arthur, did not mean to *rompere le scatole* (break the boxes), as it were."(In brief, why Arthur was asked to go on <u>another</u> Crusade shortly after getting back to Rome from The Holy Land).

--*"Vedi Napoli e muori."* "See Naples and die." The Los Angeles translation: Venice Beach at sunset off the coast reminds me of The Bay of Naples at sunset. Exceedingly similar atmosphere. Did 'what comes around go around' on organized crime, or are you <u>now</u> happy with the Los Angeles Police Department? Naples is an international city of organized crime throughout Europe. Be thankful. You would be if you knew the truth.

* * *

Lust, gluttony, manipulation, jealousy, vindictiveness, detachment, and vulgarity: The Seven Deadly sins of The United States of America, a nation surviving despite these sins since George Washington's conquest of The East Coast, and The Spanish Missionaries' decision to open their heart to God despite their culpability in all seven, a California looking to meld past with present for a graceful future as long-standing seamstresses in many of the wars bearing witness to General George Washington.

--"Their baccalaureate address was the charge for a quest to find a roll of toilet paper to wipe their ass. They finally found a roll, now they want their rights!"

***

If this indeed is a strange phenomenon, with an enigmatic side to trouble-shooting, that has beset, well, then it is similar to doing study on the moon of lunar life with the feeling something is out there. On the belief that man's problems may be environmental, not necessarily due to what we normally attribute agitation between races and religions to even sexes as well, which are social and economic factors of perception, then rich and poor alike function as a society internationally where we literally all are in the same shoes regarding this phenomenon. It's an enigmatic mystery. It brings to mind the lesser-known stories of religious knowledge, astronomy, our often ignorance on the school of physics, and even both medical and environmental equations of a scientific <u>and</u> psychological nature. These are many things seemingly setting us on different intellectual paths that even modern worship in an organized body of believers does not bring together.

Is there one God, as stated in Scripture, actually creating an affectation of this phenomenon in a good vs. evil struggle of the universe, through a court proceeding of Christ the King and the Virgin Mary, as both a female divinity and queen-figure, whom alike intervene on our behalf personally or through angels? Or, is this a pantheon of several gods, demigods, and prophets who despite similar powers and goals, might even come to war in Heaven (an expression used in a certain verse of the New Testament letters)? This latter question itself, alluding to the reason many Christians struggle with the validity of mythology, can add extra weight upon an already-described search requiring multi-tasked academic skill.

Many take the low road, avoiding said topics all together to stick to a simple diet of literal Scriptural interpretation and choice, to not take any chances on survival, carrying on their family name biologically, or risks to their status in society. In The DaVinci Code by American author Dan Brown of New England, a group of secret people with mystical rituals who had tapped into hidden secrets involving these fields is discussed as The Priory of Scion, dating back to the Middle Ages. Scholars of Dante Alighieri's timeless work, The Divine Comedy, discuss the dual existence of both a modern day of the peak of enlightenment and a tradition with roots in antiquity that finds us struggling with the puzzle of conflict and natural disaster, at every point in mankind's history, discoveries and natural wonders of old being revealed every step of the way equally through the same similar paths of knowing one's Creator, in his Master Plan and Universal Court of various celestial levels.

This leads us in this current struggle to a place in which we often feel time is running out to end the potential destruction of the world by our own wars or other doing. In this struggle which seemingly complicates further the more we specialize in one field from religion to physics (with as much diversity as we can add through extra reading while continuing focused, single-field research), one ponders which aforementioned ram of the world of knowledge actually bridges together all of them the best to create a centered, fortuitous area amongst us all from which we can nourish, to continue our own personal battlefronts vs. evil. Is there really a countering evil phenomenon to our "lunar study," or is this simply a weakness caused by the inherently divided pursuit of knowledge and a weakening effect of the fragmentation? How do we continue, therefore, that is to say with what methodologies, to empower to break off these very academic chains? Overwhelming, time-consuming, and at times a fruitless effort of challenge repeatedly following another.

Clearly to answer the latter one must begin a quest to answer initial questions (look at the linguistic relation between those two words, 'quest' and 'question,' as if almost alluding to conditions of the very atmosphere in antiquity in which language developed initially!) in order to fortify bridges: Already in place between physics and the environment and religion and psychology, one piece of classic literature and another, as well as purportedly a section of the mythological pantheon and The Court of Christ and Mary. For a true Christian or Jewish believer, is it not robbing Peter to pay Paul, nay sin, to bridge a supposed pantheon of Greek temple construction times with the mother-son team of ancient Jerusalem and the God of Old Testament magnitude? A formidably challenging question all its own in which we recall warnings in Old and New Testaments alike to have a fear of our Creator every so often amidst our true nature of having fear to quest without Him.

Indeed, in the cold recesses of the atheistic world from which few walk out and little light of the believers shines in, it seems as if a literal cat and mouse game has begun not only with the unity Presidential leadership attempts to symbolically inspire amongst our fellow Americans, but with the beauty of the institution of life-long marriage itself. Are tangents from The Church par for the course in actually doing what Christ's court would have you do, that is, continue to challenge oneself and others by taking the Gospel to a mysterious locale, an atheist's pocket, why, a modern-day Priory? How does one reinitiate a quest for said bridges amidst said challenges, or are the bridges, why, the strength to build such, amidst the reward from Above of the strength to overrule the pantheon oneself for championing Mary in said cause? Is this not what Dan Brown has one of his principle characters doing scientifically in <u>The Lost Symbol</u>, finding a situation where thoughts are not only unifying all but systematically creating some sort of Biblical hero or demi-god from the thinker? This, amidst yet another secret society's network in The Masons, who purportedly have planted their influence in the historical foundation of Colonial America? Does one have only a fictional realm to speculate a bridge between Brown's thought-provoking idea content and actual Scripture for current applications personally associated with one's walk in everyday life? The very nature of modern fiction has us leave by nature yet a question looming at all times, while answers lay mystically within.

--"Normally no. Normally, no, not a one-horse country. When it gets like that just go through your exercise routine regardless."

--"So. Nice place you've picked for our business lunch with our visitor from Milan. It's on me. What else do you do besides interpreting from time to time?"
--"I'm working on my Doctorate."
--"Oh? Italian? UCLA?"
--"No, Domestic Studies at L'Universita' di Mamma Rea Speranza!"
--"The Show Must Go On."
--"Where one evil politician is buried, another rises. Ever since the first Irishman buried a potato underneath a full moon to end the famine with good luck!"

1st Classmate: "That's great. I don't see how you remember these things about those years."
2nd Classmate: "I guess you would have to quote someone more important in college to carry on this American transition regarding that."

--We can hear you from the West Coast: "You have a different perspective of colloquial Ireland of antiquity than I do."

--"Now, when you study Neapolitan culture and language, don't over-explain 'Cazzima' to Americans: there are actually some who will attempt to apply the philosophy here."

--"Those that learn from others."

--"For a more enrichening cultural experience."

--"Now don't say it."

--"Look, Professor, no joint."

--"Why did you say it!?"

--"To point out the universality of the liberals here and the Leftist Italian University constituency."

--"Then I'll forgive you, this Happy Halloween 2010, the Year of our Lord Christ the Messiah."

--"I seem to have a mental block of what was said when witnessing in college."

--"You witnessed to liberals, that's why: good for you!"

--"What was it like, because I rented off-campus and can work? One way or another, I met Oliver Twist!"

--"I bet that helps you twenty years after getting through the Recession!"

--"Oh, I can top that: If you want milk with sausage and eggs but need to eat Kosher, drink the whole glass of milk when the plate is finished."

--"Why don't you just rob Peter for an extended version of the Gentiles' portion of God's covenant for Jews and Gentiles alike?"

--"How much is this for a 1-hour office session?"

--"100$ by Harvard time."

--"Charge Paul at Medical."

--"I'll be glad to write the chap prolifically! So will that about wrap it up on the Jewish-Gentile complex? Laughter is the best medicine."

--"I kind of feel I've been held up."

--"The premise originally was you had a problem with stuck-up people you were meeting on their way down. My premise is, Catholic you are, if this is the issue here and now, you're going up. Do not be disillusioned."

--"Even though Paul isn't the best there is."

--"We'll save my impression of the difference in manners between his and the Blue Cross Salesman for another time, _n'est pas_(not so?")

--"_C'est vrai_(True)!"

--"I know what you mean. The French Foreign Legion is looking for a couple of these people."

--"Yes, let's not give them any ideas to quarantine the Historic and Scenic District during flu season, so we can keep credit for the extra exercise we get while Daddy More-Bucks revamps real estate as we know it."

--"People change."

--"There are some particular groups that have a timeless quality even though they've made different decisions."

--"Well, I'm in a bit of a hurry."

--"The Dean's Banquet?"

--"No, the damn Medical Faculty Awards Dinner. Dr. Nicoli, for Christ's sake, is being presented, 'Most Supportive Faculty in a Theological Base!' I've got to present this dinosaur with the last bone for the skeleton!"

--"So you mean to say, off the subject of this Alcoholic Democrat psychiatrist, legalizing marijuana is a Pelozzi power play to slow down Rep. Jerry Lewis?"

--"It's not Barbara Boxer. She's actually very legitimate. It's not Diane Feinstein. You know her track record."

--"Yes. I gave her a vote for her grandmotherly touch."

--"What would you do if the doctor said not to smoke pot, they sold it, and now they were employed, but you weren't?

--"I would start over to get into Jerry Seinfeld's field!"

--"For right now, psychologists in California have done their job. They've taught us this isn't an eternal Punch and Judy routine where we keep smacking each other."

--"Great. Harvard has laid down the law with a true Democratic feel on equality and its laws."

--"Since Christianity and Christian Contemporary History are not taught."

--"Well, yes, it's a public school system that needs support."

--"Oh, true, it's not as if Catholic school funds grow on eucalyptus and orange trees, that's for sure, so I know this leads this way."

--"Well, God bless you, Doctor."

--"Thank you. God bless you too."

--"And that Dr....he's just going to have an extra glass tonight, but no, no addictive pattern in his psychological profile?"

--"You know who my doctor is?"

--"Who?"

--"Stephen King!"

--"Touche'!"

--"Stephen King's recommendation: 'Make the cash to keep the career going by selling them chunks of dried-out cheddar cheese, and tell them it's crack.'"

--"Got it, 'Careers Class,' the Party class. What are you bringing for a dish to the class' reunion?"

--"I'm bringing, 'F.B.I. Background Check and Fingerprint Pie!' So do you see why I am going to stay?"

--"Are you going to play 'spin the bottle'?"

--"No. I actually don't like the new rule where the dudes can kiss the other dudes."

--"Which way does it go when it does not digress?"

"It goes to the coffee pot. Imagine how many times the witch and the vampire went to the food."

--"Yes, when their done with the critique, they can try running for your job."

--"Yes, I'm taking a hiatus in this scene of The Writers' Strike to chart the wind for the other smokers, who are firing. We are going to get the Aristocracy out of this one, on the front lines. The Bastille has been stormed as if this were historic Paris during George Bush, Jr.'s second term."

--"Oh, c'mon, Rumors? We don't discriminate against schizophrenics. Stud! Try the Hustler if the Church isn't working out. Really. Catholic Democrats aren't playing games. It's just you. Got nothin' to do with that from Shreveport on out. Infant Baptism. Spoken. You don't have to do it over? But what did you do in the Episcopal Church? Won't go any further than the Priesthood: See? Loyalty beyond your imagination."

--"Oh, the Emperor wears no clothes period, Mass or not. And I'll keep the mask, fine. It's the family's—I'm not related to Octavio Paz!"

--"Thanks, Primo!"

--"'Primo' has stolen second. Stolen third. Stolen Home. Stolen Period. You can get in line for seconds, that's all they serve at Primo's version of the Family Reunion."

--"Seconds?"

--"Eat the main course from Villa's Family Cookbook. 'Primo' can cook too!"

--"Well, we're all brothers and sisters in Christ, Primo!"

--"My blood cousins want to keep both of their occupations, so don't lose me on this. One is with the military family we are and the other is with the Lutheran Church. Sure we'll throw Primo's recipe in the family cookbook. When you say 'Take the log out of your eye,' try not to act like the coach from Washington State!"

--"Oh, sorry, they must have a hard time understanding an Officer's son and his interpretation of 'Native Californian' by proxy."

--"Two rules: Not at election time and don't steal the family blonde's recipes."

--"I got it—'Primo' is going to Confession for stealing the game ball of the cross-town rivalry, and will say a Hail Mary and Act of Contrition, but must have the Irish priest's signature on it. The Certificate for being a Catholic now is signed by the priest who is retired, that took the other Confession, and this is the new priesthood in the line-up since Villa's descendant on Primo's mother-in-law's side has moved on from being a Man of the Cloth, who handled the exorcism so he can tell the difference between you and 'Primo' at any possible wedding in your family!"

--"I gather that is the gene's opinion."

--"Whether or not the subject is Dual Citizenship, the matriarchs of family past will see you on *Dia De Los Muertos*."

--"That's why we say, 'God Bless The Cook and Protect the Eaters!'"

--"I see. The family wants more Republican loyalty to fight the cartels than General Primo."

--"Oh, I'll take the seconds now that I've been drafted by Primo—on tobacco!"

--"I see the Catholic Church can't beat him!"

--"He's not an expert in Canonical Law, but he'll give it the ol' college try on that Pentateuch."

--"On that note, Jesus will have a word with Zaccheus at his home."

--"You mean someone else besides me had problems with the Jewish-Gentile complex?"

--"One patient at a time on the front lines for tobacco addiction, twenty cigarettes per pack, and get some sleep. Maybe a Primo impersonator with a German Pope named Benedict the XVIth isn't the topic!"

--"A psychiatrist is in session with a stable patient and claims, "Now that medication is not the issue, I have to tell you there will be complications with biological children. Do you want surgery as preventive maintenance?"

--"The patient quips, 'Dr. I thought that's what you'd been doing in these sessions for the past five years since I started seeing you. I haven't been on a single date the entire time!'"

--"You must be about to burst."

--"You mean like you did in our first session with my parents? This medication is enough to back up the whole offensive line of the team!"

--"It's a tranquilizer in slow-release form there!"

--"It's a solvent, a detergent, a sperm-equalizer."

--"Maybe you've just started producing less sperm with age."

--"My grandfather had eleven, you have three. You would have the gonads to say that, now that your thirteen-year old daughter has her driver's license and a new car!"

--"You gettin' tough?!"

--"Not as tough as it has been to start my career again with the rap you gave me!"

--"You should show more gratitude for the patience and assistance!"

--"You should have studied dentistry at this rate."

--"Well, don't drag your tail between your legs!"

--"Pardon me, doctor, a brief reminder, I am the patient who is not the dragon Satan."

--"That's right, you're Catholic."

--"Are you going to compartamentalize this notepaper into origami boxes?"

--"What's the moral of the story, team?"

--"When California doesn't like dick anymore, Dick parties!"

--"I mean are you kiddin', the one who got away? Half of these women made the college track team!"

--"What, while Pancho was makin' homemade soft tacos?"

--"You should learn to step up!"

--"Well, gentleman, I don't like the expression, but here it is...'President Obama and the American Economy players!' See, he's The Temptation's man!"

So, lecturing, jokes, and satire aside, after common daily theology's thoughts brought to mind, we hear a report on a Los Angeles station that Madison, Wisconsin students are protesting their governor's end to collective bargaining for Unions. After four beers from a Wisconsin company are gone for the night, I'm reminding myself that, off the platform of every moral discussion contemporary Episcopal and Catholic thought have posed, I choose not to comment. Women over forty years old and men with schizo-affective disorder in the state of California have been tied to a rise in autistic children's births (whether the Church sees eye to eye with science as is claimed, that we as Catholics are not anything but aligned with progress in medicine, or man is out of control unlike Christ who runs ahead of science). Furthermore, I originally have been portrayed as an intellectual.

Though it more moral dialogue in contemporary American Catholicism and not an academic one (so as to better influence a younger generation's reason alone), I assure you I see again as I did before a poetry of Antiquity that bridges all modern gaps. These Californians in this category of childbirth are in a Purgatory wherein science is Virgil, the parent-to-be is Dante, or theirs stated within. Perhaps an analogous scrutiny of Mandelbaum's Translation (of what is known as "The California Dante," a version of Alighieri's original work) could deduce something theological for those wondering, "Why our generation? During a war? With our economic issues?" Yes, I know a new version of the poetry of theology which has a stark reality to its credit deserved. I'll try to interpret Canto at a time.

Purgatorio CANTO 1

1-3 Talents in the sense of artistic or monetary, what is the difference for the Baby-Boomer on the cusp of the era? That many are unsung, and owed credit that could have been reinvested in the American economy exponentially. This is a generation of Californian that has seen an era, leaving in its mind artists from many generations of America (in all branches between music and film) behind, to embark upon journeys forward from a sea of the state's events, tragedies and crimes, during a war toppling their American right to rear a child with economic legitimacy, a cruel sea not only in this sense, but even under a darker cloud in modern medicine's bad news.

4-6 In Dante, the Second Kingdom refers to after-life: In analogy, a second kingdom symbolizes the proper academic and medical research of future offspring, given one's family trees and behavioral environment where medicine could have an answer. Citizens of prayer young and old await a new clarity, considered Heaven on Earth compared to the state in its prior situation of sin; yet-to-be-cleansed for a large, historic part of the population.

7-9 Here Dante (the proverbial future parent) tells his leadership on the topic of poetry, as well as the analogy of the times, his discipline's creation is coming up from the garbage pail of the time period. Analogous to Hell for the real Dante, who knows what Calliope symbolizes as someone who has stood through generations, on all topics, as a poet and muse capable of bearing light on one and all, from grandparents down to Millennial and/or Medieval youth. A warrior poet, not, as Dante, a poet-statesman.

Whether Hell's dead realm talks of the current periodic disavowing of The Divine Comedy by contemporary strategists in fields looking for less than a traditional framework to the future, or talks of said wake of crisis from which surviving Californians seek to recall less (so as to keep their footprints covered), we get the impression messages rise periodically since this writing of the 14th century to take on new meanings.

8-12 "… and may Calliope rise somewhat here
Accompanying my singing with that music
Whose power struck the poor Pierides
So forcefully that they despaired of pardon."

Avoiding the analogy this was any more confrontational or important a generation's identity, as perhaps not always treated as such in light of other egos and the reiteration now ensuing,

17-18 "…as I had left behind the air of death that had afflicted both sight and breast."

Indeed, a timeframe laying proclamation to less marriage and much counseling, both mind and heart of those on the poet's vessel challenged.

23-24 "…and saw four stars
Not seen before except by first people."

Perhaps alluding, in this sense, to the four Gospels, if a mythological bridge does less.

40-41 "Who are you—who…
Were able to escape the eternal prison?"

46 "The laws of the abyss, have they been broken?"

55 "…but, through his folly, little time was left
Before he did—he was so close to it."

The eternal prison speaks of pockets in the time frame of 40ish folk's moments from childhood to present, where now presented with medicine's query and angle on childrearing, if an early start did not get the wind to its sails, it is now the final evening of biological reproduction being within the realm of their very ages.

62 "the only road I could have taken was the road I took."

Why, as if this were snatched from the geneticist's mouth following a private counseling to a patient in said category, who has recounted some of the timeframe's translations.

64 "I showed him the people of <u>perdition</u>"
Which is described as (1) utter destruction; and (2) eternal damnation.

65 "now I intend to show him those spirits who, in your care, are bent on expiation."

The spirits, as well bent on Dante's care in this original version, according to his guide sent by the "lady sent from Heaven," venture forth into a California where the vast majority, judging by our survival skills intact, are in our historically worst season of State debt ever: utter destruction, as law and order has prevailed (despite its own issues) in case the latter definition be brought forthwith in wartime.

66 "…are bent on expiation."

We are bent on atonement, not at the hands of an elderly upper class, necessarily to maintain their standard, morals, and interpretation of American history, though from <u>Californio</u> to Korean, Irish to Tex-Mex, with many fish in the pond to feed in said waters, this is a generation that smacks of the instincts of a domestic guard animal on issues such as marijuana's legalization, scientific subjects, entertainment savvy, European literati communication, and the birth of their generation's baby idea for promulgation to all four corners of the State, the environmental corporation. This is all before we get to Catholic priorities for all of America, never dropped in a fumble for Diocesan control, after an era marred by lawsuits but not the sun's judgment, within which the inherent beauty of a new generation of <u>Californio</u> put on a work uniform!

67 "To tell you how I led him would take long;
It is a power descending from above that helps guide him here, to see and hear you.
Now it may please you to approve his coming; he goes in search of liberty—so precious
As he who gives his life for it must know."

This is personal, as I've spoken Italian since I was eighteen years old, turning 48 this year, with the ringing of the country's World Cup victory in my '82 year abroad, under Pope John Paul II, in my ears, echoes of Neapolitan voices accompanying to this day in the California kitchen. I went to my first Episcopal Confession in 2000 A.D., absolved of private matters, to shortly after go to rehab for a wine I still drink now, as the war hit our country the year after: I, in alignment with Christ, was stuck in a sober living home in Latino parts of L.A., due to vindictive hospital housing authorities not giving a rat's ass about my education or welfare, happy to have caught and cleaned a non anti-Semitic, leaving me with parolees to see on T.V. the events of September 11th.

In time I gave six more years of my life to Episcopal good will, before Mary's wrath split the historic American and British Church in two, and I became a Catholic in December 2008, now agreeing with their paperwork on moral fiber, but shocked at the repeat of economic sins in yet another church, and the air of Gainsbourough's 18th Century Bath, England, a snooty timeframe in which he painted the Blue Boy, and as a pen-artist as well, I have to say, there's no such thing as a Let Go, Let God organized religion party where the wine pours for one and all like it does just South of the Vatican, every day at noon and night meals, to keep the Neapolitan culture a vibrant force in a war-torn world. I am of humble and good standing with the California Church, and have not dated for 10 years, after a break-up where I vowed to do it right the next time, when a new version of both statewide recession and American economic woes smacked the docks of New York City like hard winter water, tumbling auto and real estate magnates alike. My will was to give my life, as a U.S. Air Force Veteran couple's only heir seeking love in the process, to Christian, not L.A., fiction, screen, or journalistic writing, in the name of Yahweh and Mary's only son, Jesus Ben-Joseph the Messiah. This is the nature of it, biologically sound to produce or not in a California culture which strives for more layman support for the simplest of music orders from public television to play in the home, if public radio cannot steer for street-smarts. That's what I mean, 30 years of speaking Italian fluently as a Hungarian-American, by, "I've been around."

76 "Eternal edicts are not broken for us; this man's alive, and I'm not bound by Minos; but I am from the circle where the chaste eyes of Marcia are; and she still prays to you, o holy breast, to keep her as your own: for her love, then, incline to us."

No, God's rights for his devout are eternal edicts, not Catholic and Monastic orders alone, as I am not bound by contemporary thought in their realm in order to continue my thoughts of reproducing: so many more secular women at the store smile at me amidst these platitudes, born by a distraught haul and a loose-tongued trail from man's Confessional in some parts, why sometimes, practically even an eavesdropping (so the Priesthood has notes before the Saturday's Vigil of Absolution, if one can call it such while the professor smokes the rest of this modern-day, down-to-Earth carnival of community control).

In this hard condensation of terms, Cato the younger, the husband of Marcia, is defeated by Caesar in an attempt by conflicting armies to reestablish Rome without corruption. He committed suicide, refusing Caesar's Rome, and Marcia is said to guard the infernal circle of Purgatory for the penitent. In contemporary terms, one wonders if she is a person nowadays who would honor absolutions of previous sins on California's front, to usher in a rejuvenated age of memory and progress exempt from the eternal verdict said female can place, like the Hanoi Jane of our age and prosperity, like a dagger in the back of the modern California patriot. Will mentality rejuvenate along more appropriate lines of stability? <u>Much rests here at our own infernal circle</u>.

By the next time I pick up my pen to continue this interpretation of California through the telescope of *Purgatorio's* prose, Usama Bin-Laden is dead. I wonder what spirit infiltrated the room and twisted my thoughts in said direction, to see the poetry in my life. On with the interpretation:

100-106 "This solitary island, all around
Its very base, there where the breakers pound,

Bears rushes on its soft and muddy ground.
There is no other plant that lives below:
Hardens-and breaks beneath the wave's harsh blows.
That done, do not return by this same pass;
The sun, which rises now, will show you how this hillside can be climbed more easily."

Said generation of Californian, from early mid-1960's to present, are the rushes on this solitary island, or portion of World History, where the breakers

208

of the Technological Era have pounded. We will not return to discussions of medicine past or present with a generation that has led as such, by sacrifice, through this stretch of narrow times. Modern medicine by all means, after this has been said and done, has research tools and lab knowledge that can climb the hillside more easily in its pursuit of the peak of understanding.

118 "We made our way across the lonely plain,
Like one returning to a lost pathway,
Who, till he finds it, seems to move in vain."

Simply, well, one thinks of the American way before these issues arose, a lonely plain indeed for more than one reason, amidst an age of less marriage (amongst other things). As one strolls by those with a control of their own vibrancy in a normal marriage, complete with stroller and angelic ride within, yes, to find said peaceful pathway carrying aforementioned facts can seem like moving in vain.

CANTO II

This is about Ante-Purgatory, where dawn on the shore of the island mountain is morn for those women in their forties and fifties in the Golden State (as well as men under treatment for schizophrenia, at various levels of how treatment has taken, though many in the same economic rough patch) and their challenge with contemplating a biological childbirth of their own with a stronger possibility of autism in the youth. These are the arriving souls in the boat of the helmsman angel thus far in California's own contemporary Ante-Purgatory. Purgatory, and in continuum, until the end of our times and awaiting of judgment by the Heavenly Court (Court in the sense of a regality more than a judicious body to the true believer).

Rather than discuss Canto's IV through XXXIII in said perspective, with my analogies of the prose to the situation, simply be frequented with the day and age when the University's medical center may not even return a phone call regarding what they can do, as far as genome mapping is concerned to predetermine an outcome. And, also, it is an age when parents encourage, in the face of an often skeptical priesthood, the kind of faith that got a military family such as ours through this much time and space, without going over every problem at the top whilst we work in our garden of plants, trees, and birds' nests.

To say America does not see the mental health treatment recipient as a parent, after all the media's propaganda and a complete diversion from the discussions of equality and autonomy in said war's timeframe, is an understatement. Three terms (Bushes' to Obama's) in the Oval office have avoided commentary on the topic, unlike the Carters in their lives, only to afford the patients painstaking reminders of Social Security not lasting all of their lives, while laying dormant on the often-chilling side of what actually happened in a generation of transition from one set of psychotropic medications to another. As jobs are scarce, depression and obesiosis high, smoking a more real threat, perhaps the helmsman angel is Romanticism itself, which looks to the ancient world for inspiration out of an often disavowing with a purportedly enlightened age. Said definition of old-fashioned, in an academic sense, for myself can be applicable, as the Brightside of the technological age, from medication side-effects to cell phone, cancer-causing theories, rarely rears its head in the morn without a solemn vow to repeat the class. Amidst this perspective of analogy for a translation of Dante Alighieri's The Divine Comedy lays neither the desire to prophesy nor belittle any reader's request to interpret said work for their own diving purposes. I, from here noting this congruity in the times, care for more to do with the constant renewal twould be necessary to carry on a new generation. I, perhaps, not the only one of a beautiful mind, so much as a strong heart serving the Lord at every corner in the damndest of ways. Yes, amidst the Earl Gray Arguments, with homosexuality, conspiratorial socializations, why, rubbernecking on the road of life, is the silent campaign of medicine. And keep in mind, on the Catholic end, it is not that the lawsuits preclude the commenting, but on the subject of the intersection of the interpretation of behavioral norms plus spoken tongue, and jungle politics, one has ushered self and others into an era where one looks sanctimonious at best for concluding the purposed balance for amoral discussion. "Look, Ma, no ties! Not a single tie in the Parish! Is our neighbor with the long hair and the neat toys at the pulpit today? Maybe we can go fishing!"

An Episcopal Female of the Cloth and a Catholic nun were out on the town and stopped by the café`. Amidst the purported cultural debauchery, one said to the other at the same time: "I didn't start it!"

Then the Episcopalian said, "No, it was our choir director—his list is a mile long!"

"You think that's bad here she goes, Miss Cross Country, running the whole mile!"

"How do you suppose we were not en-burdened with a hormonal imbalance?"

"Our parents had other than to make us stoop to purchase the Earl Gray!"

"So, it's what goes in the Temples?"

"Other than doughnuts, candies, menudo, chorizo, gorditas and frijoles, yes!"

"Well, then why are Italians our same way?"

"Because she ran the mile in record time, and opted for the Lipton's!"

"Hey, whose that heavy in the corner?"

"Him? He's anti-Semitic! Starts it up all over again! Likes to watch her run by!"

"I thought the torch-bearer was a Greek mule!"

"This is as far as he's gotten—they don't bother him, they bother these, so he still gets to jog with her!"

"How do you solve on academics?"

"You don't! They're running a business, not a library!"

"Oh, come on! There's not going to be some common sense meeting point between the two!?"

"The Cross as Public Eye and The Cantina as the home-nest! It's a Northern Italian border town to some, nearby snowy peaks."

"SO this is why they go out in droves?"

"And why some travel alone..."

"Why?"

"Because they can't be seen connected to said agenda on either end when they speak a foreign tongue as well, and sports is the filibuster!"

"Ludicrous! Not in our America!"

"Your Left is stupid, ours is connected! So that's why I'm a nun, to meet the Latinos halfway since they sat there that long while they thought this was a place to discuss film!"

"I should have ordered an Espresso!"

"No, it's a dichotomy between Europe and America—the clowns come in from the circus when you do that!"

"Just like to hear my sermons!"

"You now see why the Priesthood misses the Marx Brothers!"

"Oh, you want to hear the crap our Bishop came up with? Once bitten twice shy!"

"So, this region is full of hot air on the subject of culture, courting and childbirth?"

"Look at it from my perspective: It's an unusually warm and smoggy Spring day, the apartment has to air out after breakfast on the stove and although bathed, it's a three-quarter mile walk to get an ice coffee so gasoline is saved, the coffee mug they brought does not have a burning cross on it, their clothes are clean, smog checks passed, degrees in lieu of dominoes awarded and the cost for the opinion is $1.95."

"That used to be 25 cents in the Depression."

"Let the historians arm-wrestle over the subtle beauty of it all, wash your hands with Holy Water, and spread the Gospel for either faith you want...they'll still show here at night like clockwork for a cool evening breeze."

"I think decisiveness and differences shouldn't be stuffed in the corner for a time-out."

"They swept and mopped the corner where the chessboards go—put whatever you want in there."

* * *

--"*SO*, you movin' back to Naples with the cash?"

--"*If* this economy doesn't pick up, I'm moving to the bushes by the library!"

--"*I* know where you can get a job!"

--"Where?"

--"The colloquial Lutheran's Scamboree! Can you pass a drug test?"

--"Well, if you tested me, you would catch every pot-smoker in town, because then you would be able to hire me and keep an eye on me, and they would have to come up with a whole new hustle rumor-wise!"

--"All right, just be polite to the phlebotomist, because she hasn't lived here long!"

--"Deal! I'll fill out an application, leave someone besides the pharmacy and the Italian restaurants with my Social Security number, and see if we get a little bingo!"

--"Well, there's health hazards involved."

--"I'll try my best."

--"*Just* let the Vets who know your Dad see if they can land a shot on ya' at the dunk-tank!"

--"How deep is it now?"

--"Getting deeper all the time!"

--"Look at that!"

--"What?"

--"Dallas stuck to the original story and has their foot on the gas every time there's a play-off game!"

--"Commercials for true Californians!"

--"And borders for all."

* * *

--"Do you always handle the heat with that expression, or is this your first crush on an intellectual?"

THE FORMULA FOR ITALIAN POLITICS
BOXER +OBAMA= PELOZZI x MUSSOLINI

GIULIANI

Because the Queen is in Heaven, not on the meek's time-clock, and if that's American problems, it must've been Girls Night Out if you're unemployed! Be thankful she thinks American men have a half an Italian inventor's chance to solve the Lord's riddle!

--"Hey, what's goin' on—the rest of the Gentiles' side of the fence?"

--"Oh, Burt got caught with a quick left for braggin' about his kid outside at Chuck E. Cheese's, other than that just fast cars."

--"Who was it?"

--"One of my roommates, but they sped up too fast for me to catch anything but the Alma Mater license plate holder!"

"Hey, you know Serrano cloud signs? What's Chief sayin'?"

--"Don't all pile in your trucks and cart the boat to the lake at once, there's some work at home to be done!"

"Now that we've helped find the camel that one of Hop-Along's men stole like a Judas in the middle of a community work ritual, maybe the unemployment line will stick to the game plan!"

-"No, Jesus doesn't like revenge."

--"Well, what about that Killa' Judgment Day schedule? I mean all that, right back atcha', after 'End Times'?"

--"Well, you better owe me one, false prophet that I am, because I heard in Heaven they get to be the guys, and we're the girls!"

--"You mean I get to be on the receivin' end of that alimony check."

--"Yea, you—and the rest of those with anathemas!"

--"I thought that crock o' ham was absolved?! I mean, that's what they told me."

--"Oh, you can trust them all you want, but I'm not gonna letcha' live it down while I'm drinkin' with you! Yes, $5, not $1, would be more like it for a ride home from the bar, Bud!"

--"And the Final Heckle from the Jews on Rosh Hoshanah, as Uncle Steve called them to not worry about Yom Kippur, even though they got him labeled gay..."

--"So? Not impressed. Know you a long time too, jack-ass! Not like we had a fetish."

--"Oh, in his dreams! The scank!"

--"Yea, wolf down another while you are at that extra beer, Uncle Steve, before the skunk you are goes out to the patio to smoke one with his brother-in-law!"

--"Both fags...We are too Law School material!"

--"Hey, make'em pay up, Dude, after 18 years of our emergency training in case of erectile dysfunction!"

--"So, let me get this straight, you rent to your daughter's ex-boyfriend? You know how many ex-girlfriends I, Uncle Steve, have?"

--"Letcha' fill that one in yourself!"

--"I've got a harem of'em!"

--"Don't say that too loud, the Muslims'll hear you, and their harems apparently are serious business!"

--"And mine isn't? Lordy, I didn't know they were so adamant for my conversion to Roman Catholicism and had voted me the Bishop of the Mentally Ill!"

--"After sent to the Papa-cy, without a doubt., for his stamp of approval."

--"Hey, what about the mother-daughter sober living home? They won't give the deposit back if she shakes up the Till of Wisdom, snake and all, and you have to get it out of your system in your private room!"

--"It's an experiment in candidate Kerry's potential for the Oval Office, Purple Heart and all! See? The Color Purple is the textbook. It's either that or the cartels in Mexico, buster!"

--"Oh, vacuuming, rose gardens, spiders, wasps, bathroom and dishes on the inside—Injun Stephen rides again!"

--"German roots?"

--"Yes, the migration they were in the first war—ass-backwards, forward march, to the money we go!"

--"So is "Don't ask, Don't tell" more about masturbation and pre-marital sex, now that we have female military that is a sure slayer, and so Cheney dragged a red herring about his daughter's lesbian loyalties, because Sean Penn slowed down the whole damn enlistment for five years over a stoner role?"

--"That's a good question—was she bothering you?"

--"The Period in the Military is a tradition: Start with my mother, she served 34 years under Colonels who smoked, and thus quitting is a torture beginning with young pup John McCain's career. Trust me, he was a Sad-Sack reader, at the deck-swabbing since he had to have a Hostess cupcake while I did it for Mom. Her and I, the female cadet, I don't know whose strings these are, but puts me in a league with the witch who killed the elf when Harry Potter escaped, even though you have to help him through until he's bald like his dad!"

--"Don't get wrapped around an axle about it."

--"It's not an axel, it's a map, don't act like a hungry dad!"

--"Well, don't swing at me. Hey, I'm kinda interested after all in Catholicism—how did that conversion class go? Interesting people?"

--"All black coffee drinkers. True discipline more than denial, just a couple of elbows in line when I would pick up the silver spoon to mix in my Sweet N' Low!"

--"Oh, as a New Yorker, I have clarity. They said that's from the Neapolitans since they're playing the victim at the same time."

--"President Obama and Jesse Jackson both have their strong points spiritually, but they are under the misconception that our retired military officers from World War II, Korea, and Viet Nam are a modern day version of Orphan Annie's Daddy Warbucks! Not only are Officers' second children seeing their American Dream emptied out like a sanitary napkin basket every night on a media needing Hollywood root-rot reform, many struggle much as Vets. Apparently, behind the scenes of the White House, his family genealogy is more important than America's future.

--"Ah, yes, Arnold would like you to cover the center of the stage while he has an affair? Can you do that, Santa Barbara?"

--"Come on, people, the East Coast is going to pack the House tonight, let's see some real-time enthusiasm!"

--"Here, feed the Pony Express. Bishop, don't just give them another document, and a new carrot to the horse!"

--"Did you hear the scoop on the evangelical parish in Texas?"

--"No."

--"The guy who speaks in tongues is gay. Yea, I heard every California boy has gotten more than his full of Bush phenomena, and the Lakers are going to try this one again next year!"

--"Well, who let them in the Catholic Church anyway?"

--"The priesthood needed the dirt on someone who got stoned, in Confession. These ministers work with Seventh Day Adventists at the Medical School and high schools. Between the gossip and the scholarships possible they did all they could to get him in the Church, but put him on stage!"

--"Poor guy. You'd think he got high with L. Ron Hubbard."

--"Yes, Dennis Prager, college. In an Air Force brat neighborhood. With a cast of thousands on every campus from around the Golden State. When the subject changes to psychiatry, and patient's efforts to get off Social Security and S.S.I., keep in mind—from his psychologist's point of view, after seeing the patient's etiquette, it becomes apparent we had to wrap up some of these degrees in a blanket and give them a bottle. Hence, a generation unto its own, an Inferno for Film Industry folk fighting bigger demons.

--"The pharmacy, for the position of store clerk, thinks I'm over-qualified. They won't hire. This clientele is a combination of the cast from Oliver Twist, Landon Donovan fans and Who's Who in The Wags! Meanwhile, back to Uncle Dick in Oklahoma—'You gonna put salt and pepper on those eggs? And hash browns? Then you don't get to go to try-outs! Go to your room! And Stephen, you're sleepin' in the sixteen –wheel camper!' And A.F.S.—the cousin with no tongue, plus the girl from camp with the braces, jumping up and down at J.F.K.

Airport with Dad, screaming 'Soltesz is home! Soltesz is home!" Well Cousin Butch and wife Kathy from Connecticut had a point—'maybe you're a leader.' Now that half the high school gang that went along has one hair on the chest each, for Father, Son, and Holy Spirit?"

--"Oh what's up now, Sushka, wanna see it?!"

--"What happened?"

--"They tried to make him look like a nasty Christmas elf, but they love their chips, beer and Arnold Schwarzenneggar flicks after work. They like to hold the thought on what happened the first time though, and on the flip-side, half the counties in California have taken their crack at female spies on the blank slate's code."

--"Who are they?"

A) The Family

B) The Jews

C) The Hometown

D) The Neighborhood

HINT: The older sister is in all four categories.

--"You make me look self-centered."

--"You must be doing something that proceeds my action which determines whether the magic trick works on the audience or not."

--"One thing leads to another."

--"The entomology of that expression roots in an age when the British Olympic Track Team could change the face of the Games."

--"Yes, I think my sister was born a decade later to the day."

--"And verbally abusive."

--"And you're Doctoral Theology. I'll let you give out-of-state and L.A. family cues on how to keep me having mysterious occurrences when the subject of writing comes up, then we can get disciplined by Dad after we pray over the meal at the table. It's as if you walked right out of the chapter of The Eye Of The Needle."

--"I've never seen a parish put one Christian through that when the subject is a Lenten fast."

--"Maybe it's the new Altar Guild—a little less steam than the old one."

--"Here, you stop giggling, poking, teasing and farting, and I'll try not to act like the unconscious creative process evolves around me."

--"Upp. Okay, Rabbi."

--"The rabbi is gay. Really."

--"I think they should have two parishes: the writer's and the rabbi's."

216

--"Alex, what's after the category on Roman history?"

--"One on Colloquial American Catholic Historical Figures. Plus Various Kitchen Herbs and Spices."

--"Mr. Steve, just what can you tell the Duke and Duchess about the Match: Anglican vs. Episcopal?"

--"I was an older acolyte at a healing service which the London Rector and his wife were running for His American Seniors there for gall-bladder surgery and the likes, and what the 50-year old student was striking up a conversation on archeology in California with me for I know not, when a guest minister arrived, all of us close up to the Altar, a 250 lb. former leader of A.A. counseling in the military who exclaimed, "I feel the gays should have their rights." The next Sabbath, the third minister\former cop, a smoker, gave a monolithic one hour sermon of old war stories taking us to policing in A.I.D.S.-stricken Africa. The wife of the British minister's last line to me since I had seen him was, "Get a job!" We then got the sudden news the Brit had barely been rushed to the hospital on time to survive a staff-spider bite! Was this, I asked, Keystone Cops or Keystone Clergy?!"

--"You thought it was a legitimate exorcism? That kid's been spinning since he converted!"

--"Well, Mass gets a little deep. Like it goes to the bottom of the soul, where the Grateful Dead fan he met years ago is connected through the experience of receiving the Sacrament! Different than Episcopal. In this scene it's as if my demons are at a card table playing dominoes."

--"Well, you're in sturdy shape, so says the doc, despite tobacco: as a former Episcopal acolyte, what seems to be the central issue?"

--"Keep it low, but they have cut across half the chow lines in the food chain to keep the subject science and finances in the Parish. As a lad of only 200 lbs., presumably gay since the crow en route for the hostel apparently got shot out of the sky, my work is endless! You can't splash enough Holy Water on a British minister's hands to cleanse them of what he must do: Administer Communion to people who know me from childhood, not my stint on the bunk of a hostel, and believe me, I'm not Gulliver! This is Where the Wild Things Are! They said I spoke Italian in my sleep to ward off the Devil! I have no idea who I gave my resume to that said I was a Hungarian-American lad, conversant in Italian and Spanish, who had worked at a youth hostel for international travelers of low-budget, but apparently they read between the lines with much disdain and the crap hit the fan! Not only did the Church split in two, but we had several funerals back-to-back and I was unemployed for three years, discovering that my writing was therapeutic, and encouraged by the British minister to continue, took pen to canvas one day, drawing Madame Bruni a full two-three years before she announced her engagement to the French president! Whatever Surge that was in Iraq is over!"

--"Can you draw the Duchess?"

--"No, I'm not going to jinx this one over all that baggage, I tell you! That's how I became Catholic. But here comes the Bishop, with policies and priesthood both like the Great White Buffalo compared to the poor commoner. Onto another church, I carried the cross Earth gave me, and thought of the girlfriend often who liked to discuss the environment. I guess that's why several services of Jewish and Christian both down the line, she took me on a hike, I remember, and they didn't. WE found 1,000's of Ladybugs in the shade. You get me off the subject of my own again, Sweetheart. So, in case some people were curious, I attended Jewish temple on Saturday several times, but could not, logistically, make the Torah Study in the evening from a German-American sober living home. The ghost of Hitler? After the social and economic errors of the Episcopal religion 1970-2000, it is a very real experience. So this war is about the Christians cleansing their souls. Their ancestors died in World War II (or lived, like my uncle, after combat duty), for God's first people, and they forgot. From Kushner to Kohn, rabbis can be elusive like priests. In fact, American.

So, no A.ID.S. foreign or domestic, despite ethnocentricity on romance laid upon polyglots, and where does this discussion lead? Is Laura Bush ignorant? No, Laura Bush has been championed by my two votes. Other than condoms, it has been known in certain circles, that without discussing children or abortion, a certain subculture, academically (not as if they're from the Dept. that creates Public Relations executives), desires more intimacy. And that's when the American male is in chains, of slavery to Christ, in what should be a discussion of medical records, since he is a scientist impassioned by an athlete, not a shrink impassioned by a sociologist. The military doesn't want to take the jump in modern parentry, and Federal Funds alone support the scientist's lab as he queries the proper control factors for the dialogue. That's two votes to win a war, not to teach me about hygiene in medical facilities—I've been in and out of hospitals for various things(not A.I.D.S. however) as an earth-tending Texan descendant since I was eight—that's the childhood heart surgery, and now I work in California dry heat and smog. Texas, keep your air quality, and your head.

Affairs, other sins, earthquakes—California is a miracle in motion if you do not get wrapped around an axle about Latino things. Mixed ethnicities do get to date Whites, and so do more old-stone Mexicans (of 25 generations present, since the first buffalo was sacrificed on a Coahuila altar by the Medicine Man Bush thinks this is a gentleman's agreement of the magic of marriage with the Pope, and Disney is growing internationally. Like a wart on a linguist's ass. The doctor removed it. Doctor God. And He wants to see everyone's medical records at the Pearly Gates, because after this narcissism internationally, Saint Peter has a change-up for you before entering Heaven as the child you are, if you think romance up there doesn't continue as it does down here. When I see a theologian with my language skills, and a psychiatrist with my social skills, I'll let them take a shot at the books and the baggage of the playground I had, before getting straight A's in K-6 years. LE FIN

--"Are you talking the neighborhood's menu for poon?"

--"Anything you're having, big guy—do you need a buck for gas?"

--"What is this female bartender doing?"

--"She's taking photos for the downtown business center's magazine editor."

--"Any complaints?"

--"The price of the beer—I could afford a few rolls of film with that. The shoes...she shouldn't go there at Halloween, some of these folk wanted less of Kansas, so that's why they moved here, and the beer nuts...the kid in the bar needs a burger now after that tease."

--"Hey, she shouldn't be talking to her ex!"

--"Shouldn't be talking to her ex? Are you kidding? You don't have too many choices: You one-night stand her, you'll probably be collecting Unemployment, quickly moving to S.S.I. You stick with her for two years, and her ex's Mom will win the bets. She dated him for three months. He's A.D.H."

--"The poor spoiled little shit. That nice of clothes, but not The Billionaire Boys' Club!"

--"Need some more glue for those models...the catwalk has new wood paneling and he not only doesn't read People, that's quite a demonic phenomena when you bring up the Bible."

--"Maybe the whole city is just orgasmed out, and that's why the newcomers are getting the jobs!"

--"Are you sure, 'Esteban,' you aren't sending out the wrong signals?"

--"It's not like my head looks like a satellite dish—let's keep it in bounds!"

--"And this other guy...he's looking over your shoulder at your card-hand on chicks until it's time to talk Bible!"

--"Get me a golf bag! Any bag will do...I've been to that guy's Country Club!"

--"What else did happen between then and now...I refuse to believe this is the whole problem at the top!"

--"Hey, look...the musical being put on by the theatre house."

--"Can't wait to see the staging, Esteban. Who's the score by?"

--"Her ex's Jr.High choir director."

--"DO you have a borderline personality?"

--"My grandmother was born there—two miles from the border. The family's been giving birth to American citizens for five generations...can you move on with it yet?"

--"You mean all that plus a buck for gas? No wonder the cab company is foreign-owned—they pay rent! That's it—tired of going for rebounds. You'll have to find a new Center—I'm retired."

--"Another beer, Le Bron, or should I pull the truck around?"

--"It's your buck. Frame it."

--"Oh, uh, that's enough abuses of the word 'family' in the whole bar tonight, how 'bout we let this soldier sleep in my wallet?"

--"Hey, did you hear, Ted's wife, 1/8 Mexican."

--"Oh, Ted's sister, 1/32 Mexican, by proximity."

--"Proximity?"

--"Yea, her seat on the tour bus!"

--"The tour bus?"

--"Yea, ever notice how only White people in this state have seen the Grand Canyon?"

--"Oh, I know—It's not a Black Hole—it's the last 'I' dotted by Ted's sister, the screenplay writer!"

--"So the mind plays tricks: Life goes on."

--"It's inexplicable. I used to know this word, but I have to look it back up... Are you trying to tell me I have been through that much pressure? It should mean 'unexplainable.' So when I looked it up, yes, that's the truism. Now, let's try that word, off the subject of overdone psychoanalysis' effects, and see if that is a fact that is true. Here they have 'an undoubted or self-evident truth, esp. one too obvious or unimportant to mention.' Looks like we created the word for some breathing room on fair play in sarcasm. That's Webster's New Collegiate Dictionary. Good, not one for sarcasm, maybe being an intellectual is not quite the Christian calling based on attitudes we should bear in order to emulate Christ Jesus. Because Webster's was worth prayer time to purchase. Thank you. Not wasted time, albeit a bit different of a discovery. In other words, I'm correct."

--"In mid-1981, near twenty Americans stayed overnight in a nice New York City hotel, still reserved by the American Field Service to this day I do not know, yet the Lyden's unique view of The Empire State Building somehow guaranteed me a genuine year abroad in Italy. The year began rather well.

San Gimignano, in Tuscany, was the location of our intensive one-week study in Italian language via "The Baby Method," where nothing but Italian was spoken to all students, wherever they be from in the world, while listening and writing in the classroom. We were given a piece of our identities as global citizens in the sharing of the rooms, the walks through San Gimignano, the classes, the large dining hall meals, and the camaraderie at night.

--"The train ride south to Naples was long, with intermittent stops, and when I arrived to meet my host family I was definitely in a new location, with a new scenario, yet inside the central connection of Orientation held tightly, and I was not too much at a loss for words. As well, the woman greeting me was direct, to let me know she was to be a friend of sorts: "You call me, 'Mamma,' and you call him 'Babbo,' which is Mom and Dad in Italian.

Their apartment was intimate and exciting, traditional and furnished, and all this time, for good reasons, my host-brother Fabio was resolute that though three years younger at fifteen, he shared responsibility for my well-being. They had a share of quiet where they lived, though yes, the city's streets were close outside. Spaghetti Col Frutta del Mare was the first Neapolitan dish served to me, a pasta and marinara sauce sprinkled throughout with the shells of miniature-sized clams.

Soon my A.F.S.-Italy regional advisor called to let me know our schedule for enrolling me in a high school nearby. This is where the story of my stay became difficult, as my first real challenge abroad was somewhat over my head.

I sat with my advisor in the office of the school's chief administrator/principal for near half an hour straight, as they went back and forth in rapid Italian, When we left, the regional advisor, Pippo, was discouraged. Unexpectedly I was being asked to provide some documentation of my rights as an exchange student to now enroll in the school! Hmmmm......

Without too much to say in gruesome details, I was led on for a couple of months, during which I was only allowed to participate when the students had their sports break to play basketball and soccer in between study hours, if I was on campus. My advisor did his best, though in these ensuing months I was only allowed two weeks time inside class, before being told, "Out!" again. It was Christmas Break that a Senior female student discovered the principal was a member of an Italian political party which had an anti-American nature. She then went through the school halls and had various students sign a petition protesting my denial of enrollment, which she took with a full group to the District Central School Office to present. What a dedicated friend!

Things remained the same, so it seemed, and to get my mind off it all, the male students organized a basketball team and placed me at starting guard. We did not play in the country-wide Youth Olympics in Italy called Giocchi della Gioventu`(because of the principal), though our blessing was to take home a First Place trophy and a gold-fashioned medal for each player, with a scene of the Bay of Naples stamped on the front of the medal! I have to this day the medal hanging on my apartment wall by my college degree, a favorite painting, and my Roman Catholic Church Confirmation certificate as of Easter 2008.

After the tournament and our celebration in one of the city's many pizzerias, I was allowed by the host family to travel Italy and stay with the host families of American and foreign friends both before our end of year reunion in Cormier, near the French-Italian border. My biological parents in California, often sticklers for regulations, were just silently happy I had an award-winning host family, a Naples police officer, his wife, and their son, who took me in and shared their support, home, kitchen, humor, and dialect with me. It turned out to be the kind of year that good schools in America didn't complain about, so I thank A.F.S. Through many tough times in over thirty years since as an American in a global community, Italy's friendship has come back to often raise my mood to one of joy.

When on the phone from America to Naples some five months after my return from Naples, my host brother let me know a little secret…the parents of the students at my Italian high school had grouped, because of other failures by the principal in his disciplining of students and carrying out of academic affairs, and forced the Central Office to oust him and replace him with someone new, all at the beginning of their new school year without me! They got their courage up after Italy won the '81-'82 World Cup in a summer of fantastic soccer, and got a better principal.

The story of what led up to our medal is heartening. I would wake after 5:30 A.M., when the family would get the officer to work and the son a bowl of coffee with biscotti in his stomach, before trekking 2 ½ urban street miles to his Technical Institute, a pre-engineering high school(he later wound up in politics). I would have a bowl of the same, change into my street clothes, and hoof by foot uphill nearly a mile to arrive at the street to the school gates. After a weekend, meaning Saturday and Sunday nights only in Italy (six-day work-weeks!), I would bow my head silently amongst the passersby walking toward the gate. One and all, we witnessed a milieu of condoms and heroine needles trashed on the ground after night's use in parked cars of young Neapolitan lovers, and addicts, who parked in this remote place of the city for a branch of the Camorra's fest (the Neapolitan Mob operating beneath the Church's nose). We saw no street people except when venturing to see what the school did on weekend nights.

There was a coffee bar at the base of one of the tall apartment buildings in this section of the city, named Parco San Paolo, the location of Gli Azurri's—The Blue—outdoor stadium, the international playing center of the city's famous soccer team. I momentously, with my own feet to brag about, would pass the slid-back metal gate, say hello to the whole crowd, and wait an hour until after the first class in the fog to pick up a game on their simple asphalt court at break-time. Often soccer with a cheap, miniature orange ball would take priority, and defense in said sport is an art of a sacrifice for to learn against young Neapolitan minds on asphalt.

222

But the war was in basketball, without a ref or yard-duty: I took my knocks endlessly, but somehow didn't end a semester-long scoring drive, throughout all the aforementioned bureaucratic mischief. The romantic male lead in the school soap opera (which only has Naples life, if I can lower that at all to help you understand, not a prom and queen—too many Italian women would bear the crowns) sectioned off a group of five of us with stamina who had tournament potential to practice as a consistent team. He was studying to become an accountant.

We wiped out our first team in tournament play at Campi Flegrei, the outdoor men's club court named for its location by the train station of same name, which was the nation's stop before heading 14 hours south to Sicily from the port city. The second two teams were scared to show up, even though we were playing no-hang-up contact ball, and just taking our shots. In the final, a team of sturdy, private Classical high school men showed up, and we tossed the ball in the air. I watched from my guard position our taking of the toss, only to see a very weak shot taken, grasping nothing but air (as if the romantic lead had been at the school gates all night in his car with his girlfriend, newspaper covering his car windows and all). We knew we had to double back non-stop all night after that on defense (we did), I grabbed the ball on offense the whole game after that, driving to the hoop and dishing off to various unfaltering supporters for outside shots. It commenced what added up to our four on their five.

Whatever the score was going into the third quarter, the opposition was not going to stand for our hustle, and gave me alone a quick uppercut in the shorts while reaching for an offensive board. We finished the game after getting our second wind with a heavy, controlling full-court sprint to walk away with it 56-26 (I put in 28 points in what I thought, amidst the back-and-forth play all game, to be a constant catch-up effort on our part). The stadium's section of the town kept its pride amidst a city-wide offer to keep their own brand on the game. I had an offer from an expansion team at the base division of their 5-tier pro-league, a true amateur attempt to see what a group of less enthusiastic Italians a few years older could do if they played me at Center, standing 6'2" myself, the only other player on the team that tall. We lost all our games that season, except the first, in which I pumped in a last-minute outside jump-shot, back in my youth, when my game was hard to rock if I had support. It is more remembered by ball fans as the year Magic Johnson and the Lakers won the '82 championship against McHale and Bird of the Celtics, a game in which my cheering for him rocked the apartment complex before a good Neapolitan lunch. I witnessed his shots on international T.V. with an Italian-language broadcast, a year in which for the first time he was awarded MVP of the NBA. Athletic fever had overcome a young Italy in Pope John Paul II's day, and they went on to their victory as a country against Germany in the World Cup of soccer, also beating Brazilian favorite with the great player Falcrau,

3-2, in the pivot point of the Cup. World Cup is an event I rarely miss after this, now cheering again for an ever-improving American team.

Yes, hard to believe, this is what my Christian calling was in the country, other than introducing the Gospels to a street-smart crowd moving on with its young adulthood by going out into the city from Parco San Paolo at night, to pizzerias, Pino Danielle concerts, films, etc., men and women, brothers and sisters, all strongly encouraging to speak Italian and their dialect to a decent level of fluency by the end of the 10-month's time. I often think of them while having a mini-cigar is a Southern Cal rather resilient to smoking, and dangerous as well, but not the same ambience as this city, known for its cuisine, and infamous for its international crime syndicate.

The one year I was there, '82, happened to be the peak of the Secondigliano Clan Wars, Mafia city infighting which produced more homicides of such than any other year of the War's timeframe in Italian history. Amazing how sunny it is in day and what an exchange student never finds out about for years following his own Bible-thumping! Some thought this was Country Club money getting me abroad, because our neighborhood in America, though my family was on a Major's, not a Colonel's salary. Besides extra study time for grades, I crammed harder than most for tests, did well on interviews, held a job in a downtown restaurant, lettered in a sport other than basketball, in addition to many extracurricular activities, and won a scholarship through additional effort. Many of the other exchange students were from California money and otherwise: at times I couldn't help but see it in their words, though I got on with my trip abroad. Napoli is known for the survivability of its culture and its protection of the art of cooking amidst attempted conquests, especially seafood. Southern Italy, the Deep South, sound familiar? Neapolitans face prejudice from their Northern neighbors too…mentality-wise, nation-to-nation, you would think it was the Duke of the House of Bourbon on sojourn for his court in Naples who sold us Louisiana! General rule: That's a Roman tomato, and that's a flour product, plus, if it shakes, rattles, rolls and utters an animal noise, you can cook them all together on an overcast seaport day, if you've got a pint to wash it down with that doesn't end before or after the meal. That's a cable T.V. That company carries Naples soccer on it while you're washin' your mugs, ugly or female, and if the latter, you can squirm out of the kitchen when you want to watch the game, unless you're La Mamma, in which case your younger children can steal the olders' birthrights if they can carry the T.V. into the kitchen without dropping the box when tripping over the cable.

Where do you put it in the kitchen? You don't: You stand for Parco San Paolo, while the pasta boils! What's the American adage? "That's tough shit!" Yea, you pinched my muscle when I got back from Naples and spit that out of your vocabulary! Here, I'll hold it, Mamma, until your husband is done with his New York best-seller in translation. Then we'll watch the penalty kicks. The author of the best seller is Apostle Paolo! Three hours south of Rome, under the full moon at sea. Drop it there! And swim....Fascita 'ma magna! (Make me eat!). That's what the Priesthood means by "getting in the mud" with your proclamation of the Gospels at certain times in your life, such as the Christ displayed for us himself, all the way to the upturning of the tables in the market. Hoops is a business to some, to others it's a Bible. But, no international saints down the line, no one considers it just a game, and as a ritual, it has a movement and a school, college or otherwise, as well as, in my belief, a Master of Divine Nature who constantly seeks to save us, not just rubberstamps us a good sport on the Judgment Day. Oh, and no, I'm retired from the sport, tournament and Italian semi-pro both, since quite some time (no more martial arts since the 1980's either, and only six months at that). I've never played it anywhere since what I call the Tattoo Generation of Hoops became widespread, and I don't have one on my body now. I think that's how both Pope John Paul II and Pope Benedict XVI would put it to you, if this ever became an issue for Roman opinion. So that's not a dark cloud, neither the spirit of Herod, as the priest\pastor will say in his sermon from time to time, that's the smog in Naples: drifted over to keep me company, buddy, in a small American city where nobody speaks Italian!

Yes we think of Italy more as a soccer court than basketball, and yes, soccer we as exchange students were challenged at first thing as a team together, all our countries combined; Turkey, the U.S., Malaysia, Ghana, Venezuela, and Western European nations represented, all combined, our first week on Italian soil, verse the residents of Andrea Boccelli's home region in San GImignano, Tuscany. You would think it was anything but an international fare, but the Catholics settled it for all on their home turf, fairly, safely, and legally. The exchange students mustered one team verse the Florence, or Fiorentine, locals in the mountain town, not far from where the bell-towers of ancient and castle-like realm ring at saint's dawn. The students were assisted by a lone Sicilian, who made a social statement that day on behalf of his region by out-dribbling the opposition and scoring at will on the side of us students, most of who cowered back and roughed our way through an otherwise defensive struggle. Three-four Florence natives would challenge the Sicilian dribbler at once, being out-muscled by foot control and left wide open for his shots on goal, him scoring over four.

At three quarters way through the game, we moved up field, the students, and as if for good behavior backing him up on defense, he prodded me to march up the field with him and passed me the ball to the far right side of the field, after a furious display of mastery over the dribbled ball amongst numbers mid-field. I did what an American would do in said situation: I parked the passed ball downfield with my foot, and shot it clean from the right side of the field 25' to the left side of the goal box, a long but clean and hard-driven shot that moved past a stunned Italian goalkeeper who could not believe we had doubled back from defense this quickly.

We swept the game well, far from even the regional tournament-level play we witnessed on our school grounds year-round to follow, though our foreign victory to keep, with a native dribbler complementing our foreign defense against his own.

To win at the beginning of our year was not unlike drawing blood from an angry bull: we learned much about soccer in our respective host-regions over the course of the year while studying everything from science to literature to religion, championing Italy one and all as she led her way through an international World Cup field by the end. Our Sicilian friend was not around, yet seasoned players such as he drove the ball amidst a foreign offensive flurry from all fronts, to vouch for Italy as champ in said year of '82. An introduction as an on-field enthusiast of soccer fever an American does not forget, to remember even up until the late 2000's, when he cheers his own national team into the Knockout round amidst new-found international players. Donovan. Dempsey. Soccer is open: They play for huge teams globally, and might unite to win one without their foreign support. Soccer, the sport of a huge population of the globe, as much as other sports seizes the limelight on our own American home turf, not just on T.V., but in our school system. Is it possible for us to win a World Cup? The more Americans mature together in later years, the mightier the tug on the foreign cord we yank.

More St. Patrick's Day

"The Sacramento State Boys Choir, now performing their command of musical expertise, all ambivalencies intended for state issues, with their 'specialty' in the diminuendo."

"No, are you kidding me? That's my sister's version of corporate L.A. Those guys are driving to work in a Yellow Submarine with wheels, worked on by Vern the mechanic. Be careful you don't get hepatitis at Denny's in Bakersfield!"

"Yes, McDonald's in Santa Monica proper, where the migrants, a little gruff in the morn, just order their Egg McMuffin. To the McDonald's outside Camp Pendleton, where off-duty troops monitor the military brat crowd getting their noon Big Mac, before a violent element moves into the tables. This is why humor is not the City of San Diego's forte`. That last guttural shot had the only part of gravity not only firmly solidified with the ground, but still with ass lying detached mid-air, floating into outer space, as Sehau rolled to his death. Before we get to this problem with others getting in words when visiting Huntington Beach from out-of-town for a tourist day (Huntington, just like Laguna Beach, where if you don't shop to her standards, the regal and sexually-experienced local has a right to comment), let's pray…Hence Laura Bush's worldwide comments on A.I.D.S.

This is so loose that the parents together for almost fifty years don't get fatigue, but an all-new psychosomatic reaction comes manifest into a medical community by the time they get done crossing cultures at the Art Festival. Manners, please, your California graces, one and all. And 'punt' on that invitation to be Protestant in this state again. I did not realize we were reliving this state's particular phase of the War On Drugs over and over again with that famous motto, "That's the attitude that gets the sales!" I mean believe it, she's got an apartment in a beach community and she thinks she's Neapolitan!

"So, why do people become writers?"

--"Because they want an all-new cast of the memories inside and out, on what tradition represents here and there."

--"Isn't this enigma of conversion a theological explanation?"

--"You would have to dust your sandals off from the mountain tops to take the quote on that in proper context, and let the avalanche of disgust ensue."

--"So this is a problem discussing history in this time frame?"

--"It has amounted to Alka Seltzer, lots of it, on those we feel from our time frame do half for this country what Washington did!"

--"So other than The Revelation of St. John the Divine and events of comparison, this war amounts to..."

--"A caveat from The Communion of Saints in the sky now in a less-than-couth era of America."

--"Kinda leaves your ass in a sling."

--"That's how it's gone. In one aspect my support through blood drives has been offered, in another do not remove the tourniquet on what it means to have a good time that the Priesthood placed on at First Communion!"

--"So nothing really happening for quite some time in Southern California, I take it?"

--"The physician introduced the Mammy's, an awards ceremony for his favorite female patients."

--"Kind of a bit morbid of a generation of beer-drinking buddy, I see?"

--"The last one thought he was Robert DuVall."

--"That ought to end this latest in Off the Road, a response to contemporary enthusiasts of Kerouac who thought the Golden State had undying qualities."

--"How did you get this caffeine addiction, anyway? Was coffee at noon in some outdoor places that exciting?"

--"No, that's the past, this is current: I'm back home and now everyone is out of the nest from youth, but not so much as reminiscent types, as 'business and network'-oriented. Unemployment in a town this historical makes this apartment look like the room for the switchboard to a nation-wide 'cups' dialogue complete with strings!"

--"You're going out of your mind."

--"I'm attempting to quit smoking and reviewing over the Klingers of all-time who had me smoking long in the first place, with their rock-hard compassion!"

--"I see; Here comes the nicotine craving when you're trying to quit, and you hear two voices: "Is there an H?" "There are three." "Would you like to solve?" "No, I'll spin again!"

--"Jesus, I did not realize Grandma and Grandpa could tear that far into the dialogue with a couple of t-bone steaks. I see. Twenty-two years of high school counseling behind the scenes! When I can smell the scented votive candle again, we'll have another long talk!"

--"Again, the discovery of new, hidden dimensions by the physics and philosophy world, even thought the language is other than Spanish. We're still connecting on a plane of Chicanesimo when we light up a cigarette!"

--"Must've been a long semester!"

--"It's the C.I.A.'s version of my portfolio, that's what's killing the comeback."

--"Don't drink coffee with strangers if you're the friendly Catholic type!"

--"Maybe it was the counter, not with the barrista, but the Left infiltrating the Employment Development Department and their version of my e-mail address!"

--"No, that's the C.I.A. again—You're last name's Hungarian, you once thought of changing your name, you were on bad terms with your father, a U.S. military officer, he thought Cedar-Sanai was a good move, you had countless conversations with and dealt cigarettes to both Jewish and Black Street people from Los Angeles while at Cedar-Sanai, collaborated with a foreign-born psychiatrist, had a window and light-switch overlooking the French restaurant, 9/11 occurred, you got bailed out by your Dad, but they're not convinced. It's their backburner pride!"

--"That's like a salam' passed so far up my Italian-English speaking ass I can't believe it! I'm not an alcoholic!"

--"Don't jinx me, neither am I! Light the candle, again, quick Dude, this is my flashback trigger!"

--"In Dr Burn's 'Crystal Ball Effect,' collegians of the past amidst depression is a theme who are asking for extra cake and extra ground-grown vegetables, as there are that many words sucked out of our present-day mouths on alcohol spirits as well as tobacco, which I have tried to leave behind. If this state made sense spirit-wise, we are having a wave on the grave of high school and college past, plus waving off the hygiene of the new-born collegiate in society and society's synopsis (including advertising) on our lives with senior parents, so well, Onward, Celtics. You show to Mass. Even though your weather for a friendship on this planet is not the compassion Mary and co. are known for in Western society. Maybe you can pull all that out and show you did not come to this country to act like a jack-ass to other families that have served this country for 200+ years! You and your hired guards on the team, as if the White boy in me has to use you as his filter in the Big Time, when you cross the country and kick the goal but can't wipe your ass when a Hungarian of youth shows you dance steps on the court, starting your own rumors in White society and then keeping it to Confession in such delight of your politics and numbers!

That's not how cards were played when the West was won, but if this is still your pride in olde-country media art, I'll bow at Mass, and you learn to properly reciprocate again when I entertain, since you can't entertain without my nature tied down to the tetherball pole! Go on! Shoot! You want another fan on the West Coast who got dicked by the Lakers regime—see what you're shooting is like when you carry my Italian semi-pro legacy in as large a city as Boston or L.A: all demons aside post-exorcism since this isn't, 'If you can't beat 'em, join 'em,' it's 'You've got a Parish with its head up its ass, lazy at home, too, and I need a job!'"

The Sister and Her Boyfriend Got The Jokes Too

So, yes, on the topic of quitting smoking for the mentally-ill, the voices, the non-ceasing temptations, the number of things, that go wrong on other topics of life at the point of qutting: It was said a more veritable definition of the situation in a psychological context was needed than 'simply Satan', not that I'm recommending a new title for an L.A. sit-com. SO, lo and behold, I went to Mass this morning and heard appropriate readings from Scripture and a most valuable sermon, seemingly addressing the removal of this very nail from my 'coffin' as a smoker.

In the Old Testament, Yahweh keeps his promise to the Jews by raising them from their grave and giving them a new location of dwelling as the tribes of survivors they are. In the New Testament, Lazarus, thought by some to be asleep, is resurrected forth from his tomb by the Lord. And in the sermon, we must not be afraid of dying, nor of what has to die in us so as that we may be closer to God and holiness. Which explains why a smoker has a fight over the negative input now stored inside that must die when he becomes a quitter of tobacco, and lo and behold a new chemistry will modulate over time, a new smell, away from the tobacco death, a former smoker cleaned and cleaned up, environment and all, so that they are not holy at Mass and a negative at home.

Consistency psychologically as a Christian will improve when simple medical steps are taken to walk the jungle like the non-smoker you used to be! SO, explain some more about the schizophrenia and smoking in religious terms? The voices? Yes, for brief, hot seconds you are in a pit in Purgatory with the demonic side of your Earthly enemy and you must be patient and wait for the Lord to come. How do they know to preach so prevalently? English is a living language. That's the Holy Spirit.

--"And yes, after the party is over Mom said, you eventually have to understand the Parish's paradigm."

--"Fine, just don't get awarded a snack every time I light up!"

--"Yea, help you with what that issue was really about. Wanted to discuss acid rain at lunch. Felt his daughter should switch her career to Environmental Science. Thinks we're splitting hairs with a difference between behavior and natural environment. Says this is just semantics!"

--"Okay. Thank you for the advanced counsel. Waiting for Gandalf to return to the family myself."

--"Did you need a cherry on top of that hot fudge academic discrimination, or you getting kinda' full?"

--"Is more silly than pushy."

--"They'll drink your beers at the tap, and when you drive home don't forget where we're at on water running down hill as well as water under the bridge."

--"Can you actually drive that thing safely?"

--"Not without The Good Sam Club pushing!"

--"Yes, with as popular as The Photography Club is in this town, I didn't know smoking cessation kits came complete with coverage by those wandering about a Paparazzi career."

--"When is yours not the longest nose in town? When they know you better than you know yourself."

--"That's funny. I had been told I was isolating."

--"About as much as Madonna on tour!"

--"Madonna. Good comparison. I wish I had hung in there with the Church choir."

--"Don't make all those career moves at once, bub!"

--"Yea, that's the hidden beauty behind Campus Crusaders For Christ that you don't get for years down the line—by the time the professor is done collecting on info about the true nature of the situation from Mom and Dad, 'The Beast' is the metaphor that may apply."

--"Don't go back to earn your Navy Cross—I mean your Master's!"

--"Betcha' I could do it without smoking."

--"Uh, as far as this whole subject is concerned, family-wise, I think the other competitors are done with The Pit and the Pendulum and For Whom the Bell Tolls."

--"No, I'm done with dessert. I did not realize you were writing Sherlock Holmes and Dr. Watson in The Street-Smart Potluck Mystery."

--"You should talk. It's been a quasi-version of Opus Dei meets Malakh for guac and chips as far as other talents' lives are concerned for the past 10 years of this tobacco problem!"

--"Yes, was that the 'borrow the Bible' play, the 'new Bible' play, or an appearance and dissapearance by Omar Moreno of the Pittsburg Pirates at that last Church?"

--"No, this is a literal cake-walk in a high society town, not a baseball season opener complete with stolen bases!"

--"Meaning?"

--"'It's not your cake' is the motto."

--"I need a beer."

--"Run to the store. Mother won't see you. She's engrossed in that new Harlequin book, The Last Dagger."

--"Hmmm. A bit of a ritual on the competition and its field?"

--"Was above your head. Interview With a Vampire séances carrying as much weight as a Dantesque torch through time."

--"Was a surprise to many that this planet has only one mood."

--"You didn't know Freud, not Lewis, was the Honorary Dean of your Alma Mater?"

--"Your John Hancock for missing class excuse notes is worth more than H.G. Wells!"

--"You mean it?"

--"Sis, I think time is of the essence if we're going to save this town, and it's not about your finances alone!"

--"What do you mean, you want a favor, Obama fan? I'm not picking up a conversation with every cast member of Angels in America, and you're interrupting me digesting my turkey frank! C'mon, a little family loyalty! It's not as if you're in that knock-down, drag out Blind Man's Bluff game you got in with the Full Bird Colonel's daughter where she spanked your ass with her eyes closed!"

--"Oh, I guess if that's how I look to you and how I objectify girls' butts, there is something besides my penis that is going to have to swell in the neighborhood."

"Winter Swell."

--"And I'm now a veteran of every city on the West Coast that got hit, while you're still infatuated with Dan Brown!"

--"Infatuated?! You look like the exhaust pipe from Chitty-Chitty Bang-Bang!"

--"Please. Mom must have fed you the Extenze Dad's been keeping, not the Pamprin. It's having said side-effect and I'll get my beauty sleep if you do!"

--"Hey, that's the phone!"

--"Answer it!"

--"O.K. Hello?! No, Bill, Bill...scoot, buddy scoot. This is his career. He's not gonna keep smokin' while you make hand puppet silhouettes against the bathroom wall while takin' a dump at your Dad the divorcee's house. Our Mom was divorced too! No, not from a California boy, William. From a Tex-Mex. Your powers are in your dreams, there's the Bush ranch, go milk a few cows for him, punk! (Click)."

--"Stephen C. Soltesz, please."

--"This is he. May I ask who's caling?"

--"Agent Forthright with the Federal Bureau of Investigation."

--"Wow! How exciting? What is the issue?!"

'--"Stephen we understand that for years on your voluntary forms you were marking 'Mexican-American,' and then they show that mysteriously at the beginning of the second Bush term, you began to change your answer to 'White.' Would you like to comment?"

--"Um, I was trying to get a job, that's all."

--"Stephen, you expect me to swallow that? Knowing you used to be a bilingual substitute teacher with a rocky past? In this day and age of controversy in the Hispanic community, and your prior study in Central Mexico?"

--"Yes, I was unemployed a long time. I took a guess."

--"Stephen, don't plan on leaving the country in the near future, and we're keeping our eyes and ears open, understand?"

--"Uh, yes, Agent Forthright."

--"Get back to your <u>chorizo con juevos</u>!(Click)."

A Northern Californian girl and her gay male friend were catching up with each other at the store. He says to her, "I saw the perfect man for you, sweetheart."

"Really? You catch where he's from?"

"Southern California. Now you want the good news or the bad news?"

"Just lay it on me!", she claims.

"He's got an ass to die for, but his head's stuck up there!"

"Where did you spot'im?"

"He was laying mulch with the landscaping crew at the doctor's residence next door."

"What, besides his ass, made you think he had any potential whatsoever?"

"The first guy on the job, not that it actually occurred to me, ripped a big hole in the mulch bag, picked up the twenty-pound bag, held it upside down, and began pouring it excessively on the dirt everywhere! The Spanish-speaking super shot a couple yells to kill, and he dropped the bag. Then he picks up the bag, puts it in the wheel-barrow, moves the wheelbarrow to more fresh dirt, and starts bending over and getting small handfuls out to spread quickly over the rough patches, and the Super calmed down."

"So why do you say he has his head stuck up there"'

"Because when he was through, he tried to translate the word 'extrapolate' into Spanish when he asked the Super what was next!"

"What's wrong with that?"

"Nobody speaks Spanish conversationally like that in this state! The Super shot him a look and told them to make it an early day, saying he would call when there was more work! That's why I think you've got to stop smoking those Natural Spirits on your break after the morning brunch shift at the café`! The element we're attracting in these parts—it's getting to be a Native American satire!" –

"Mercy! I hope he doesn't walk into a downtown shop and get the sandal treatment when he's only half-awake!"

--"Not to mention what he claims in the letter of explanation\absolution is the 'inefficacy of the professorial staff to give him a taste of what all-American achievement will look like in the near future."

--"Yes, this would be the retort from afar of your browbeater-lecture series on World War II, which I know he's experienced in some way, shape, or form before, but....GIVE'IM THE HORNS AGAIN, DICK!"

--"Mom, look, a lucky penny on the ground!"

--"Don't pick it up!"

--"Why not Mom, we always pick up the lucky pennies!"

--"Yes, normally, son, though it is considered taboo for the back to bend over too much during Pot Season."

--"The Alma Mater:
"Not only will the ball not roll
On the good times
We will change the way it does
They do
Loma Linda
Thine own be True!"

--"I don't know, Dudette, I don't know."

--"Oh, keep going, he lives in Redlands, according to word of mouth, you'll find it, The Prophet Dude's temple if you recognize me as one of his followers."

--"Can he walk backward all the way to Santa Cruz County, without turning around, and stand his ground protectorally, as we have? Consider my courtship, not his, fair lass!"

--"Ciao! No, no, we're going through a phase apparently this Spring: Please, keep her company and don't worry about your rep. It's not important to the Lord. He's the one with the bird's-eye view of the potential rise in egg sales by the men of the village if I ask someone new out after all this time. Is similar to the problem people have with marijuana and paranoia, but more of a sticky-heat anxiety producer. Please be cordial."

--"What the hell happened? What the hell is going on in this place?"

--"Listen, I understand—let the married men in the country tell you what's going on, try not to get wrapped around an axle about the rest."

--"Oh, great, if that's the weekend trophy on this topic, somebody polish it! I mean, Jesus, a beer-drinker's job in this town isn't revelry and romance, it's casting out the evil spirits so people can sleep and go about their business while I practice the faith: It's like being a shaman in this town, not a Californian."

--(At breakfast in the cafe.)

--"That man is a wolf! He's a wolf!"

--"No, don't call it that, because Eagle Scouts start as Wolf Scouts as a rank when they're Cubs, and then they grow to become strong environmentalists, in theory, if not just down-to-earth recyclers. So, find more appropriate, less reverse-sexist, language choice. OR better yet, for kind of a post-Girl Scout project, have a 'slut' campaign. Label all the guys sluts, and Let Go Let God on your own female hygiene (or lack thereof); kind of a counterespionage tactic!"

--"Is that it!? Are you through?"

--"Yes, I'm done with my grapefruit if you are—I like the way you brought your own spoon from the dorms, instead of picking up a plastic one from the utensils display. I live off campus. My spoons have been stolen, otherwise I'd do the same thing!"

--"And Colorado, you <u>suck</u> at 'cups', how the crow flies, American tradition, and spirituality in the modern day. You suck! Did you get enough aspirin to get that priceless career off the ground? Was that comfortable enough a pad on your economic base to relocate, or do you want to sell another gold-filled tooth pulled from a Californian's mouth to buy the ski-bunny a ring, now that you're <u>Who's Who in Saunas and Jacuzzis at High Altitudes</u>?"

--"Yes, Dr. Burns, who wrote the best-seller on clinical depression, described the 'crystal ball' effect, where some people act as if they're all together to monitor depression, a false method. I call it 'cups,' because I'm convinced that as the physician said, I'm a healthy, young man, not growing old, and that some people in this country, amongst other errors, are just <u>stuck here</u>, but working hard, pulling the whole past forward in a silly manner with stubbornness instead of a refined study of history. Say, what happened in California from the time of the Spanish oppression of the Serrano to the Mexican Revolution, and why is discrimination a factor to many today in downtown California small towns, as well as reverse-racism from the Hispanic community, because my Tex-Mex relatives have better manners often enough, real life problems, and work ethic?

You're a teacher, if that's the words—seriously, we've sweat more in our family's history, if that's the subject. Now pick that baton up for the relay to get immigration laws, from someone whose parents are a successful mixed-ethnicity, military marriage! Thank your flag for standing on its own sometimes, it's so beautiful and you do have a moment when you look at it in the sunlight now, don't you, one and all! Come up with better language for your fish talk, from flounder to large-mouth bass, so that more can afford any fish at your aristocratic prices in this sledgehammer tension! Thank you for the tilapia! We needed something tasty we could afford for Lent! But I will eat beef, or turkey, on Friday if this economic slander keeps growing on my very own Landlord, who throws it over the plate! It's less expensive!"

--SO, no, I have three cowboy hats (one for working in the sun, one for in town in the day, and one special black leather one. I supported the mail-order company during wartime), but I'm not a poet—you're an obese American politician—my immediate family served 56 years in the military combined while raising me—I'm obedient, not so capitalistic I talk out my ass: as my father says, You don't have equal rights for writers—this our Mohawk challenge if the country is in this state, this is Cooper's Calling! When you let us educate in peace, you can dictate, but you're an excuse box—squeeze it, another medicine comes out for Seniors! And Rome, don't invite them to be our Achilles heel with their economic problems. More credit in other parts of the country for California Catholic sermons is thought. Since it's an often out-of-control state with large numbers to take the pulpit in on the Sabbath.

You wanted a Great American Smoke-Out, a Heart Association, a Cancer Society, and money for that commercial's air-time! Okay, I wanted to quit, plus a sermon from a theologian who had soul, and I got it. You know, I believe I'm starting to hear the real animals' inner voice in this discussion—They said, "We'll be your friend—our hygiene is different too, but our spirituality is the same." So I told you a joke—a rat and a baby possum killed each other in a fight—both died? Yes your mother's backyard is a polite place for a burial to settle the feuds in the animal kingdom. She is so nice. "Thank you for helping her! We'll be your friend!"

SO Dad disposed of both critters for Downtown to pick up, sealing them in zip-locks to avoid germs, since the neighbors all have policies like that, and we've all been there for awhile. No, are you really mad at your neighbor? Naples-style? Tired of soft-speaking Teddy Roosevelt? Don't get hunted by his ghost in my town—he doesn't like logic, he likes scents. And Ulysses S. Grant is here too. You can go to Lincoln, but apologize every time before he gives you a new lesson, and he likes all people. That's our Lincoln Shrine west of the Mississippi, God will address your standing-in-battle and whether or not you can serve with the quarters you come from now. With female angels present.

--"Yes, people, dangerous country, the Southland—no real way to trouble-shoot problems when you're stuck in a real estate sand-trap. Pure poison. Networks and networks of men with wheels whose bookmark for what this is about is the brick they threw the first time you asked by saying, "Hello, how are you?" What their dysfunction was, since they act like narcissists if you prefer wisdom of California, goes back to their own personal version of Pagan religious comebacks, even though it's not historical Paganism to say. And some of this new breed, since Dad came in '72, was not only born in that timeframe but served. It's not the same—they need favors on neighborhood politics—it took us years to get our California neighborhood in order, whereas in other states, they are down with Frosty the Snowman. Well, wise up—it's a Judgment Day, not a time clock.

--"Yes, my new one, after <u>extrapolation</u>, as far as expressions people used that don't speak Spanish from the crib, is <u>Gerente de la Yarda</u>, or Crew Leader, not talking of jail, talking of the land. My Mom's landscaper was Saturnino—"Little Saturn" to the Serrano! Hate the film industry—where's the Laundromat, since this is a turf-war over the shorts the Major handed down to me from his Naval Academy Chest of Drawers, wool-manufactured on the East Coast—want a status report from the front lines plus quarters for the machines. Laundromat fell to the enemy? Who was that in the backyard within '50 feet of Mom's laundry room door? Hector? "The pesticide man, Stephen, don't smoke on the patio!" A spider crawled into my apartment, far from my Mamacita's home, and I did not see him. He was hungry. I was confused, with my instincts for serenity, as if I was being molested by a fat person who hated God but loved White men, and then there it was—3\4" in diameter, by my candlelight, crawling to the right of my lap-top, in between the three-foot space between that and my computer's mouse. I missed him with my sandal. I sprayed: the whole room, the whole room like Hector would if he were low-tech, but still knew he was alive, as I could sense it still wanted to kiss me. Everywhere. Under the chest of drawers, the heater (a Williams model), and the T.V.'s chest in the corner, as well as all over the cable and Internet wires. Then I moved the red trash-bin. I grabbed my sandal again and am the new owner of a fresh kill of a venomous spider and a Pabst Blue Ribbon. That morning beer gave me the wits in battle to save my life. That's Chicanismo for those of mixed hue. That's what 'lit the campfire' as I write to you this day before Palm Sunday, 2011. That spider was near my bed at 4:30 A.M.—I woke and ran to my beer, without knowing what I had dreamt because this was Reality in motion, for two hours of the spider licking his chops before his death.

--"Oh, okay, my cousin from Texas who served was in charge of planting The Stone that they Romanced in Latin America to create a diversion, and doubled back undiscovered to lead another battalion of different men and women into Central America."

--"How is it going in the 1920's apartment building? You like your room?"

--"Oh, great, you can't hear much between the walls, though we have more smells for the Home Team to experience before the couples try a different position."

--"Okay, we had some interesting stumpers on the quizzes from the people of different faiths, so we can't actually say the smoke has cleared from the gray area to move on, but we'll be able to stick it out, as the parents suggested, with the religion of our Baptism and eventually start getting along with our own again."

--"So you think the original prayer circles of California past that never broke are trying to tell us something?"

--"I think we had a bit of a crowd, several fires, a trip to Pompeii that was fun, and a Twenty-Year Storm that hit the coast, before somebody said, 'Look on the Bright Side—I think it's Saturday Night!'"

--"Can we stay up, Dad?"

--"I don't see why not. Just no bickering."

* * *

--"Love is blind, they're pullin' the wool over our eyes, they..."

--"Seeing 20\13 with these Progressive Lens glasses, Huck—get over it."

I had a very difficult life event exam starting with a year of education in The Bay Area which emotionally was quite a range. People were informative, a brick wall on the perception of the individual with my forte's, sometimes perceiving things that, in reality, are solutions or expressive resolution that doesn't surface because another wave hits. There is a retaining wall on the pathway away from the Italian restaurant waterfront in downtown Sausalito. What is the metaphor? They couldn't afford to understand what this felt like to me, only see and order certain items—I'm an in-path as much as an intellectual. The psychiatrist is the same way—he sees a monotonous, often intimidating person going through smoking cessation. What is invisible to many is the relationship with Jesus' suffering as a means of remembering the Crucifixion to honor his sacrifice, and the love of The Virgin Mary I am connected to for healing my rifts. In-paths—I internalize too much criticism, not saying a word, but feeling it like someone who loves his craft. The doctor does not seem to understand completely the challenges of the street-smart academic in the lower class.

Nature is the priority topic: My mother and I saw a two-year-old fox, reddish with some pepper in its coat, for the first time in our own Redlands backyard. She went to get my father, but the young thing had scampered up the hill and taken cover in the entrance from the huge growth of Vinca Minor planted there around a large pyrocantha bush. It's a reflection of me. When I came home to Southern California, more educated but with difficulties, I did not adapt to the way I had successfully adapted to Naples, but continued my education. The environment for the seeds planted, in metaphor; is a job processing tortillas at a tough taco factory. This is why schizophrenia is connected to autism. Southern California has its tricks as far as collective culture in a competitive environment. Some impossible sand-traps, if you really have experience to recognize when a normal drive is all of a sudden, a twist in the game unforeseen. Children are born in the same environment as always in which I was diagnosed, the counties surrounding and including the Greater Los Angeles area. In the Greater Bay Area campus' natural setting, the deer would do the same thing on my way to lunch--run to a camouflaged setting in the brush of the low hills. The environment makes metaphors with its setting and its creatures.

The psychiatrist doesn't study this: He\she studies medicine. There is a portion of Old Testament Scripture that talks of "jumping like a stag." The doctor sees me after I do that in an urban setting. It is not the schizophrenic, not the child, it is the tough tacos. Hence the rise in autistic births. New discoveries may advent more adaptability in the future. The psychiatric, not Chicano, paradigm must expand, not draw a connecting line from the dot of economics to the dot of medicine alone. Is that out of the pan and into the fire? In current terms, I don't crucify the chorizo con juevos but sometimes I call it 'making a sacrifice!'

I finally told family and friends at an Italian restaurant the anecdote of what started it when I got back from the Greater Bay Area one night upstairs, watching T.V. with my father, when we all were gathered some twenty years after at said restaurant. This time they smiled and clapped, and my sister said, "You have a good memory." I looked at him, in my rebellion of early childhood treatment by my fellow youth in this environment and claimed, "Dad, I'm a Chicano." The window was open to the whole neighborhood of Air Force families, in their respective dwellings, at night as the room suddenly became somewhat of an outdoor theatre. With all American recalcitrance he bellowed, "A CHICANO?! A CHICANO?! YOU'RE NOT A CHICANO!" So much for Santa Ana's winds that night: the proverbial General died of pneumonia again when he was forced to carry all that through the air! On a final note, no, it's not easy to study the environment under the tutelage of men with certain career paths in the military taking precedence. That's why I mentioned that he was stationed in a huge underground silo during Viet Nam! So when he came out of all that, he wanted a garden, not an avocado grove.

--"I think you're missing the obvious. Between the collective staffs of the pharmacy, the psychiatrist, the physician, the dentist, and the optometrist, these enterprising medical Americans constitute enough to make a semi-pro baseball team—maybe they're making some side money in the league, to give the clients a challenge in getting a job, and the guys in the streets and supermarkets are the crowd. If the apartment is too small, sorry you don't like the seat. Maybe if you can't field a team, you can sing The National Anthem at their next game, like that one autistic kid did for the NBA!"

--"And you want the insurance company to ref the game?"

--"Well, I'll tell you the same promise of a mitzvah, angel that I am, that I once similarly promised a former teacher of God's ways—As you look out over all of California's night-lights from this peak in the desert, I can promise you all this if you follow me!"

--"Help you with this one, as psychiatrists, try not to study the study habit, with your big ass blocking the patient's rear-view!"

--Mom, I cannot <u>dead-lift</u> your Mexican ego off me, I cannot <u>dead-lift it off!</u> I would like to move on with your discipline to finish the Italian portion of the book, <u>cabeza de Calabasas</u> (head the size of a squash)! You're worse than Garcia. R-i-g-h-t, mother, r-i-g-h-t-t-t-t-t, it's a gross attachment to the talent of the upcoming generation of Latinos from 3 to 33 years of age, by proxy of your friends' children and their offspring. Are we racing or fighting over the Pope's last pretzel? The professor didn't say a word to the whole class today, and then he drew a big <u>cow</u> on the board. He then said the exam was to write an essay on what we see. Most saw burgers, carne asada, carnitas, tacos, beef enchiladas and oso de res—I saw milk. They think of feeding, I think of the number of tits you need to quench their thirst!"

A Message For Easter

So, you broke? Banking issues reported you to the credit bureau? Potential new employers check with the credit bureau? Low on food and drink, gas and cleaning supplies? Politicians angry? Basketball team in a slump? Crime wave? Health issues? Miss your cousin abroad in Iraq, lost touch with old friends, the new town-folk racing around as life is passing you by? Internet and e-mail problems? Can't find a good show on T.V. when you finally sit down to turn it on? Not lacking on inspiration, but can't afford to buy new supplies? The market playing your work like a pawn? Low on gas, high on anxiety and the month is only half over before the rent is due? Quiet birthdays and New Year's Eve celebrations? You've got a gift but can't stuff their stockings? Everybody listens, but only if you don't move from this position? Easter is a time when our resurrected Lord remembers the suffering He went through on his path to the Cross. Be not unlike the Biblical character who carried a flogged Christ's cross for him part-way before the Crucifixion. That's why this tone is set as such for you. That's why you're important to Him, the Redemption of Mankind, despite your sin, and what America's version of that Cross means in the present day and age.

* * *

--"Your teeth, now that you''ve quit smoking, are starting to look more 'dental.' Yes, more 'dentally-correct.' I hear you will not be discussing any more Texas politics at the Korean dentist's chair!"

-"Kind of gives the impression that life, after being an exchange student to Benigni's Italy, has overtones of a Leslie Nielsen-type spoof on the <u>Eye of the Needle</u>."

--"Holding your breath at the beauty in the surrounding area outside of the Trojan Stadium in Pasadena?"

--"No, have to open the windows to let a little more American chutzpah in the stuffy apartment."

(Meanwhile, back in Redlands....)

--"Sir, Inspector You Never Really Knew Her is here from the school. Wants a few words with you about The Alumni Section at the football game the year preceding."

--"No, send him away. We're not interested."

--Sir, he knew you would say that, upon which I was to seize the opportunity to ask you if this was a denial stage of an alcohol problem, and if you were off your medication again. As well, here is the card of a fellow Alum who works at the Rehab Clinic—says they can get you in this week. That is provided you sign the paperwork guaranteeing no courtroom arbitration and they can have your parents' phone number again, in case you have problems with treatment—appears they've misplaced it."

--"Tell them she's in therapy after back surgery, otherwise up and about, and he's out of town on a Church affair, as well that they can catch up to me around town when they are not busy with school functions!"

--"Yes, sir, Do you want the card?"

--"Let me see."

--"Here."

--"She's not from my immediate vicinity and I never had a class with her, though have heard the name. Just tell him to hold on to it."

--"Very well, sir."

* * *

--"We've got beef comin' out the ears!"

--"Then clean the beef, not your ears, and serve from one ear or the other!"

--"Now, I made a complaint about the English I had to honor in order to proclaim these people hard-working professionals."

--"What did they say?"

--"They said, 'We were discussing nomenclature." "I said, 'I thought the subject was to assimilate.' They said, "It was enunciate, dictate, and accentuate,' and I said, 'What about the norms, like hoto, puto, and maricon, you monolingual S.O.B.'s!' Well, the argument ended there, we closed the meeting, they drove home in a truck with their buddy, and I drove home in what you bought me!"

--"And you all have gas in such—I cooked dinner for your father, talked to my brother in Louisiana, and watched the news on Japan on T.V."

-"Are you trying to tell me I'm lucky to have these jerks?"

--"Look, drop it, help me out at the house tomorrow."

--"What's for eats?"

--"It's a surprise!"

* * *

--"The American family is an endangered species!"

--"Oh, Jeez, the family values are endangered? Got news for, in case you didn't hear about George Bush's historical treaty with Chief Little Asshole, so is the American penis endangered!"

--"Man, been that long?"

--"I'd be a millionaire if I had charted different flight patterns of full moons in the sky since the last time I had a cuddle-bunny, by proxy of the American Astronomeres' Association!"

--"What do you do for an encore, Quixote?"

--"How about Spain's first World Cup, jack-ass?"

--"Yes, I see your point—took them long enough, too!"

--"Want'cha to think about it!"

--"Wow, yes, I mean, since the…treaty: interesting economic drop-off and other events!"

--"Was not our first rodeo—but our last PI!"

--"Shit! Shit! What are we going to do now?"

--"Quit smoking."

--"What about the environmental movement? The Italians in the Olde Country? Our other European allies? The Asian friends that like a cigarette? The Treaty?"

--"Annul it. That's over twenty years of many Americans' time with these foreign and domestic companies."

--"What if we substitute addictions?"

--"Hop back up on the Clydesdale's saddle, Hoss!"

--"But Bush wasn't a drinker!"

--"I say annul the treaty, hop on, and see if Ron Paul has better luck at Indian Bingo!"

--"Dear Brother-In-Law,

Happy Easter 2011! From one religion to another, how's Los Angeles treating you? Hey, tell that crazy Italian law partner of yours to watch the news: That is President Barack Obama from Chicago on the outside perimeter. Want you two to bet on me, your children's uncle, outside downtown Naples. Tell him this is no longer psychiatry as usual, in fact by rabbinical law it's considered one long Mitzvah since I don't call the boys. I don't know how much more motivational empowerment I can do for both of your sons' blossoming careers! What's that, you're going to place a safe bet on Downtown Freddy Brown, regardless of my writing career or the President's? You do that, that's Jewish wisdom, if depositions are occupying your reading time on flights! That's the only problem with this state, whether or not you prefer the weather to New Jersey: They think 90 per cent of the times in terms of the goose and the duck, but on the subject of the eagle, the seagull, and the stork, they like the New Messiah.

Okay, Fannie May and Freddie Mac, meet Jews For Jesus and A.F.S. See? You have to help them with this one: 'That was fact, that was fiction, this was work, this was masturbation, now how many fingers were held up?' I didn't know both the Catholic Church and the school district were going to ask for a play-by-play for the duration of the recession. I bet that's why you love trying cases in San Francisco, eh bud?! You miss the way Feinstein hits that curve, despite the rest of the Party! Mom thinks just the opposite—this place is a Catholic's dream. From Fedoras to Coppolas to keep the head warm—next time skip the law partner's home—and class and buy your father-in-law a Russian top hat for snowy weather! Does it give the Rumanian in you the jitters? All right, all right, it's hard enough to get business. You'd have to give the Pope himself the speech on that, not a California writer. Before we get any deeper into a dialogue on where we would be if Jim Kerry had won, since it's not the point. On our side of the State it's what if the Black fellow had been Bishop, not Jay John Bruno.

Peace on Earth to you and yours. Perfect timing in history for him, when some Americans don't know the difference between straight and gay. Am one to talk? What, my cousin? That's not a fair question, she's as pretty as my sister in my opinion. Come on, come on, pick a better club for par on the course. Golden Bagel! Get goin'—keep flying over the cuckoo's nest to your case in the Bay. Go!"

--"Don't get a sweet tooth."

--"Oh, you've got to be kidding me with the porkers competing on the subject in this town! Back off, Indian! To the fence, Koufax, and hold out your mitt, since you're the only one who can pitch this game!"

--"C'mon now, buddy-boy…patience. Great scouting virtue. Remember, 'tis the Lord's day, not yours, now that you're dedicated!"

--"Yea?! You can tell we're nearby the lake where the animals swim once the demons are cast into them!"

* * *

--"What was that about?"

--"You would have to be a red-head to understand."

--"Good point, good point. That's the hair where the rationale ends, and the enigma enters."

--"You shouldn't use all the rationale in the world to figure out any of them!"

--"Call me American, period. And keep the code we got if you want Hungarians like my father and late uncle to serve the flag!"

--"Don't bullshit me!"

--"I won't. You're the hair color that doesn't dye their hair black or blonde, you got integrity!"

--"Got to stand your ground in a cross-fire. I don't have much choice with the nails and frill they have these days!"

--"No, triple it, triple it, on the number of guys who fish for a complement regardless of the truth, just so they can say they can kick the ass of the son of a Neapolitan cop. Triple it. And that's the exponent. And serve them the Triple Club Sandwich at Denny's, when I want my glass of wine for the day. Facendo L'Americano, or pulling a token American maneuver on someone in Italy, when actually in Rome. You'd think we were writing Reynaldo Goes to Manchester. Better yet? The number of children's authors is stacking up so thick, it's as if they read Stephen Gets An A at the reading room for K-6 around here. Pretty simple. Except for the fire across the street: that was my balls backing up and it blowing out my ass!

--"Kate Middleton's dress? Her dress?"

--"Are you kidding, this is making me hungry, when can we grab a new plate to celebrate the reception in front of the T.V.?"

--"My menu is as immaterial as the comment about Kate's weight by the papers. They apparently aren't aware Californians are taking no prisoners on the subject of which country is fatter!"

--"You mean that's the synopsis? The California girls outweigh the British Belles, 250 to 121? No wonder the NFL may not field teams next year!"

--"God forbid!"

--"How 'bout them Lakers?"

--"Bryant and Blake have shed some weight, Gasol and Bynam gaining an extra meal before the game, and Brown\Fischer combo abstaining on the tug-of-war over concession snacks with the bench."

--"I think there's other issues."

--"Yes, I agree, I hope California boys don't have a temper-tantrum on a demographic scale if Middleton agrees to marry and stay off the availability list."

--"You're thinking of British locals."

--"No, I'm thinking it must be the ale drank by both, because I know some Americans that can squeeze into the category applicable."

--"The ale's Kosher in my book."

--"Yes, before it's called the 'K' word or the 'alement.'"

--"You know as far as this California thing, are you sure love is not blind with you and your compadres?"

--"And that's why I've been chosen to lead and selected to wear the glasses—the vast mass can't learn too much."

--"Yes, CLUE. It was Mrs. Gardens in the hot-tub with the tongs, to flip the bar-b-qued pork loins!"

--"Say, what is your opinion of the bride?"

--"Getting more beautiful all the time."

--"Enough to lift the curse?"

--"Off of Ireland and Southern Cal both? Would have to be a nuptials blessing indeed!"

--"Oh, come now, they don't treat you like dogs!"

--"On the contrary—what's worse, there's a supper dish thief in proximity. And it's not our sin!"

--"That's the law."

--"Was the air sucked out of both valleys on the subject, I'm sure."

--"Until the next fire in the North, with the currents of the Santa Anas due South."

--"But stuffy in the apartment?"

--"Oh, after all that 'internal, external', what's the difference?"

--"The Lord has his reasons."

--"Yes, so we don't get another 'Love at First Sight' issue."

--"God forbid!"

--"Remember the last time? Tourists booked later flights so they could see the Christmas Lights Parade at Disneyland."

--"Won't hold it against you then."

--"Oh, don't get caught letting me off the hook."

--"You callin' me an asshole?"

--"Haven't heard the word for years—mind if I look it up?"

--"Which one? That's alotta White Pages in the phone book."

--"Just call 4-1-1."

--"You know whom they call it 4-1-1 in honor of, don't you? The first Chinese panda in the Berkeley Zoo!"

--"That's...that's a 'dog' to say that! So that's how the expression came about!"

--"At least we can see straight in the heat."

--"You know what I have to say to that?"

--"No—what?"

--"Just tell'em what city wants a call now."

--"What was this, Kate Middleton's sister decorating a room at the Palace for a fest with disco balls?"

--"Maybe she studied decorating in New York City."

--"Well hell, I was going to say she was a British exchange student to an American high school at Prom Time!"

--"Come now, jack-ass! Let the girls have their disco balls! It's a Royal Wedding!"

--"All right, all right, so every Italian-English interpreter isn't John Travolta!"

--"Won't hold it against you again!"

Microcosm/Macrocosm

Well, here we were, Western intellectuals in our "station"(as a Hindu would say for their position in the world of worship) of California, when the earthquakes in Japan and storms of Hawaii, Santa Cruz, and Santa Monica, respectively, threw us for a loop. Yes, while ruing over the worldview in the context of Rome and the possible abuses of power of Italian president Berlusconi, we had this cataclysmic interruption, where, as my Hungarian-American father says, "We lost our ass!" In recoiling from our forced efforts at constructing a paradigm of modern thought to more appropriately support Western countries in transitions of time, we almost got spanked ourselves for our own abuses of power. We had conditioned ourselves as believers in Christ, and the Catholic teaching that it's not about power, to see ourselves as conduits of empathy, when strikingly, championship basketball became the side of the boat we clung to as the event of the Royal Wedding of Prince William and Kate Middleton, a Roman Catholic, took the stage.

It was more of a jolt than a shift of the paradigm, the crumbling of the great thinker's prefaces of the Church of England juxtaposed to Catholicism, as God, bored of the mere exhaustion of the contrast and comparison in the monotony of time's academic tone and follies, creates the spectacular union of two young deserving faces from said Church's roots! So, in this light, in an effort not, as mother would say, to shift the blame, the struggling thinkers, in more of an imaginary than present unity, finally were told, "It's not the T.V. or the Media, it's your heads!" So this way we went after looking in the heavens to await with anticipation the coming of this new cloud on the horizon from Great Britain on which to rest ourselves, a buoy all its own in a storm of world strife. Would the couple's good fortune be a blessing to come of stability and actualization for many in the Western systems? Or was this a hallucination of Quixotesque proportions of a Catholic stuck in his own American version of the ritual of the cult of Saint Patrick, never to free up his ropes for sustainability's pull?

In this sense the vines of time wrapped around the contemporary thinkers who positioned in the sunlit garden of The Golden State like a simple yet somewhat sturdy trellis, as now I as disciple knew the difference in magnitude of the metaphor of my gardening in the backyard and the statue of Saint Francis of Assisi in the same garden. We had been conditioned to think less of ourselves and perceive our fortune in survival, yet adamantly knew it wasn't a failure to perceive some of our own bondage, whether imposed or, in said situation, discovered and beset. To lift from the muck of the mote and dry for entrance into the castle through this entrance required no steed free to buck us off, yet a reminder of our innate swimming skills, this time the swim in the full armor of God's word! Thusly the low Church and its challenges looked in the dark recesses of the spiritual kingdom at what connections were asked for since leaving Rome as a youth, maybe not hell-ward bound, but smell the abscess of Purgatory all the same, wondering what on Earth had predisposed them to such Benedictine penance in effect since demonstrations of loyalty to the American flag.

In sense the Italian had understood the American better than the American Catholic since moving forth from The Earl Gray Arguments of Episcopal past, instructed along the way that the Anglican Church in America was not the Britain of Europe, and the British pub in town was not the Church, neither do all benefit from drinking there. The incongruities were uncomfortable, and the cultural fragmentation astoundingly silly and presumptuous, as if the day and age of culmination of efforts to escape Paz'' labyrinth would not surface, but the garden would have its variety of Christians locally all the same. The thinkers let go of each other's hands in this metaphorical prayer circle in the catacombs of modern

American historic events, praying for work, a prosperous turn of events, and romance anew this early morning before the wedding.

Retention of society from awarding credit to the efficacy of the grape-stompers in creating a wine process was perhaps the very reason Pompeii erupted—God wanted to see it on the table if this was a wedding, Cannes proper or not, and thusly under the eyes of Caesar even the debt-ridden, as <u>bon-vivants</u>, challenged those opposed to the championing of their rights to raise their glasses for said occasion. Thus a glass might rise at one A.M. Pacific Time after another day at the races, looking not so much as to the times over with, but to see what they brought if the variety in the garden required a thinker who could volunteer.

Yes, the volunteers for The New World Order Project all surmised the change-in-tactics to cultivate the economy had its reasons, if not its regality. "Land of Leprechauns"—I know the feeling, after having seen the advent of newness in America's version of The Anglican Communion, Catholic for a price to pay now. That's what I always liked about the Brits—they talk about women and the word dignity is in their vocabulary, even to describe, why, how to treat your own ilk. Contemporary Italy scares me for this reason, to have a dark cloud hanging over the prospect of whether they've changed all that about themselves or not. The leaders thrived on the challenges of controversy, comfortably since a generation of their leadership was in retirement. And that, after tendering, was the cut of beef served.

Yes, we as Americans would be wise to not make it a habit to issue commonplace caveats, especially when we're at war together in loyalty to the Stars and Stripes, forever by Biblical interpretation's decree of eternity through faith in Christ, for a caveat is not a toy one puts into the car while going to work.

I've heard of crashing the wedding, but crashing the beer run? Yes I suppose I'm Texan enough to observe the difference between there and California! Not that I know the crowds of the four different shifts of the twenty-four hour pharmacy like the back of my hand: It's amazing what loyalties mixed between generic beer companies tells you about the truth of street-life. <u>The Barack Obama Ground Game: The Documentary</u>. Just add beer, instant Catholic. Is that right, Loma Linda University, is that right?

"Yes, Ambien is good for sleep—it only makes you drowsy in the morn if you're on tough love terms with this year's enrollment. Hence the beer! 'Bedside manner?' This is Redlands Cable T.V., not the Coveting Channel, and thank you, nurse, it was a good visit with the doctor today! Yes, don't put down Anne Rice's work for the sensitive male patients—it's just the weather. They enjoy the cult, please continue the inspiration the office brings. If the specialist with the accent is wondering how I got to the country, tell him to stretch the Catholic Communion metaphor a bit—I <u>have</u> sailed with an Irishman! Hungarian on board who speaks Spanish, technically speaking between the part of the Isles and Eastern Europe proper, as the warships are charted! No, London, never been rough waters for a cruise. The Pope taught me to do water-tricks—that is the rope from the look-out basket to the Irish boat as The Royal Navy asks, "What's for dinner?"

So no, nothing really bothering me about royalty in this world, though Brits can be a little difficult to reach. After not fitting in at the C.S. Lewis Foundation, I decided to take a course on how to get along better. The textbook was entitled, The Irish-Americans! No, I don't miss the trip to Budapest all that much—it was nice to see the street acrobats do back-flips, no trampolines, though the other folk in the train station don't look Hungarian, and they're the ones I ran into that didn't speak English or Italian! And Australia—another dialect hard to follow. CLUE: It was Mel Gibson in the bathroom of the A.A. meeting house with the strings from the tea-bags! Yes, why, right in downtown Tinsel Town! With a note attached to the body, "Don't go over the media's head!" Clue #2: Edward James Olmos put a beer can on the hook for bait to go fishing for the ones who needed a lesson on why they don't belong! On that note, we can move on with my louvre, Around the World In Eighty Beers!

--"Dad, Mon, did you hear the good news?"

--"No, what son?"

--"I got accepted into Loma Linda's new private business school, 'Tude Psychiatric Services!"

--"Wow, how did you do that?"

--"Just by being myself, having good manners at the store, and not going back to the bars until the undercover cops were caught up on their casework."

--"Undercover cops?!"

--"Yeah, so the town doesn't get too slick, and the doctor can take any patient he wants! Like me!"

--"That's fantastic. I'll go pick up a pizza. Stay here tonight!"

--"Sure. Thanks!"

--"Stephen, what does the doctor think was the original cause of the schizophrenia?"

--"A domestic discussion-oriented schism."

--"Oh, come on, you had this problem in Naples."

--"No, I did not."

--"Well, you had problems in Naples, didn't you?"

--"Do you want to apply your education to understand problems in Naples? Or do you want closure on the topic?"

--"Oh, that's okay."

--"Then it's a domestic-oriented schism."

--"I thought you could take the boy out of the country but not the country out of the boy!"

--"You can't. Despite the family's true ancestry, Ia so Napuletan'!"

--"You mean honorary.?"

--"It's not a degree-awarding reception."

--"So you didn't get busy with school."

--"Oh, sure. I graduated *Molto Piu` Furbo* in my class!"

--"What's that mean?"

--"Magna Cum Urban Anyway."

--"Yes, it's all fun and games, but the foreigners seeking permanent residence in The Golden State have to remember that Ramo in <u>Island of the Blue Dolphins, Theatre Adaptation</u> is not played by them growing up in California. They need a fourth grader more indigenous to the Southern Californian section of the Pacific Rim Basin who's accent teachers here attribute more to the speed with which he reads out-loud at that age, not his Tex-Mex roots!"

--"It's a Beautiful Mind."

--"It's a Hungarian baby's birthmark on his ass!"

--"Well, I didn't mean to get aristocratic, but I've never been slapped in a bar and have moved on since this is The Killjoy Era of Hollywood!"

--"You mean Russell Crowe doesn't get the Alliance for the Mentally Ill Trophy?"

--"You're sister says you need the Stanley Cup!"

--"I don't play hockey!"

--"No, the cup Stanley wears on the pub crawls!"

The Girl That Got Away

(Because her boyfriend was going to hit her but you made a quick phone call to Starsky's new partner at 911). Now that the athletes call you a woman. But... they moved far away from your part of town with their wives and kids, and the cats moved in the neighborhood. Time to learn something new if you're tired of drinking with dogs that fart when they hold beer. I love the magician who pulled the rabbit out of the hat, began to put him to sleep with hypnotism, then woke him up, and did this over and over several times. That's my resume, that's the police chief's top-hat from the Smiley Era. Happy Easter, dog. That's the new position on the roster, your police department.

A Gear-Stripping Systematic Race

--So if politicians in Congress are wondering what this looks like now on psychiatry, politics, and economics, the mentally-ill patients who are still employable, take their medication, and pray for their country are President Obama's and George Bush, Jr.'s Lucky Charms. And you thought it was Patrick Kennedy! Let's pay a little more attention in there to the routine one goes through to fill a prescription and be administered a medication in order to apply and receive a job, to have a good work experience, how hard it is to date someone with a college degree with what we take in (since our bank accounts cannot go over $2,000 while on Social Security) and the discrimination those willing to work go through. All the while, painstakingly trying to catch up to normal Americans by doing so, and how much of an annual salary it actually takes to make a break from $845.00 a month support while below the poverty level. Any guesses (Estimated by a county leader of AMI at $39,000 a year!)? How many jobs for college-educated patients in California do you have in the ballpark? We're not just starting out. I've had over 20 jobs that the Social Security System knows about since I report my wages and pay back my over-payments. I don't think it's time for a promotion. I think it's time to thank our military, retired and active duty, for leading us to the slaying of Osama Bin-Laden. As well, I believe we should continue to say our prayers for the Commander-In-Chief of any party in office during wartime.

--"What happened, Superman?"
--"Well, I thought the three-mile run would do it for the overall event, but the psychiatrist whooped me on the beer-curls!"

--"So, this taste in your mouth when you try to quit smoking...No ideas yet?"
--"It stems from this discussion we, as Catholics, had about Tolkien and his writings."
--"Tolkien? What does this have to do with the present? I thought the discussion was Tolle`'s philosophy of the now, and Dr. Merkel's 'existential vacuum'?"
--"Well, all tying in to Tolkien, a pipe smoker, and our question as contemporaries, 'Why could J.R.R. smoke and we can't, without dying?"
--"Oh. Well, yes, I suppose a subconscious thought, when all is said and done. Well?"
--"Appears for magical reasons this is a moot point, even though mythology has prevailed. We don't have an American Middle Class anymore, but we have a Middle Earth that's on a roll."
--"Following, yet...the taste...this is a serious disturbance when the will-power to quit is there."
--"We abused the privilege and the dwarves are having us kiss the magical dragon's asshole if we want to quit!"
--"Where? I see no creatures whatsoever!"
--"Then why don't you have a maiden? Why don't you have a job? Perhaps you're not in America...since you missed the Continent so much and the fairies granted your wish, for respecting their boundaries."

--"That's a terrible explanation!"

--"It's better than Bocelli's---that it flies in your apartment landing on your tongue!"

--"Why can't you just get a job and smoke some more?"

--"Because the dwarves aren't going to fight the campaign again they already won just so you can gloat."

--"Since when is cancer always the option, even if this mythology has prevailed?"

--"Because those were the war-wounds to you and yours, who do you think you are, and why push it—quit and win the dwarves' allegiance! Just learn to handle it when they dish it out! You don't subdue the dragon alone!"

--"I'm speechless."

--"No, c'mon, let's hear it one more time for a good laugh, light up another, circle the room, and say, 'Size of an Eagle!' in your Fruit of the Looms!"

--"Oh, shut up."

--"No, bub, there's light on the other side of the underwater cavern after kissing a foreign smoker!"

Cinco De Mayo 2011

-So, this diagnosis is because of hallucinations associated with 'Chicanismo'?"

--"A mixed ethnicity's bag, it was simple to end the dialogue at the ripe old age of twenty-two of the village it is taking to raise him, and open the dialogue of the mestizo within, though an assortment of discomforts were expressed by the United Nations located in suburban California. Disneyland did well that year. So, too, did the Protestants make headway with the Jews, the Democrats with the University, and the Catholics with the Governor."

--"God doesn't change: Man does."

--"Let's rethink that one."

--"Why aren't you working?"

--"I was assed for being a professional."

--"You mean 'dissed'."

--"I prefer the Olde English, as in 'I've assed another Earl in battlefield strategy'."

--"I think that's enough Senior abuse."

--"Dad picked it!"

--"But he's your father!"

--"No, don't let go! Don't let go! We can make it—promise—wait a week or two!"

--"Okay, I'll rudder some more, but catch something before the new moon appears!"

* * *

It's not verboten to communicate on the topic of mental illness, reproduction, and marriage, or economic problems, but solve one of these deep issues at a time—my lifetime. Very well I am a Knight, I will do two jobs for contemporary psychiatry and first, I will find out about Catholic birth control, so The Lord's emotions through the saint I am may carry forth for the sake of more souls inducted into The Communion of Saints through proper witnessing*, and at the same time pursue the money ($2,000) for genome mapping (I'm being eaten at this time by fellow Americans, the disease, the enemy, tobacco, and modern psychology).

*Current Catholic priesthood members sometimes call for an evangelical Catholic Bible (Father Robert Barron), or, as well, point out the offensiveness of testimony, not deeds and manner, of which we don't seem to care of as much as music (my hometown minister). I write in mind I have an editor to pay.

Art And Evangelization, An Apostolic Succession All Its Own

O.K., we have two Christ's here: One, a Lord of the Church international, and two, a "How the West Was Won," rugged Jesus for tight spots and challenges, tradition in America, action, indeed a savvy, Charismatic Lord of Jerusalem who is here in America with us, and our fellow patriots, who says if this is about the environment, assimilation, a moral issue of conscience, a border conflict and a claim to stake, Indian curses of old for the abuse of rustlers and Spaniards alike, then this is where He hangs his hat on the subject of that voice that just came to you in your 1920's apartment. To make a long story short, on the eerie odor of usurping both political and economic from which we are healing in our own deeply rooted Depression (since we remind The Father of The Men who won the Big One), the silent response escaping my body in thought of The Holy Spirit, like a whisper, was, 'Bring me more followers and I will solve the issue."

On the Continent today, the completed Basilica was blessed with oil on the Altar, by Pope Benedict the XVIth, in Spain. Mass was held in Catalan' The Litany of Saints became a musical ritual connecting American, and other Catholics worldwide, as one Church. The thought: Art and evangelization, by the sight of the Basilica and in memory of Paul's incarceration in pre-Church Rome, are tied historically by the Apostolic Succession and the inspiration it brings to the constituency who are so thusly brought into the Arts.

Eco-Psychology, Cheese, Cheesy, Chicanismo

Changing the topic back to session studying how modern American, not Freudian, psychology can repair this internal system, several points are made. Schizophrenia is a point in life where several vampiric drains attach. This is the Freudian past. It is believed, since only subconsciously is there a framework for contact with vampiric drains, that shifting the blame of current issues from the self will re-induce the Freudian role model, when, quite recently, eco-psychology is clearly a better paradigm. Let's see what I can see on the Internet on the topic. Also, my own humorous thought is it depends which paradigm you're in as to how much power you have and over what aspect of good vs. evil. Hence how can eco-psychology fly with wings spread, avoiding the mundane and rudimentarily historical, or connect properly this time to ameliorate the internal now? Traditional vs. Modern, yes, and now, the bridge between the true is shaky with enemies looking on until the crossing to realization and actualization. Yet factors, both of these, in a continued walk with God, even in one's forties.

Eco-psychology is the study of the relationship between the human psyche and the natural environment. Each person has innate love for the environment to be uncovered. Eco-therapy uses nature to promote healing and growth. It incorporates biophilia, our relationship (innate and genetically rooted) into relationships with animals and other living organisms.

It's amazing what Californians do with cheese that the Italians don't. "Cheese" the word, and its derivative, "Cheesy," neither current Golden State definition ("Good Sex" and "dubious behavior,") found in Webster's Collegiate, last edition I bought, talking about a business purchase now, not State traditions. This is what Brits mean by American English when they comment. "Cheesy." Might see it as good as the high school ending English class with a call to enunciate, "scheduling a Don Quixote reading in Spanish" for those who know why our parents tell us to say Tex-Mex and not Chicano, residing in California to go on occasion through this education system's horse manure. I actually can address all 49 other states with one ounce of belief that these 1)Are things particular to California; and (2) preach "eternity," in that which only will be the effect delivered by said ounce. Since I'm going through this horse manure one more time. I feel as if I've been hired as a backstage hand to handle the horses for John Wayne!

As far as how Chicanismo, the academic discipline, is handled by the Sociology Department, they've got a Latina Mother of Invention, not a Tilma of the Virgin Mary. When they prayed to it, she gave them the idea for a two-fused stick of dynamite, to light one end in Texas and one in California, symbolizing Chicano unity, and now they've tossed a similar model over the San Diego border. It's not Montezuma, it's the Ph.D. on the subject in the Dept.! He has published his dissertation on Frankfurt, entitled, <u>Chicanismo and Hermaneutics; Global Cultures</u>, thought he now has to rely on Federal funds for research, since University things are tight, and they finally cut back his pay increase when they caught him with a joint at a private party, even though he's got <u>tenure</u>! He's now begun a speaking engagement\lecture series tour in the VW van throughout the University of California year, braving fire season in the dry summer, cruising down the I10 at 55 MPH, and with the spirit of the Santa Anas and the tenacity of the Gregorio Cortez mare, has succeeded in brow-beating on various campuses otherwise productive Black, White, Indian, and Asian students alike of a generation he has yet to name! As far as the opposite pole, it's turning into "Sir Arthur Conan Doyle Writes Redlands." Believe me, it's as far as how strings are pulled and what's going on in the middle of all this to substantiate "Who's Who in American Colleges and Universities."

--"That's the invisible string, you've been a guest assistant for a magic trick performed by David Copperfield!"

--"How to Translate Hungarian: 'Grilling the Kielbasa': 'Cutting Through the Red Tape.' 'Cooking the Goulash in the Crockpot': 'Singing Every Song and Staying the Whole Mass.' 'Bell Pepper': The vegetable in the Garden of Eden nearby the Tree of Knowledge of Good and Evil."

<u>The Cross and the Cantina</u>

--"Did you play college ball?"
--"Are you kidding?! I didn't get away for a single sporting event!"
--"Don't complain—they didn't get us to a single play-off until we got out of college."
--"Walk-on?"
--"I think I walked a little farther than the guy they gave the free car keys to at that!"
--"How did they miss you in the draft?"
--"Our coaches? Our division? I just hid in the keg and didn't come out so I wouldn't have to deal with that until they were done!"
--"You're a big guy—how'd you squeeze into the keg?"
--"You must be from a small, dry town. I'm from the Milwaukee area, but I..."
--"Prefer British ale?"
--"Yea. How can you tell?"
--"You wouldn't even know what Milwaukee's Best tastes like if you weren't saving your money for the Pub on the weekend!"

--"What makes you think I can't afford British ale all the time?"

--"Because you know Milwaukee's Best has been around awhile!"

--"Oh, I beg to differ. I'm younger than you think I am. I'm in my early thirties. I'm an actor."

--"Yes, I know, I saw the movie. You don't have to worry about your next role and got paid the same for the cameo that I got paid for doing demographic research here for Milwaukee's Best this year!"

--"Did you work your way up in your industry like I did?"

--"No, I auditioned for the job after getting my M.B.A. and got told to start at management."

--"M.B.A. Must be pretty easy after that."

--"Are you kidding? There were 35,000 applications for demographic research in Southern California, why several Pepperdine, USC, and your other private schools."

--"Wow! How did you nail the interview?"

--"I scooted in the bar stool when I got up to exit afterwards!"

--"Faye Dunaway."

--"Faye Dunaway?"

--"With the 'gentlemen and scholars' M.O.!"

--"No, after you."

--"That is after me, Coach, I was born in '63!"

--That's 'Kamana uana pray," for future reference.

--"So, class, as research into environmentally-friendly cleaning products in the Hygiene Specialist field approaches the theoretical, since this is EnSci300 now, we look at the ground breaker's beliefs. Long held succinctly from academe and modern science, those who sought a career from the ground up (as opposed to the then-remedial EnSci College and Faculty) the era after service in the Korean War reached adamantly conclusive thought. Principally, for your acquisition of advanced knowledge' sake, the straightforward theory has it with those who are actually full of the most crap handling the bulk of the planetary peacekeeping, avoiding all metaphor at said time in verboten manner. Hence, a complete bifurcation of global culture along military and civilian discourses' lives. Though restricted by law from said bar for reasons of recruitment, it has resulted in a somewhat antithetically-healthy nepotism for nearly the duration of the timeframe embarking from the days of the Constitution's writing. Any questions? Yes, in the back."

--"In the context of said hermeneutical bifurcation, do we have direct correlations theory-wise within the civilian diversity paradigm?"

--"Assuredly. The Academy of Motion Picture Arts and Sciences.

The Off-Shoot Joys Of Patriotism

I overheard a woman say to a friend, a woman taught that in a town with a reputable K-12 system on one hand (and a bustling, inspirational business community on another) that she had the authority to point the finger at unmarried men regarding human nature, though more Jungian thought was needed. What was needed was to get the Holy Spirit breathing through slanted, still, and digressive collective consciousness. People knew this was not a Mom and Pop town in a large state of a modern nation, that many had grown here to obtain full-flight goals with the wingspan's strength developed here, yet some were sat on as if they had gotten several pregnant, when they had not even seen a private date during the first two years of the new President's term.

It's when the eye-ware people have is great for driving for distances alone as a respectful leader, yet exercise as adults is not a continuation of adulthood but group therapy. One finger pointed when some are trying to think their way out of a recession, taught they have to come up with their own finances to support their lives, and suddenly the consciousness of the past decides to rest on these men as if we have a House of Representatives for high school graduates, and if you're not wed then you cannot carry the vote to get a job when there is important community work to do! If you did not get a seat on the Capitalist Committee after the initial run-off, that was life. There had to be a means to an end other than political and economical amidst a bona fide social issue and, last but not least, social fabric of the Holy Spirit's initial outpouring of love. It made one examine how much ecumenical, moral barter was done on the topic of the un-wed's freedom: certainly only connected by an opinion that, typical to the very sand that the apostles dusted off their sandals with to spread the Good News in more receptive parts, had a different sum total in the other cultural wombs elsewhere.

You mean again, after 25 years of struggle and input into various avenues of society, along Christian guidelines and ropes not the least void of ominous temptation from several sources with potential to create debility in differing parts of the soul, a "You Cannot Be A Prophet In Your Own Hometown" Biblical reading at Mass on self-actualization is coming to more meaningful fruition? The only glass cleaner's job in town he could get was for the wine cellar business, complete with an original Margerita pizza for the low price of six dollars. Other than that, this was a bad dream: People in high percentage no longer cared for marriage across America and spent much time alone. To look further into this it needed, as a collective voice taking a pit-stop on the caravan, as the men carried the beams holding up our cherished Grandmother's canopy while we marched through the abstract terrain of domestic America at war, we needed either more (1) strength (2) a less-obstructed route; (3) something to protect from the sun overhead; or (4) a miracle to transition these men with more breathing room if inside all connections <u>had</u> to be kept as sits on the beginning and end of the twenty-year hourglass. The pot that is stirred when a modern-day boom of infants sparks the matriarch's pride. I wondered how to handle childbirth and conception on my own side with this flooding of collective thought connected for the sake of other's esteem, amidst this kind of weight on starting my own business, remembering today is…the hope that my father on his 77th birthday does not fold the progress of the plan from within when on some days he is miraculous support. This is a struggling son—the father needs his breaks. What if he, too, heard the same collective, though did not have to personalize it since the wife had respect? What a sin all its own to disrespect a forefather of the twenty-year supposed cycle! So in this morning, as my Catholic parish did nothing but drive the caravan, we were to worship as a family at his Anglican church. Amidst the wrong way to interpret *caballero*, from my mother's side, and artist from both sides, by a parish otherwise fluid in their ritual of conversions such as I, I doubled back hoping for a correct interpretation of American in accord with some semblance of esprit from the ally Great Britain had become. Aware Anglican ministers were crossing over to Roman Catholic in this time frame, I was somewhat ashamed at the mask worn on the moral front by my fellow Americans of all denominations when quite clearly the topic had been the strength to carry on while God received our prayers for American troops.

And somewhat strengthened I was by the fact that we had, despite our simple, not truthful but simple tendency to discriminate, thus far protected our right to worship where we please. A sexist portrayed? I rested aware of my own discord with some of my fellow males, while being helpful in a family that had evolved to offer female and male combatant support alike, no, a seed of faith that more could look at the new U.S. military after twenty years of this (and domestic discussion springing forth throughout said timeframe) in just that beam of light's reflection.

--"These guys need a psychologist."

--"How about I take the option since no progress is being made in society?"

--"This is crazy."

--"Methodology. Since my real problem is money, but the country as a whole is rather lugubrious if the subjects aren't athletics and toys, and it is attached to me in my desires to see less debt in the whole Western world, period."

--"What is your point, man?"

--"I believe we have a bit of a fetish in the region whilst we're exercising our rights to the augmentation of our vocabularies, and I'd like to concur with someone who has better statistics on the outbreak than the mayor. The Apostolic secession will hold court."

--"Where is the Queen?"

--"She's preparing for a proper wedding."

--"Isn't that the country our ancestors left behind?"

--"Good to see you're not daydreaming. Continue with the esprit` of what other letters have been posted in scarlet, and don't believe it—you don't have a tail, there's nothing between your legs from buttocks exuding period, and the subject is cash. Until we stop flushing the history books down the water-closet."

--"It's called a _gabinetto_, a loo, and a _cuarto de bano,_ and I want you to take a piss in all three countries and crawl to the sink, wash your hands, and forget about ethnocentricness."

--"I'll call the Alliance For the Mentally Ill, since if this is the rate we're going at, I'll have a job preachin' Episcopal side by the time we're through with these history buffs' speech on The Tower Of Babylon archeological dig trophy. You couldn't get a straight roll of craps if you're that smart outside of Sandusky, Ohio!"

--"Please continue to wear a suit and tie to Mass, whether or not you're the new Messiah, Jesus Christ of the apartment scene, or Jesus Christ of Mexico excursions.

Sincerely,

The Rector

--"C'mon, give me the binoculars!"

--"Can't you hear him? He's the one making that noise."

--"What's he doing?"

--"He's telegraphing the other pot smokers on the farm by tapping his SKOAL can on a wooden log. You know what he's telling them?"

--"What the hell could this be? Has he seen us?"

--"No. It said, 'Hey Buddy!'"

--"You mean all that log-tapping for 'Hey Buddy'?"

--"What, you thought we were the slow ones in this field?!"

--"Did you bring the weasel-food?"

--"What, you kiddin'? You start making a trail with those Gummy Bears to get him to come off the bag, and we're going to have a moose run."

--"Student 1: "Northern California where the girls are warm…?"

--"Professor?'

--"I believe I would like to help you with the amount of less verbal flack you take in a long-term relationship context, within the exterior perimeter of the Southern California metropolis, and dismiss you for the day."

__"You're not going to 'rob the cradle,' are you?"

--"Good, let's keep that 20-20 vision in the driver's seat, though after Catholic Church lawsuits, disappearing children, child molestation, infant mortality, abortion, autism and other childhood diseases, you might keep the cradle a little more sacred as a symbol, and use metaphors that don't inspire Satan to come up with unique ideas. I know, that Mormon real estate in Texas on metaphors. Would make any Californian wonder. Still, most have toned it down."

It mistook me for a drug dealer not an Eagle Scout, when I wore my Coppola, and the dead are still rising from the college scene in the early 90's. Yes, stay current, no baggage. Carry the luggage at LAX as a porter. Snap out of it. I need a CD player that doesn't break two weeks after buying it from Walmart, not a California Stop ticket. You know how I know California history repeated itself from 20 years ago, with my college Sophomore agenda-now-national knowledge on the media? Because we brought up the environmental issue 20 years ago before the war. Pot then too. Pot in junior high. Look at the State's budget between then and now, and who's prospered in the time frame. I'll be the needle in the haystack on approximately the same economic terms who feels four out of five people's manners have improved in the State since I first compared them with a European country and decided we had a 'tude, changing gears on how I thought of America. It was not any propaganda I saw, biography of a media liberal, or literature I read in particular. It was behavior. I faced discrimination in Italy as well, from other California students, not just Italians. You can place a bet on "no common sense" or "no street-smarts" because of eye-ware, but please, don't get diabetes, it's not that personal (And no, Dad and I are not working on plans for Bush of names on a Black List for Californians who have thrown on weight unacceptable to Disabled American Veterans, in case a Republican takes office again). There is another way to move on with it. Teach Sarah Palin how to handle North Korea and other world view issues, before my symptoms of paranoia have a new cycle, keep her waist-line where it's at, and let the stars dance in the sky. When the Big Dipper is upside-down, it means we did not cut spending on the East Coast. When it's right side up, it means you're lying on a blanket, with the waves crashing, with someone who can turn your life upside down and filter out the ageing oils in your skin.

--"Why did the cop go on furlough?"
--"Because he did not have the instincts to survive amongst the business animals and the subject was lunch downtown."

So Anne Rice has left the Catholic Church. I see this with clarity, much clarity. "Anne had The Passion, though became disgruntled..." I believe that's the part where someone evangelized to the quarterback of the Philadelphia Eagles, as far as recent moments in time. How did news travel on this from Philadelphia to Louisiana? Wasn't the quarterback, as he and other NFL players during the off-season were, on the German Autobonn? What, in a truck? No, actually, no trucks allowed on the Autobonn, only tanks and BMW's going home now. It's like America—you need a beer, you ask your mother. However, there's beer mysteriously everywhere, which eliminates need for delivery. Once they did think of talking to Italy about the sale of some trucks to Angela Merkel, though, surprisingly enough, Italy's only reply was, "We don't have delivery trucks either: only tourist busses that go down the stretch of freeway along the Amalfi Coast." SO, yes, the news to Anne from the quarterback: TO make a long story short, who is the best truckin' nation in the world? In sum, without excessive lecturing, to quote the unknown American beer drinker, "Keep truckin' for the Lord!"

--"Steve, you passed Judgment. You're in Heaven. Congratulations, Stephen, I'm Eli Whitney, and you know what I invented. And, Steve, we have a lot of festing to do, I've been dying to meet you, and don't ask how we fest, because there's no wine in Heaven."

--"And Stephen, I'm the Mother of Invention, I heard you and many like to produce artworks. You're all my children, and, Stephen, I apologize to you for over-eating. Now make way the way for the coming of the Lord: Entering Heaven was both a joy, but very difficult, because it was not about my rich cuisine's traditions to inspire, it was about learning our spiritual bodies. Earthlings are in transition, and Heaven's agriculture, well, can't begin to tell you....it's Christ's parable from the sky today, and you'll prepare, but without food for eternal dreams played out with the Heavenly Court: actual regal fellowship, but you can show your bottom to Mary, when you exit the throne of God, she's not the Mother of Britain, she's the Mother of Sacrifice For Mankind, so very alike, but with different humor. She wants to see where that last shot went into your butt to heal it, and has sovereignty over every baby's bottom since the beginning of mankind. When the Earthly skit is over, you won't be able to answer Her when asked by an angel, 'Do you have a baby's bottom or a man's butt?"...You'd be struck in awe by the light emanating from the throne of God Almighty."

--"As far as collegiate pursuits go, a sermon that a member of the priesthood gave (as opposed to a liberal thinker at the café`) about the blind leading the blind seemed to apply. As far as the pursuit of a higher degree to 'expand my opportunities for the workforce,' I'll put it in more concise terms: I'm about to go to Disneyland and pay the tab for forty days and forty nights since this has been a Mickey Mouse fare. I bet ol' Mickey would...it does not work as such. That Disneyland is about new beginnings, and college is about history and tradition! I think the 55-year-old Indian sculpture salesman who told me to focus on the connections in play was rather devious in retrospect. Now that we've got a connection academically with every single dirty apartment—leaving behind the roommate, an un-showered-in-the-morning-thinking, my milk-drinking, Absolved from dishes-doing, beer-drinking, hole-in-the-wall-punching Californian who was here over a 10-year time span-- I'm going to forego the new challenge that always exists as life is full of challenges(according to the first Bible-thumping devil-woman—see Isaiah—of a rich Protestant I met right after the 12th grade). It's not like they don't make them like that anymore now that people of all races have now adopted or married into Protestant faiths, to continue the ecumenical dialogue at the same stereotypical standard of scrutinizing the White male with short hair and glasses. If this many people are getting the same lottery draw from my chemistry and these are the End Days, what does that stand to say and how do you turn over a new leaf with that many maggots underneath?"

--"Hey, Mr. Mouse, meet me at the Art Festival: I've got news! Now the Italians, after a briefer definition of this phenomena for once, were not their talkative selves and simply say, *"Madonna Mia!"* You'd think the Alumni Association's inventory, while serving food and beverages, is where they pick up with this eclipse of The Bright Side; now that you have a degree, after said Olympian hygiene event. You think the Southern Italians have flies? I'll keep the medal in a private men's tournament, the Eagle rank, the Baccalaureate, and convert to Catholicism, get cable T.V. and watch the Pro-Life March at Washington D.C. on EWTN!

So, do you understand, in a more humorous tone, what it mean when The Nice Guy (not to be confused with The Little Guy) becomes a Knight of Columbus Internationals, after confessing the past as a partner in crime during years culminating in the L.A. Riot, and being absolved? Thank God I took the Italians advice: I thought of not a mustard seed, but a clove of garlic, to protect me from the little vamps! Yes, the College Years, hosted by the great modern and efficacious patriarchs, the Jews, and co-starring with an award unique in all the pluralistic academy, The Asian-American Students Union. The handbook at the campus bookstore was entitled <ins>From Diversity to Dishwashing Liquid: 10 Sure Steps to Getting Your First Degree.</ins>"

So, yes, fellow Gentiles, some of these people are new to this country, as well, for reason of Apostle Paul's bit of heroism when he informed us(Jew he was), this covenant is also for the Punch N' Judy Gentile scene, not to mention, The Messianic Marx Brothers. So, I know many abroad sacrificing for the U.S..A. may not always understand what I will tell you as far as why military service was not an option for me, even though not gay myself, but this was a continuous war story for me!

Yes, I hear it, the Officer's (like my father and his fellows from 'Nam and Korea) blindness, and the lack of a Commander in Chief plan. 9\11 was a cheap shot. I feel our own version, as the country with the mountains and plains, will never lose its soul since the birth at the hands of discovery, Puritan ocean trials, Indian wars, British and American bloodshed, and the Declaration of Independence, and has been heard by the healing hand of God on a world he does not see as several nations, with all his wrath poured on. SO in the middle of this build-up, I was reared by a woman who served as a civilian secretary for the U.S. Armed Forces for thirty-four years, held at GS-6 until retirement, and now lets me know that is what to come and how to vote. What is inseparable from this essay's discussion is that yes, it was difficult for me despite being fairly intelligent amongst my peers to <u>ever</u> have the control of a man who would lead others. I don't get disciplined with a whip, I get disciplined by pale green eyes born in humidity: Yes, to me this was an irrevocable shot from behind when the opponent could not see, and you don't mess with Texas, as this sentiment was returned directly and equally.

As she is ten years older since 2001, I have heard more of the ills and strong points of the nation my great-grandmother and grandmother came to their crossing into during the 1890's, the latter giving birth to eleven(none twins) and living to a keen age as a devout believer in Christ. Yes, their eyes, her and her mother's alike, get like a smoking gun even in the milder California climate when I'm given instructions for work, and now, that I've had the chance to carry on after three years of unemployment as an enumerator for the U.S. Census Bureau, I have seen our nation's Alamo, and am alive to fight in conscience at our San Jacinto of the future: These latter troops have not left the field from Afghanistan and Iraq, though this aftermath in the dust is ours to reap with open arms, as much as a fool can be, for all amongst a world that may continue having inexplicable changes until they, nations one and all, realize America is their friend.

So that is the nature of it, before another political party came in to prevent us from striking again with this country's name from the depth of Earth's pride. And yes, poor as some Americans are now, Christ is your brother and has come to dine with you, do not fight Him, or the America(ns) that sit as stands(and yes, we've all gotten in a fight with our brother in this country). I do not recommend revolution, though die if you must for Christ. This was not a Jerusalem revolt, this was a back-stab from the crowd and a crucifixion, a sacrifice, for our propensity as humans of original sin to do the same.

--"Is that you, Master?"

--"Yes, is this your art display?"

--"Yes."

--"So, artist no—not studying martial arts? Ever study?"

--"Oh, long time ago. But now Doctor, not Master, wants me to take medication. Not good to have psychological issues and study: should focus on problem and just take knocks."

--"What happens when you get a wife?"

--"Oh, she has to do her own cooking. Has to grow up too!"

--"What's wrong with this picture?"

--"Lines hard to see—should draw them in darker for contrast."

--"Why always have to be contrast? Why not blend?"

--"Because viewer in search of something, like artist—can't make it too hard for target audience."

--"So, that's why you need glasses: Could not find it first time, but sure were looking for it!"

--"Oh, Master, I went to apply for a job or two, and you know what happened?"

--"What, forgot resume?"

--"No, their eyes welled up with tears and they said, 'We don't want to talk with you!"

--"So, Dad, the subject today was 'stigma'."

--"Oh, okay, before we get to that, Honey?! Honey—where's the popcorn machine?"

--"Sweetheart, I gladly will retrieve it for Thou and you—Cure whatever it is that they're going to name this disease by taking care of <u>yourself</u>, if you've got the same taste in your mouth I do, and it's not the sweet-smellin' tobacco or the beer!"

--"Yes, sport, look on the bright side, if you got the same bed-sheets that you did at 18: You must be one of the most fertile turtles in The Valley!"

--"Oh, honey, he doesn't see a psychiatrist: he sees clean while others are lookin' in!"

--"Oh, you know why his last girlfriend left it with, 'You're see-through!'? Because she's been to every bar in town!"

--"Hey, son, did you see that girl with the same thing as you at lunch today?"

--"Yes."

--"How's she doing?"

--"She had surgery. She's o.k., she survived it."

--"A female thing?"

--"No she had some of the bone under her scalp removed for the marrow to put in her knee, in straps for going' 15 rounds with three different Psyche=Techs!"

--"You know, I can't believe the hospitals get to go that far!"

--"She said the prize money she got from The Patients' Pool was worth it—she don't really need to worry about work even if she was ready again."

--"How long was she strapped down for?"

--"Until the doctor was done flirtin'!"

--"You mean they still got that same womanizin' S.O.B. headin' the staff after all this time?"

--"Well you didn't hear about the young upstart they had hired to prepare him for retirement, 'bout the time we went through this with her from years ago?"

--"No, what a motorcycle accident?"

--"He tried to slip her the medication and the tongue at the same time while she was strapped down, and she bit'em!"

--"Speakin' of which, aren't you supposed to take your evening pill?"

--"I told you I get a shot now. Plus I did not bring my overnight bag."

--"A shot?"

--"Yes, two of 'em. Right in the derrier!"

--"Ouch!"

--"Oh, he's good at it. Don't think twice about it!"

--"Well, you know what I'm thinkin' twice about that must be gettin' a kick out of this, if <u>this girl</u> is still unwindin' with same crowd every time they release her...?"

--"What?"

--"Yea, who?"

--"Not 'who': 'what': 'Oink!'!"

--"Oink!!"

--"Oink!"

--"Oink!"

--"O.K., popcorn's ready!"

--"What are we going to watch this time?"

--"A healthy cop movie, with lots of cussin' and gunfire!"

--"No, I don't like cussin', guys. Pick something else!"

--"Let's see—<u>Girl Interrupted</u>, <u>Fatal Beauty</u> with Whoopie Goldberg, and we got that French one with the blonde holdin' the pair of scissors on the cover of the video cassette box."

--"Oh, all right, much as I'm tired of hearin' about New York City, put it on!"

--"You wanna listen to music instead, honey?"

--"Oh, hell, if they had a real radio station anywhere near this State, I'd be dangerous! Let it go!"

--"Yes, that's true—you know why they think he's the tweeker-sketcher-wierdo? Because the underwear's tight and his resume's long! It's been hangin' out to dry on every street corner since the first P.O.W.-M.I.A. flag went up after 'Nam—they're not done with what the sacrifice of lives did for fillin' in black books!"

--O.K., O.K., put it in the VHS player!"

--"It's stuck!"

--"Let your Dad handle it."

--"Hand me the remote, son."

--"Here."

--"No, the other one."

--"While he's toolin' around with that, that one guy Bill called."

--"Screw Bill. That's the dude who used to moan around the girls every time we went out in public until it was time for his macho routine and his ride home."

--"Oh, you.."

--"No, honey, you drop it this time. There, it's startin'. Son, you wanna beer?"

--"Sure."

--"And you?"

--"Some Chardonnay with a little ice is fine."

--"And I'll take that glass of red wine I been waitin' for."

Who's Behind All This?!

--"There we go, team, if most Italians from San Francisco to New York and back to Sicily have relatives in the Boot, what would you call a Southern Californian military Vet's son who's fluent in Italian?"

--"The Shoehorn."

--"CO-starring, 'It doesn't translate: with 'What means P?,' and the musical soundtrack by 'We're Havin' Problems." OR take the other one of Andrea Boccelli's small markets. City of four million. CD stands for Cristiano Democratico. It's the most popular political party affiliated with the Church. And that's an 'Hungaro!': It means, 'Hungarian.' See how Litsz got labeled another Mephistofoles? He's only fourteen hours from Italy central. Still like the film Rome, Open City: Except for the nationality that freed the Jews—a little discrimination. On the other hand, watch the most modern Italian move in the 21st century world: The hand-off to the American brotha'! See? Gotcha. Their quick hands if you don't like cards. Get another shoehorn for Las Vegas. Call it the Icehouse beer from Milwaukee. Little sex? Puerto Rico and Mexico date Dutch compared to the Lamborghini Love Triangle. The David doesn't move: Everyone else in the country has O.C.D. That's why a Polish Pope in The Reagan Years! 'And that dish, _figlio mio_, that's _Canoli alla Messican'!_' I got it, this was a paradigm shift from past to present, preparing for L.A.'s future. And that's not only Amanda Knox with a joint, but Amanda Knox's sex with the new generation of Italian male. No, I'm going to get a swine flu shot and a bucket, I haven't had this much fun since David Bowie did a sci-fi film in '82! Or we could see the porn in the Raging Bull theatre scene. What happened after the _Gianni Stacchino_ scene with the banana peel?! That's a bit of

Or we could see the porn in Raging Bull thetre scene. What happened after the _Gianni Stacchino_ scene with the banana peel?! That's a bit of overload if cleanliness is next to godliness and hygiene is not really a contemporary problem! 'Pecora.'

Or we could see the porn in the <u>Raging Bull</u> theatre scene. What happened after the <u>Gianni Stacchino</u> scene with the banana peel?! That'a'<u>Pecora</u>.' It's a Passover Seder at an Episcopal church, not and A.I.D.S. clinic in downtown Milan. The stones of the Olde Country's gutter are barking this one out if that's the American platform for Purgatory after life on planet Earth, since we decided to depose a dictator in the Middle East. The breathing room on a job in America if 'Italian spoken' is on your resume: Pizza to Teaching to Chance, Tinkers and Evers were resting on their laurels and then came real estate's issue. Where's the neighbor now that we have an Italian doctor in the neighborhood by the Country Club?! 'You're not Italian!' You're not American: That's New York Mob and that's F.B.I., 2011. That's the first non-fiction point on the board since <u>Gianni Stacchino</u>. How are you gonna spend my Mom's, since you found my Dad's wallet? No, horse manure doesn't stink like that. That's the New York-style or Chicago pizza, not a Margherita you can foot home where wine is brought up from the cantina if we're runnin' low, not another pharmacy staff member that insists we drink the whole thing if we pass when we're carded by the cashier.

I got it. I picked Bush, Mom, and McCain, L. A. picked Rocky, President Obama, and the Security Guard at the twenty-four hour pharmacy. They I got it. I picked Bush, Mom, and McCain, L.A. picked Rocky, President Obama, and the Security Guard at the twenty-four hour pharmacy. They I I called it a ground game. I called it reverse-sexism, now that I got a point from the clouds, for technically-speaking, still being straight, although they really don't care as long as the plan sticks in order to get Northern Cal off the rap for the state's debt. I'm surprised I never found a roll of toilet paper that once unwrapped, didn't have recycled crap stains on it in this region, since the side of S.C. bordering Orange and L.A. gets the food for the animals, not humans. See, we're not a romantic region, but we have cat's eyes, so don't give us the jobs from the People's Table. And Rudy Giuliani to U.C.L.A.: 'I plan to help the business community' and 'We don't offer more than a Business Minor at this time, despite the Italian Trade Commission office in downtown Westwood.' Did they say, 'We need a doctor' or 'The doctor needs a date'? Fine, the angels will comment: 'It's not necessarily B.O.: You've got a small mole in your armpit, though!'"

--I'm unemployed? My economic theory, stemming from their 'rugged individualist' credo of famous author Ayn Rand, inspirer of Alan Greenspan, is 'You take me off the team in this town work-wise, the City goes further into debt!' Whereas, with the Federal Government, the reciprocal is true. They keep going further into the red whether or not I'm in the Continental U.S.A.!

--"But this is a cluster: as we say in the Armed Forces!"

--"Bud, you have not only thrown this conversation, you have blown so much smoke that this cluster has been re-charted into The New World!"

--"Now, instead of prayer before the meal of our gathering, how about we just wish everyone 'Happy Exploring'?!"

--"Yes, you don't have stars in your eyes, but look at the constellations in your academic galaxy!"

--"Yes, that's a good question, while we're all up on this floor of The Enterprise, whose going to take over for Pickard in Star Trek: The Quest For Reality!"

--"Oh, it's 'universal'…and the subject *du jour* is 'tweeking'!"

--"Why we have come to popularize amphetamine use-slang is anyone's guess, and that's 'disintegration'—it's just tooth decay. Maybe you use the term since you last kissed him!"

--"Well, I don't know about 'disintegration'—how about keeping our teeth and just saying the food is melting in my mouth?!"

--"Should we use the knife to cut this, or the coffee and tobacco?"

--"Pass the rum!"

--"If Kissinger had this chemistry there'd be peace in the Middle East!"

--"Seriously Bud, how did the Finals go?"

--"Got my ass kicked in there by my own private Alamo!"

--"How did it go on the Financial Aid Award?"

--"Well, the research and paperwork of exhaustive measure took several all-nighters, before I discovered I'm a Straight White Male, and the counselor at Aid is a Lesbian Mexican-American!"

--"So…no money?"

--"A work-study option. Even if I change my major to Chicanismo!"

--"What's the job?"

--"Security at the medication-testing laboratory when the Animal Rights activists come around!"

--"You gonna be packin'? You get a uniform?"

--"A uniform? This is an assignment where I'm Covert For The Lord in the campus bar scene as a marijuana rights activist! Packin'—won't need to. Got a beer allowance to stay on their good side until I get information!"

--"Ah, I wish I could go back for a Master's Degree…you have all the luck!"

--"Yea. And the good news is, the campus hasn't changed a bit except there's more to do in town as a resident artist!"

--"You paint? Sculpt?"

--"I work on Classic automobile engines at the security yard, so they can re-fire with perks that don't come out of the budget for the new Law School's construction!"

The existential vacuum is the negative past, the sins committed, or lack of progression of a few in adaptation, preventing a cathartic, clean brain chemistry from rejuvenating in the present. OR is the present outside, the past internal? Is this all the depth of the soul? So the new book, <u>The Power of Now</u>.

<u>Jack Kerouac in L.A.</u>

--"Whatcha got in the grocery bag?"

--"The screenplay I'm working on."

--"What's it called and what's it about?"

--"It's called <u>Brotha Lova Bacon Barn</u>: about a Scout who rescues a live pig-mascot from a fire in a store the size of Walmart of Los Angeles, which sells nothing but bacon and bacon products!"

--"Why <u>Brotha Lova</u> Bacon Barn?"

--"NO, not <u>Brotha Lova</u>, etc.—that's just the whole name of the store, in neon lights. The store owners never appear in the movie, but they are believed to be from around the area."

--"I'd think you were a ghost who used to drive a public transit bus in the South, that is such a bad joke."

--"You would have to read the book. The book is ten times better."

--"I've heard that before. Step out of the car."

--"O.K."

--"So you can stand. Need you to, one foot in front of the other, heel to toe, begin walking a straight line until I'm done perusing the contents of the bag."

--"There's 350 pages! It's a normal-sized standard, made-for-producer screenplay!"

--"350 pages of that!?"

--"I'll start walking, unless I can just take the breathalyzer...I haven't had a drop all night."

--"You know why I pulled you over?"

--"No."

--"Because that's my screenplay....and I saw the California plates and 'I Love L.A.' sticker."

--"And...but you're in Law, not writing?!"

--"My city. I <u>own</u> it. Always wanted a piece of the film world. Now give me the screenplay, take the beers off the floor of the car, and get back home. You breathe a word of me being dirty and the next time you shop for beer, I won't be there with a uniform!"

--"Okay! Surrender. I'm not going through that again. Back to the drawing board."

--"Oh, you an artist?"

--"I've got my paintings in the trunk!"

--"Open the trunk."

--"Okay."

--"How much for that one?"

--"Cost ya' twenty."

--"Here's your twenty. Stick to painting, kid, and remember, I own this town."

--"I hope you go undercover and bust the last producer who gave us one of those rotten eggs we saw before I got this great idea."

--"Next time watch your karma, if you want to write. But this time, I gotcha."

--"I'll keep the twenty, and swing back later with a whole new painting—I swear officer."

--On the road, then, Kerouac: the signs. Read them."

--"Ciao."

--"That's the mentally-ill guy over there who likes to work, I take it."

--"The only one of them in town. He has a few simple needs and likes to do his own cooking, but other than that he'll till the earth in the hot California summer sun for a few extra bucks to buy cheap beer."

--"What's his name?"

--"Injun Stephen. Those are his quarters. Where he's kept while they run the schools."

--"Should we go in?"

--"No! The manager alone handles visitations."

--"Can we talk to him?"

--"They're from New York. They said Injun Stephen is a curious sort—he doesn't like it when he has to see the doctor and gets easily upset. Best not to disturb."

--"Is he violent?"

--"Hasn't gotten in a fight for over thirty years, but shouts at the walls a lot."

--"The cat seems to like him. Does he feed her?"

--"No. He speaks in a foreign tongue to her and she doesn't mind, because nobody bothers her and she gets to stay in the safe neighborhood. Injun Stephen chases away bad cats, raccoons, and he likes dogs too."

--"Does he have family?"

--"Mostly from outside California parts. His Dad is a disabled Vet, Mom is popular in high society, and the brother-in-law an attorney who makes big donations to the Humane Society."

--"How nice of the city to house him in this day and age. Some of the others claim they're experimented upon."

--"Takes his meds from the doctor and has only been late for three appointments in six years. Apparently got a kick out of the stories they told him of Mary Lincoln. Says it made the most sense in a long time that she moved to

Europe after the war. He was treated for something like her shortly after coming back from there."

--"She didn't come back though, did she?"

--"No, but he sure did—to till the Earth."

--"No, remember the Good Rule."

--"What's that?"

--"If the joke isn't good it's because smoking makes writers lose their focus if they've actually got a better attitude than to do so."

--"Hey Sport, what you doing with your time?"

--"Picked up a new book: <u>The California Gold Rush of The 2nd Millenium: Influx of Out-of-State Degrees and the Dichotomy Between Education Then and the Free Market Now</u>."

--"Who's it by?"

--"A certain Sociology professor from U.C.I. was raised here. Says a few who didn't serve as long as you jumped the fence 'bout when I reconnoitered to L.A. a while."

--"What's the thesis, c'mon, stop playin' around."

--"As I read it, sits as, "More Book, Less Wood," with no analogy for traditional Catholicism, in our region anyway."

--"Yea, certainly not the Catholics. Enough there for More Book <u>and</u> wood!"

--"You miss the Episcopalians, Dad?"

--"Where did you think your new leadership got the wood for the new Cathedral plans?"

--"So what was your first clue that the hostess wasn't exchange student material?"

--"The toothpicks to clean the teeth after the meal were stuck in the cheese-ball."

--"And you say you were rejected at the affair?"

--"That's not a knife through hot butter protruding through my back and making my shirt stick out, that's the tip of the blade of wild grass protruding from the mule's mouth. He was right behind me the entire time."

--"Ya workin'?"

--"No, nobody'll hire me."

--"What happened?"

--"Oh, got fired from my last job."

--"Where were you working at?'

--"That Louisiana Chicken chain."

--"What did you do? What's that about, you're a good Louisiana boy. You get along with the folk."

--"I got a little high, and they got upset I brought in my own doughnuts one day and cooked'em in the deep fry for a snack!"

--"Why are there so many weight problems in this area?"

--"You have to know the underbelly of the region to understand. It looks like a typical region of California on the back porch of media giant Los Angeles, doesn't it?"

--"Well, yes, kind of an Air Force town and retirement community in one as well."

--"It looks like modern America of norm, but actually has a post-war history more unique than any other region involved in the conflicts to such an extent."

--"What does this have to do with the diets of the inhabitants? Fast food and restaurants, home cooking and barbeques, cafeterias and food drives. Lots of people exercising it off, though. What gives?"

--"Without much known about the legacy, operation, or network, we are in the middle of the mysterious dynasty of none other than Hung Yung Phat!"

--"No!"

--"Yes. The region was offered by the military as collateral in exchange for the release of prisoners of war. People don't know that when they move here."

--"That tops that business about cities of amnesty. What's next, California?"

--"South African diamonds, Jewish and Middle-Eastern in-fighting over the market—all counter-espionage propaganda. Hung Yung Phat has complete domination."

--"Let's go to the store.

--"It's getting late! Remember what your Mom said—'The Fuzz' is out."

--"Let's not use the ol' 'barometer for a relationship' trick right when I'm stopping on a dime before taking a right on red!"

--"I'll stay here. I think it's risky."

--"I guess you did not hear the good news."

--"What?"

--"I made Who's Who in California Colleges and Universities Streetsmarts 2010 edition!"

--"All right. Let me wear your London Fog coat!"

--"We'll be back by 1:30 A.M. for a black and white movie on the Turner Channel. Promise."

--"I'm glad it's not the weekend."

--"No one said the adventure would stop in the middle of the Alumni Counselor's career transition coup."

--"Yea, I'll take this over a _coup d'etat_ any day."

--"Yes. Anymore jinxes and those aliens in Real Estate are going to signal again that there's a new apartment to rent or two."

--"THAT'S what that is in the dark sky outside the window."

--"You thought it was the Red-Eye to March Air Reserve Base?"

--"All right! There it is now."

--"Stop saying 'All right!'"

--"I don't know how to stop saying, 'All right!' if you use proper English in America these days, you get a stiff warning, a scrutiny, and possible repercussions of inventory, financial and medical!"

--"In accord. _D'accord. D'accordo. De acuerdo._"

--"Let's move out!"

--"To the Fatmobile!"

--"And there it goes, the roach-coach on 16 wheels to Rosarito!"

--"Bring us back an exchange student!"

--"Ever notice how Wall St. is always, when there, up a patented 3%?"

--"It's the girls, their protest. They refused to do more shopping until Clooney movies get more than three stars!"

--"Clooney? Nah, he's fine. Can't a guy have a little fun?"

--"What, fun, o.k., but 24 hours-7 days? Let's not drain the sap out of every family tree that has a t.v. set!"

Master's Degree, T.C.U., Faculty of Hang-ups

-- "Hang-Up 100"

-- "Hang-Up 200: Advanced Hang-Ups"

-- "Hang-Up 300: Hang-Up Crisis, Family Intervention, and Reform"

-- "Hang-Up 400: Religious Transformation and Collective Armistice Agreements

-- "Upper Division Hang-Up 100: The Job Application Process, Completing Your Degree, and Collective Armistice Benefits"

-- "Upper Division Hang-up 200: Benefits, Employment, and Sports Enthusiasm."

-- "Upper Division Hang-Up 300: Sports, Politics, The Arts and Temporary Real Estate Issues"

--"Upper Division Hang-Up 400: The Battle, The War, The War Goes On: Dead Terrorists and New Leases

--"Upper Division Hang-Up 500: The Cowboys' Play-off Hopes"
-- "Upper Division Hang-Up Final: Hang Up Your Cowboy Hat!"

--"What actually happened when there occurred an actualization in Greek mythology between philosopher and Pantheon?"

--"I believe they arrived at the point of the Christian scales, required a cross reference for purposes of the Church's own reputation, and it culminated in the matrimony of Pandora and Narcissus."

--"Now anytime you want our minds out of the gutter, just remember some of us have to park there because we're real men!"

--"You are free to your own time from the people who will not cut ties with you and your parents, but if you're too tired to roam about the country after the filibuster we understand!"

--"One world traveler rep in the ground. *C'est la vit!*"

--"Europe's got problems."

--"The fall-out from America's moral meltdown is #1 on my list!"

--"Pot calling the.."

--"The University White!"

--"Wow. Sniff-sniff. All that caused by the connection of the spirits?!"

--"Believe me, it's beef, potatoes, and vegetables, little tea, not Return to Sender Souflet!"

--"I'm surprised the postal carrier doesn't handle it for the whole building but maybe he's not insured!"

--"Can't get to bed now, might as well stay up!"

--"Here, return this 16-pak of toilet paper to the Post Office, the whole staff needs to get with the Morning Show!"

--"Those evictions are determined to lower the rent."

--"Are you kidding? It's the first time the gutter by front-street parking hasn't had $50 worth of aluminum in it once a month!"

--"Are you kidding me about some of these apartment scene partiers from Christmas past?! You couldn't quarterback them out of The Little Red Schoolhouse with a Green Bay '10 jersey!"

October 9th, 2011

The Gospel lesson given in *La Mezia Terme* of Southern Italy (perhaps one of the communities similar to those along the route of *Campania's Circumvesuviano*—the regional train—though given in *Calabria*) talks of a man bound at the wedding feast of the King for not having a wedding garment.

In one sense it is not talking of the Heavenly Banquet, but an Earthly one, where Christ was rejected by the Jews of Antiquity and forced through the temptations of Satan. He heard and sees things in the desert, whereas in this Gospel lesson the alliteration of gnashing teeth.

Similarly, people often want to know what schizophrenia is like: In this sense a Catholic one I heard nothing more than what the priesthood says (those chosen must endure with Christ through Earthly trials, though will exit them all the more faithful). As far as gnashing teeth in the darkness, perhaps it's a nocturnal animal, such as a raccoon, with its young. As far as what we see, why yes, I've seen a raccoon outside with its young, after a moment of confusion upstairs, with a tension in the air that blocked out my thinking, causing me to go outside for fresher air. As I moved backward knowing raccoons are larger and less domestic than cats, which are normally outside, the raccoon moved backward just as quickly, mimicking me! I eventually made it back upstairs.

So, as schizophrenia goes, despite confusion that can last a long time, belief in Christ will be its own harmony and grace, until the sun rises the next day, which may not be the same one on which I keep the appointment with the doctor. That does not rest us with confusion by light despite a harmonious world! Many have, and others will, continue to wake to Christ's call, forward moving from the confusion.

In a more surreal manner, as the Pope blessed The Host, I saw the faces of my former Italian friends in those of the priests at The Holy See's side, in the nature of The Holy Mystery, many of them not Catholic, though more than merely Southern Italian to me. I had no idea that during Advent, 27 years after sharing the Good News For Modern Man with already-mature youth with real-life adult challenges to their region's integrity, that I would connect to Roman Catholicism. It had never been a long-term plan, as my Episcopal baptism had been a sufficient faith in which to know the Holy Trinity. Not so much as 'all good things must come to an end,' as, judging by a raccoon's litter being over my head, I don't question the catching up in a whirlwind of myself from time to time, nor Christ's changes in my otherwise sincere life plans.

At first this was a faith with "*cose assai*", or many, many issues, and I do now take rest from normal mass in an otherwise younger Parish to see Catholics from other parts of the world, as my only Earthly father had offered to include the Catholic television station in our plan, in exchange for car-washing and any other hand I could lend him and my mother their 27 years after an exchange they saw come true, as with many parents and students. We had our problems like any traveler, though perhaps still cannot explain our fortune in light of other Americans who have difficult stays compared to ours. I see a modern outdoor Mass, not an historical Neapolitan church fortunate to be standing without scaffolds, after the Earthquake of 1980 that struck the city, bearing to light the mobile homes I saw of *The Terremotati* (Earthquake Victims), those dislocated from the quake. To say there is much of Italy I did not see is not the point: this education continues, whether or not my "fellow students at campus" agree with Pope Benedict the XVIth. I suppose I was surprised to hear of liberal Italian universities even in Rome proper, though not of the nepotism of their professorialship. The study habits of a doctor in any field, though not 100%, may be hereditary—certainly I can't blame my cavity on the dentist's son!

--"Stefano, what-a means..."

--"I do no sacc', becaus-a my watch ish in za shawp!"

--"No, Stefano, (Fabio, com'e se dice—How do you say)"

--"Stefano, what do these things say? Translate the words."

--"Yes, I know...My Way..but that's not where you took me!"

--"*Gesu Cristo mio*(my Lord Christ Almighty)!"

--"That's what I said the first time, accept Jesus into your life and then go out for a marghuerita pizza, whether or not it's a delicacy compared to American pizza, because as far as that's concerned, I've got news..."

--*"Che*? What?"

--*"Caggia` fa` a copp' o' camo`?"*

--"That doesn't translate. It's untranslatable!"

--"No, it does... It means, 'What are you doing on our table?' Or as Jesus would say, 'First take the log out of your own eye!"

--"No, he's got a point, do you cook, or live at home, or cook, but still live at home?"

--"Not I, wise guy. I'm on my own. ME. Stefano."

--"Mamma, Stefano came for a *soggiorno*(extended stay). Please let him stay for lunch."

--"Well, it's *Pasta Arrabiata*(Mad Pasta)."-

--"*Signora*, it wasn't the American budget, was it? C'mon, 'Mad Pasta,'—I'll work it out at the table, *Signora*."

--"Then wash your hands, everyone!"

--"Stefano, is the glass half-full, or half-empty?"

--"It's a clean third of a glass, and is that your new Red?...keep it pourin'!"

--"Stefano, are you sure you'll be able to walk back home in this heat? You can't take a siesta with Antonella!"

--"Signora, before we start another Dungeons and Dragons game out there between Notre Dame and Los Angeles, just don't forget: I speak fluent Amaretto and other cognacs known to interpreters. Please, if it spills, it's the tailor's turf, if it lands in the mouth, it's mine, all mine!"

--"So you won't need surgery for a scorched hernia?"

--*"Ia' so' Napuletan'! (I am Neapolitan!).*

--"A little Neapolitan, a little American."

--"C'mon, _ragazzi_, the Pope doesn't have all day, the Bible Study has begun, and _Pasta Arrabiata_ is <u>served</u>!"

--"Hey, Signora, did you see that Raiders' sack? Caught'em lookin' like Chicken Licken'!"

--"Say the prayer!"

--_<u>"Che la California tien' successo, e che Napule non soffre pelicolosamente. Nel nome del Padre, ed il Figlio, e lo Spirito Santo, con La Bendizione della Santa Maria, Amen</u>_ (That California has success, and that Naples doesn't suffer dangerously. In the name of the Father, and the Son, and the Holy Spirit, with the blessing of the Virgin Mary. Amen.").

--*Signora.* Interception Raiders. Stefano, we owe you some _gettoni_ (coins for the pay-phone)!"

--"Donate it to your favorite charity. I'll be around. That's a big plate of pasta, *Signora*, I didn't know we had incurred some wrath, I'll be polite!"

--"Well, yes, Stefano, we know Italians in Northern California!"

--"I hope this isn't another long story, *signora*, it's not as if Antonella thought I was ugly—she asked how much I would pay to sleep with her and I didn't know Italian currency at the time!"

--"Stefano, it doesn't translate—well, I'll help you—she doesn't want to sleep with *Nando,* and you actually have manners, not for your lack of wisdom!"

--"Would you prefer we talk about Patton, or my mom and dad? They served too!"

--"We want to talk about Antonella!"

--"Then read your own Bible, o, _scugnizzo_, you broke the boxes—or did the Raiders!?"

--"Do you want Papa`'s opinion?"

--"Did he need help in the shower or what, I'm clean as a whistle!"

--"Eat, Stefano, eat! Here's to Al Davis!"

--*"Chin-Chin*, everyone!"

--"Yea, Mamma, was that 'The Classic Hungarian Period,' or the classic Hungarian period?"

--"It was you, *scemo*, she doesn't want to sleep with thou!"

--"Don't claw each other over the *verdure* (the greens), folks, Jesus will take the prayers, but this is not the mouth of the wolf, this is the Pope's town! Are you here to eat, pray, and love, or to launder grades? It takes more humility than street-smarts, though now that you're out there, don't speed on the stop sign! Teaching you to drive an American street requires an American, not a mail-carrier! Don't act like the whole family can take a vacation on your vespa!"

--"It's half-full now. And there's another game. San Diego. Are you watching? Or do you need another movie?"

--"Movie why not?"

--"Antonella want to go?"

--"Everyone pay their own sin tax at the box office and you'll need a coat for when the movie's out!"

--"What happened to soccer?"

--"It's an away game, *fratellino*, c'mon, you need a movie!"

--"Antonella, I'm sorry if I didn't understand! But no, I don't miss America!"

--"Stefano, don't worry. I'm sure there's something new showing besides *Il Tempo delle Melle* with Sophie Marceau, or David Bowie, or The Godfather."

--"Yes, *signorino,* other than Benigni, new Italian actors actually aren't just here for a quick steal!"

--"And, yes, Troissa is deceased—you are not!"

--"Maybe Carla Bruni's documentary is showing—we didn't get it in America except in the inner city!"

--"That's a great idea, I hope she keeps going."

--"And learns her place!"

--"What's that supposed to mean, Stefano?"

--"That the meek inherit the Earth, according to Scripture. So, are we going to walk?"

--"It's our walk, that's for sure."

--"Stefano, she was wronged."

--"Then <u>listen</u> to what this is about, because it's not Catholic t.v. but Antonella likes it, and I'm with her. Get you coats, I'll take the dishes to the kitchen. Stay out of the *cabinetto*(bathroom), I need to dump some water."

--"Thank you, Stefano, I'll take the help this time, just don't look around the kitchen for every spice we need for *la Pascua*(Easter)."

--"I'll walk with Stefano, since they get confused, after all is said and done, as to whether he's gay or straight."

--"We didn't start anything, Mamma!"

--"Well, mother, teach him not to worry me, we don't have accords, we have Catholic tradition."

--"They're okay, they're all okay."

--"I'm going to stay here with Mamma, you kids have a good time. Stefano, can you get back to your home from the theatre?"

--"Si, Signora!"

--"Then do so. Make sure, *ragazzi*, you don't split up until you're at the stadium. Otherwise you don't go to school, you stay at home."

--"No, we don't have problems."

--"Oh, and Stefano, only on your birthday do you offer—otherwise, we have our own wallets!"

--"Si`, Maestro. I'm sorry for the *faut pas*."

--"Now do you feel better, Antonella?"

--"Yes, I'm ready to go out!"

--"Well, it's a lot of new traffic, but film is there for your enjoyment. Learn something and don't drag your tails."

--"Ciao, Mamma, is Stefano ready?"

--"I'm through! Let's walk it one more time!"

* * *

--"Yes, the Black Club Scene: No Filter. The only one you really knew was the one who wanted to do you in. The Jewish youth: Money and resumes, No Filter. Then the Classics: The choral groups of San Bernadino; Protestant and Catholic unite. All united on the chords for "Old Man River.""

--"English Lit in the U.C. System. They actually feel your career was better spent keeping up your Spanish, as World Lit. is about diversity. No Filter, no good glasses, all hermaneutics from here on out, but no penalty kick, no goal, free kick in the shins, the opposition, and the only thing "L.A." about it is the damn cigarette."

--"This is the campus library."

--"I thought I was at the hatchery on the Colorado River, that was such a tedious explanation of why <u>now</u> it's the Campus Crusaders For Christ play, now that the tennis shoes need a trip to the Laundromat. I tutor Italian."

--"I'm studying for my Master's in Nursing."

--"That was my minor in high school."

--"And now, Association of Western Colleges and Universities, known for their affectionate student population and 5-Star staffing."

--"This is the Counseling Building for career management. See the chimes made of straws, in the shapes of pyramids and triangles all attached by one string? To remind you the California Art World, analogous with ancient Rome, was not built in a day."

--"Inside the building, we have on display the 350-page chemistry thesis for a new psychotropic medication, reworked into a huge origami stork!"

--"Oh, can top that: We've seen everything in the archives from El Cid to Almodovar's, and it's still a patented White-Boy 'B'. You'd think I rapped my way through the job at the Deli bussing tables."

--"Who was the chef for the rest of the cafeteria, in the back, Erma Bombeck?"

--"Oh Jesus, after the make-outs and wipe-outs, the flip-flop store doesn't want my business, <u>and</u>, I'm surprised I'm not walking distance to my job as a concessioner at Disneyland!"

--"Maybe the glasses are a good thing. Otherwise it's Universal Studios!"

--"Are you kidding? The coed dorms in San Francisco? In retrospect, this was like being the burglar in <u>Home Alone</u>!"

--"What about the gentleman who uses the word 'Betties' 15 times in the story on the way to class, putting them all in the category he can sing <u>*Ojos Verdes*</u> for at his Latin Lover moment?"

--"That's the script if you're an assimilated Angelino."

--"Assimilated?! Lost the pole vault to reach the bar that year in the Pac-10 Finals!"

--"That's why some old-timers jump it with a Cuban cigar, not a pole!"

--"If I crap before I quit, I promise you I'll read more Lit!"

--"Oh, communication: Blow the ram's horn, just don't blow."

--"Get a breathe team to help you quit. You're obviously not the only one anchored to go back for Degree #2 in Golden State times!"

--"Hey, what about the Community College that thought a straight or gay verdict depended on their sparkling working world EOE rank?"

--"Is that what the blonde psychology professor had to say?! Wow, I know, the ringer goes from class on Maslov's hierarchy of needs to Solid Geometry, acing both, but he's got a dragon's tail that comes out in the parking lot!"

--"Was the spelling bee winner 'hieroglyph'?"

--"As atheists go, I've seen semester-long protests about parking citations, but that's county history now!"

--"How 'bout the fam's post-collegiate Academy Award nominee role? They open the window to the apartment, inserted suggestions for the new personality, and continued with their traditions in marriage."

--"But you don't have one husband named Peter amongst the whole congregation of 'em!"

--"The men of the fam? You ever seen a Texan stand by while the cock crowed three times before the Cowboys' semifinal playoff game?"

--"Aikmann. Aikmann's the problem. He thought this was as simple as a ring on the girl's finger."

--"Don't know how to break it to him that not a single member of his Bruins teams got an 'A' in Italian 6!"

--"Must've been some schedules centering around not bobbling the cafeteria trays!"

--"And pledge season. Animal rights activists outside the sororities."

--"Parking permits 10, students on foot from the dorms 1."

--"Too bad you don't have a commercial license to drive an RTD at this rate."

--"I see you know what the wind is sayin' with the ringin' of the chimes outside the Counseling Center."

--"Yea. Mind if she nibbles on a granola bar while talkin' to you about what comes next after Italian?"

--"Italian Trade Commission is five miles down the street from campus. That means new neighborhood, watch yourself, can't help you there, and remember your doctor's appointment: It's a big world—he can help you."

--"C'mon now, sports fan, 'pussy-shit' is terrible collegiate English!"

--"Sounds like <u>you</u> threw his cross-town rivalry verse Rodney Peete!"

--"I wasn't at the game! Hungry campus to feed at the Law library grounds and that was the deli schedule, before the film school rolled in with a documentary on Bangladesh!"

--"You make these Bruins look worse than Harvard Boxing!"

--"Don't like the sounds of that. They've got Benetton clothing outlets in Westwood too! Equality!"

--"Benetton?"

--"You know, orange, red, blue, green, yellow, and no black."

--"I thought you were describing a Bruin fraternity!"

--"You wanna see the lease on this off-campus apartment one more time?"

--"Didn't realize a roommate from the Austrian Consulate in Vienna was gonna co-sign."

--"You had your chance."

--"Now that the Italians, ten years later, learn to translate 'Boys Night Out' for the <u>other</u> guests at the building."

--"I heard that was <u>you</u>."

--"They act like I'm Webster's great-grandson when they need to pass a UCLA extension course."

--"All this, plus the <u>one</u> Jew on campus with a pot habit?"

--"Those doctors' night-time reading assignments of patients' profiles must be done in Jack Nicholson's suburb!"

--"Can't do shit right."

--"That's how the Left in L.A. became so popular. It's only human."

--"Who's on the secretary's desk at the Italian Department, that short lady with the stiletto in her boot from the James Bond movie?"

--"An Eddie Murphy wanna-be who speaks only English, doesn't type, but has phone time with a good studio producer."

--"Asshole Factory?"

--"No, the editorial staff of the Daily Bruin?"

--"You ever see a cop in that side of town?"

--"Only once: I thought I saw Catwoman at night on Westwood Blvd."

--"Get me a $1.50 slice of that café` pizza, no beer, I'm on foot."

--"Need new shoes?"

--"All this. After a Northern Californian school?"

--"Until the next nightly news!"

--"How many Los Angelinos walkin' this one mile of the field from the psychologist's office?"

--"Three quarters of the number of immigrants at McDonald's in Downtown Santa Monica before work in the morning."

--"So, different crowd dynamics."

--"The Bruin clothing store is a choice internship in corporate public relations."

--"And so Troy is drilling up the coast."

--"For cavities in the molars."

--"Makes Dr. Phil's clientele look like the Best of the Best."

--"And Judge Judy's neighborhood—good gas station, but stressful on the traffic's driving skills."

--"Try the low-budget hotels. It's only a little pot."

--"I knew you expected me to run that by the doctor one more time on Tuesday morn, 11:00 A.M."

--"So that's how late-rising patterns got started—it wasn't the family after all."

--"Texan for <u>breakfast cook</u>: Army corporal."

--"Westwood for <u>breakfast cook</u>: Sephardic Jew."

--"Peete clean his clock that bad?"

--"No. Same week the Wrangler wristband snapped on the time piece."

--"And is not an Italian job?"

--"110% positive. There is <u>no</u> Italian job in Los Angeles! Even with a degree!"

--"Get the feelin' the coaches' wives also enjoy the village's live theatre productions?"

--"Is that diversity, or Jerry Brown's debatable budget proposal hovering overhead?!"

--"<u>Was</u> Harvard Boxing. They can't believe the architecture on those Westwood skyrises! Beautiful work, if not East Coast proper ameliorations—just gorgeous!"

--"Urban renewal."

--"Yes, got to finish this extra reading but need two more weeks!"

--"I'm going to Google Charmin company for a 'Contact Us' address, since déjà`vu, speak of college, and we run out before Monday!"

--"Bulk mailing of the goods?"

--"Fed Ex or UPS? I'll do the research free for an internship."

--"Aren't you trying to cut in to the ACLU's campus contract? Study business law."

--"My B.M. Business Management."

--"Peurile!"

--"Only a Greek with a maid and cooking staff likes Mom and Dad's neighborhood!"

--"Hike! Google Charmin! We'll network to the Mexico City corporation's Public Relations Dept.!"

--"Quick thinking!"

--"What about Milan's? Since you're in the ballpark?"

--"Last I heard at the Dept. from the chair, they're looking for an Italian-born research scientist to discover a high-tech serrated edge between each sheet, now that the perfectionist feels it's a broader-reaching environmental topic of discussion. And they already have the girls planning the conference, the hotel arrangements, catering, and transportation to and from LAX to hear Luigi speak in Winter Quarter, the same day, why, as the Sock-Hop after the NCAA's!"

--"Yes, American Literature, dispelling popular Italian authorial myth, no longer falls apart after one sheet is torn from the book, nor when a whole chapter is used to wipe European asses!"

--"Don't drop it in the ground hole in the shower room's crapper section, where it goes if the conference facilities are medieval in tale!"

--"Too much information."

--"So was Cajuonga's Plumbing fees for a simple studio apartment water line issue."

--"Is that why the guy hired by the state in the orange vest got his kid into USC?"

--"Good economy, several guns taking shots at it, but hey, they're not the Texas Rangers!"

--"Yes, don't mind the fifth-generation Californian of Hispanic descent—they're at the private school altar, where the fact they don't speak Spanish is not the subject!"

--"Oh, say hello to Virgin Records' Christian affiliates at Kereoke belting out the theme song to this land-right's treaty!"

--"Hell, the chairs of the Medical School's Departments are out of conference—they want a raise, and it was all simply the beer! Where's that Italian speaker you used to know? Makes a good case study—has the manners for the girls in research with their Psychology M.A.'s!"

--"Could sign him for two more years, yea, cosign this decision, but don't give him enough pocket change for his own marinated jalapenos!"

--"Picky, picky, picky!"

--"Sacramento has to start somewhere."

--"How 'bout we don't get them thinking this is a bank-teller's, girls'-night-out gossip coup, and cut him a dozen tortillas, ground beef, and a jar of La Juanita?"

--"That's White!"

--"Actually, descended from Ripley's family!"

--"Was that the money for the float?"

--"Built in San Fernando Valley proper."

--"Don't get your Camel Joe towel stolen when you're in the water!"

--"I love L.A.!"

--"But is this really L.A.? OR simply a Stanford interpretation of such, belatedly?"

* * *

--"And now, in an act of true obedience to the High Church, the priesthood will champion the Bishop in the bobbing-for-the-apples, to hold the root of sin in their mouths and spit the seeds into the Protestant wind, symbolizing the historic act of them, belatedly, after 200 years of California discovery, shouldering their ecumenical brothers' burdens elsewhere than Bank of America's front door, 8:00 A.M. sharp Monday."

--"Such imagery. Like the flan, uneaten, becoming biologically alive, expanding, and nurturing itself off the calories of the workers in the protected natural setting."

--"At least Henry the Eighth groomed."

--"Oh, I would use the urinal only twice a day, on the water topic, but apparently it's my coffee habit now and I owe both Italy and Mexico back-taxes for their services serving such in _Mitzvah_ fashion to Americans."

--"That's why I drink the wine at noon, so the coffee at night does not have loose strings attached to it from daytime T.V.!"

--"Hey, what country made the catheroid tube in your large screen? I can Google them."

--"How 'bout I just clean the screen with Windex and a paper towel?"

--"Your role in the movement to nowhere."

--"Or, as we more affectionately know it, "That Happy Place!""

--"To think this started out with a study of the Spanish language, not Italian, those hypocrites!"

--"Yes, Spanish, not California History, Organized Crime, Mexican Pueblos, Texas politics, and dietary advice for those of Latino persuasion shopping in California stores themselves."

--"Oh."

--"Yes, bud, give me the _paella,_ give me the _paella!_"

283

--"Then don't kiss me!"

--"Is that what gave me schizophrenia, Loco Coco?"

--"Intellectual Snob!"

--"The curtain or the box, Chicano, come on down, I'm Monty Haul!"

--"So, the Espa~nol in you comes out. And I thought you were gay."

--"It's your campus' student selection committee, personally, I hate taking a dive even though Apostle Paul says circumcision is not the subject. I mean really, next time I'll start a small business bed-sheet washing to pay for this 'Masters', since this cloud has been pierced by the ozone known as your Tiparillo's power!"

--"It's a dirty game."

--"Run all the bulls at once. And away from the football stadium."

--"Are you criticizing my protest training?"

--"No, your wife, my cousin, what's the difference? My cousin drives the speed limit without Rat Patrol of The Greek System following her to the Medical Marijuana dispensary, and she raises children, not rewards them for prolonging high school. Your wife? I'll get my paint brush if she can remember how to sit still in a nice dress like my cousin. We have a *dicotomi`a*! Your staff, the staff at the tobacco store for the Colonels in my Mom's offices."

--"Why, I..."

--"Never mind, I'll wipe it for ya', if it was bad *pulque* to go with the rural menudo."

--"So did you get the joke, you Mormon-Mexican half-breed? You have a Californian degree in Italian, not an Italian degree in California, and your *cinturo`n*(belt) sales seem to be the main topic on Girls' Night Out, since with your cooking we constantly need a new one. 'Stay away from pasta.' Eat the pasta, and pass your gas on a long walk, not in the counselor's office when you show them your 5-generation family crest. Is that family crest, or Family Crest, a tube big enough for all of La Familia? Enough of white and brown, Laurel and Hardy—there, there is the Panama Jack outfit at the Outlet and those are the trunks they're selling. Stop swimming backstroke in my coffee pot, amigo! Expand the Chicano paradigm? Have I got a sale on fresh Jalapen~os for you, $3.00 a bag, and keep the whole cult of San Juan de Aguas Altas in the bag! Like you to drop by the bank other than when your second-cousin the teller wants to break for lunch at EL Torito!

--What do you know about L.A.?"

--"That their version of European influences is a 14-carat, 5-finger gag!"

--"Okay, that ought to be your black number on the roulette wheel!"

--"Where your balls, from Lieutenant Governor on down?! Is it laundry day? Get some red dye #5 from the tape staining your golf chorties! Captain Alimentos?! (Captain Nourishment)"

--"You think it was cold for Sanchez in New York last playoffs? That he missed the warmth of Michael Jackson's ranch?"

--"The Parish is a Wienie,
   It's Sad but True...
   But the colors of the Parish
   Are Red, White, and Blue!"

--"Jesus, Coattails Jack and Other Villa Men. There's the bar of soap, thesis lovers, don't all get in the shower at once. Or was that the Bronco's 2nd string calling the jewelry store by the studio a couple meters below the Hollywood sign, doing a heat dance since they can't play in the rain? Nice going again, jack-ass, a perfect beginning to Fall in California. Next time put your shaving cream jar and brush on display in the County History Museum! I've heard of Yukon Jack arriving from the snow-covered gold mines with the seed and White Fang pups, but don't look now, it's the leader of the writers' group at Denny's waiting out front the restaurant at 5 A.M., twelve hours before the rest convene with him."

--"No, buddy, it's the last Parish you were a member of. Their snitch had an argument with his new date. Things like this are why a hole is sunk in the City Treasury, you have to remember, he's a living legend on the subject of time-clock repeats at the age of 25. He gets overtime, but doesn't punch out until today's 12-pak is through your gut and in the commode--all twelve. Anxiety Magazine's Cover Boy for December."

--"Her only line in this 24-hour, apartment scene Film Festival is, as the sun goes up, "I'm going home now."

--"Is this a prototypical apartment scene, or a Starbucks manager special?"

--"The worst student of O.U.'s Freshman class finds the very same drive enrichening. You talk about a state with cash: must've been ranked as Sarah Palin's potential hidden electoral punch!"

--"Do we have an Oval Office, or is the last campaign manager's sister still in labor?"

--"Time to leave the park now, protestor, don't forget to pack out your beer cans and say hello to the new Euro-hostel girls!"

--"All this, just for a large can of beer."

--"You miss home?"

--"What happened to the privately-owned American liquor store?"

--"No turnin' back to Iraq. Shop where you can."

--"That cashier makes $350 a shift, will not move out, the difference between her and the manager is he's got a bed on the ground floor of a building in this town..."

--"And the bagger's Dad's have better sports cars than British motors. Pick your poison."

--"The ink on the bill will get me if the beer doesn't."

--"It's not that they don't know beer, but that's the price tag on the block of cheese, they're trying to move that too!"

--"I got it. The kid's got a clip-on I.D. for the Medical Center and suddenly all apartments' the stages. Checks his mail at 5:00 A.M. after coming home with the bed-pan's contents on the soles of his Nunn-Bushes."

--"And her curtain tells all at the front of the home when it's time for a beer run."

--"Santa's Village finest date they're not!"

--"On Butt-munch, on Mobile Home Manager, to the Community Center's parking lot cross-town we go!"

--"Did you see the University's off-campus student pull-up? He's got a 'Vote Yes on Prop 8' sticker on the air-bag compartment of his compact, and a 'Protected by Jerry Brown' decal with a sex pistol drawing on his bumper!"

--"I think the maintenance man paid his own next with fake I.D. sales."

--"What was that? Was that a football player in a wheelbarrow?"

--"It's a ghost. That's how they got injured players off the field when the campus first suited up a team."

--"No stretcher?"

--"No. Only enough for the Red Cross tent, before the Engineering Fraternity had actually completed construction on the original campus."

--"Where's it going?"

--"The Charge Nurse's roommate is shattered."

--"Strong Democrats on campus at the paper this year, I hear."

--"Try the silence, if you want to hear the truth. The only flag going up in the nearby vicinity has Ronald McDonald on it."

--"What's on the other channels?"

--"You Republicans have to make your own decisions. Me, I'm Chicano. Our own nation in-between."

--"So that's the size of the salami that got passed! You've been saying the same thing since we graduated from campus in the 90's! I can't believe you've held the Sacred Crap in that long that calls The Wild Buffalo!"

--"What's in that box?"

--"Paperwork. Look at the magic marker words."

--"CASE STUDIES ON SKELETONS, DONE BY CHAIR OF THE FACULTY OF PSYCHIATRY, '32."

--"THAT'S ALL? YOU COULD FIT A VICTROLA INTO THIS BOX!"

--"You'd think it held Jerry Garcia's Will and Testaments."

--"So, we can get rid of it? For more space?"

--"No, Rabbi, No. No dowry before it's time in this department."

--"Hey, look at the newspaper on these office lamps: FDR establishes Social Security System as Eleanor Returns Safely From Great Wall."

--"We're getting a raise next year I hear. It was a manners contest with Unemployment Office Employees."

--"Oh, keep smiling—know it's like a second major."

--"Cheech Marin, for your information, ye of White Pride, is not a day over 45, according to his birth certificate records in Fresno. He is your peer!"

--"Oh, a real Singing Cowboy. All that dressing on his pants, no ranch! The instigator of Catholic legal battles! Can he partake in Holy Communion for 45 minutes without farting?"

--"You do not know God's contract, His own commandments, with The People! We are absolved!"

--"Time to change your tire with a real Mexican Hat Dance and a tamalada!"

--"Democracy? You said it, Crusty!"

--"I do not have time for Schwarzenneggar tactics at The Community Center! We have President Obama's brethren by our side, making the karateka moves of The Swan Kata, together!"

--"Anymore piss in the wind and the North One will blow chunks!"

--"Take us to Del Taco. You'll understand all!"

--"The Party Dime, fine, The Party Dime-baggy, no!"

--"And George Lopez! Goin' on his 15th anniversary of marriage at 35 years, what do you say to that White Knight?"

--"Chicanasaurus Rex is showing at the L.A. Zoo, after her parade on Olivera! Tu Mama!"

--"That show's in Texas, Ranger-lover!"

--"You guys don't see in sun-light too well."

--"You mean with blue eyes? And glasses? Don't see well period. But the problem's at least not the cookin'."

--"I think it's more important to have an occupation."

--"Oh, yea, but look, the roles have been reversed on that in the Mexican-American world for quite some time, especially for us with the specs. Ugly Judy went pretty far, but me, why they think I'm behind the times."

--"Get your name in lights. Do Ugly Esteban."

--"Seriously now, I think you need to lose the goatee` and the _bigote_ (mustache), El Diablo!"

--"You betcha, pinche-picky-pouty-outy! Should see how crooked my aunt's finger is now at 90, as it's pointing right at your Budweiser allowance! Drive safe, you low-land heart-throb! Before we throw Eric Estrada back to you in California real estate for another 'slide show!"

--"CH-I-I-I-T. Battle of San Jacinto, my ass!"

--"You and Mexico pushing, you vegetarian!"

--"Did you see how many entrepreneurs wanted one-bedroom apartments after getting their B.A. at U.C.L.A.? What did Real Estate jack-up the prices again for, were they hoping the guys would marry each-other?!"

--"I wanted to work on an advanced degree in Beer Run Logistics too, but only got a half-tuition offer."

--"What's in between New Orleans and where they threw The Boston Tea Party?"

--"Betsy Ross' parents' home."

--"Now after Washington threw the silver dollar over the Potomac, what did Lafayette do?"

--"He retreated to where they would eventually build Fort Knox."

--"The other men followed the star in the sky that night to defend the Ross area, but there was no fight."

--"The winds came in. There was no fight in upstate New York, where Fort Knox was being planned, either."

--"What did General La Fayette do?"

--"He established terms on the shores of the St. Laurence River between the Canadians, the Americans, and the French who were painters."

--"What did Washington do?"

--"He sent a carrier pigeon telling La Fayette that Betsy was available and La Fayette hiked all the way of the Mississippi to her home. He was not married nor divorced at the time. Franklin presided upon return from France in their ceremonial friendship, before the conclusion of the war and La Fayette's return to the native French parts. Shortly after, the colors of the American flag were decided upon by Betsy and sown. The entire time, Martha protected the kitchen in the Washingtonian troops' parts, despite the freezing cold."

--"Bob, we missed you on All Souls' Day!"

--"Thank you, Stephen. Do you know anyone who works for the U.S.O.? We're doing a benefit in the sky."

--"My mother used to work for the U.S.O.! She gave me a pin to wear on my coats."

--"Coats? Are you a Hungarian cross-dresser or an animal changing seasons?"

--"I'm straight, Bob, but I keep it low."

--"You keep it low. Why Stephen?"

--"Because I speak fluent Italian. It's that time of year. The girls'll pass me around like Halloween candy!"

--"Oh. I didn't mean to get nosey about what was in the bag. Those tricks. Don't worry. That's the only Italian Dean Martin told me: 'Assoluto!'"

--"You can't teach that to a Texan, but I know, it means 'Absolved!'"

--"Stephen, please retire from tutoring Italian. Be a library assistant. We don't need all hell breaking out!"

--"That's an offer I can't refuse."

--"Then I own you a favor. What'll it be?"

--"I'll hold on to it. That's what the Major does."

--"I know the routine."

--"Who was that you were talking to in Spanish?"

--"Oh, that's my cousin."

--"He's Mexican."

--"Yes. We. We are Mexican, even though I look White."

--"What were you two saying?"

--"Don't ask. Family stuff."

--"Do you think he would date me?"

--"Date? I'm not sure you understand, if you allow me to make the translation. He's five potatoes short of a 10 lb. bag."

--"Thick?"

--"100% pork chop. But good defense on the line."

--"Right tackle, right?"

--"No. U.S. Marine Corps. Eats Air Force brats for lunch."

--"I'm surprised he doesn't eat you. I guess he doesn't eat his own."

--"Oh, it's not that, he would, in theory. We don't have a personality conflict, and don't ask why."

--"No, tell me."

--"Well, we were crawling through the backyard lookin' for snakes as kids, and he got his pant leg stuck on barb wire. He was stuck there, and I told him to trust me, I would return shortly. I brought back our Uncle, who got him out of the barb wire, and then we flipped over a wooden board in the middle of no-where, tree-fort size wall of a board right on the other side of the barb wire, and before it could move, our Uncle killed the biggest black snake you ever saw. He never got caught in barbed wire again!"

--"They're Czech, Stephen, they're Czech: They bypassed guard Nando's yield sign in downtown Naples and stormed right into the N.A.T.O. base to announce their support for America! They're Czech. Nando was served for breakfast as a giant pancake for the Officers to divide amongst themselves!"

--"Where did the air go? I took one drag off that cigar!"

--"Did we mention the support of British muster on the ground?"

--"Believe me, the Brothers Rimm…and it's not the basketball hoop, it's your beer-mug."

--"Next…Attack of the Slippers Gang. 'He drove home at 5:30 A.M. One car turned the corner before him. It was the L.A. Times Carrier. Suddenly he had to swerve to avoid the shots. As he ducked below the dashboard, the last thing he saw were slippers."

--"Barry? Didn't you used to know a Jew with that name?"

--"And now they're doing radio commercials? During wartime? Was it that great the first time we got one of these token Barrys off the ground? Too many Jews on the field."

--"No, the opposite, son—there's not too many of'em, and they're getting up there in age, even your generation."

--"Sorry I commented. I guess you do have a blind side on this if all uncles came home to raise families after World War II. Barry, sure, good vehicle for the Vets."

--"That's if you can get him to cut his hair."

--"No, really, Dad, let the civilian scholars keep their strength!"

--"But what about the hair-styling industry? Whose going to feed the legacy's children?"

--"Dad, the only thing 'had' in that legacy is my rep after that last ethno-torture. Another bilingual woman, more slander than a Congressional vote. They make more off of a Barry bet once every ten months than they do off my keeping them punching a quarterly time-clock. It's not about money. You wouldn't understand—Mom buys your shampoo!"

--"Ahhh! Listen to this! Denial!"

--"Try the Green Dye #3 and the Yellow #5 in the Pert and No Tears, mix'em together and you've got hair that stays on your head dry-dead in it's tracks until your next shampoo at a saloon."

--"Rinse it after you swim, idiot."

--"Swim? Your water? That pool? Dad, you're from the Atlantic. One swim there and your scalp is cold-treated for 50 years. This is California. For water that clean you're sending me into Ponce De Leon's battles."

--"Oh, C'mon, Bill the pool cleaner knows what he's doing."

--"Of course he does, he grew up in Barry's neighborhood! Chart the wind better next time you argue while quitting the Olde English pipe, I not only can't quit cigars but Mom's got gas. You exploded loud enough to be heard by Barry's Mom. She told her husband."

--"So? What do I care?"

--"He's a capitalist like all the rest, Dad, it's no big deal, but he runs a psychologist's practice!"

--"Oh, cut the shit! C'mon!"

--"No, no joke. Dad, put two and two together. In every other backyard in this town, the mice throw a toga. In your pool, it's like they're Twilight Zone. They don't escape the Love Seat or the steps before they're floating back-up. You're taboo!"

--"Read the paper."

--"Los Angeles Times. For subscriptions in Tagalog, dial 1-800-FIL-PEEN."

--"Go to bed."

--"Can't, Dad, I'm scared. Haven't been able to get the vampire out of there since the last Dungeons and Dragons game, and that was 30 years ago. I actually haven't played since."

--"What have you been smoking?"

--"That's twelve years ago, Dad, this has been a brief war going back and forth. Dad, I don't know what your gardener smoked when he was my age, but I can't believe he's making comparisons and then witnessing God to me. I worked that crap off in one week. They've got Peckerwoods south of the border too Dad, so before he makes Barry's dad look like a hero, could you let me know how Cousin Billy on the Hungarian side is doing on the East Coast? Last legitimate beer served me in this country."

--"He died. I didn't tell you? Passed away a while back."

--"Never married."

--"Nah, he had children. You've just never met them."

--"Thank God. There is a Prince Stephen of The Olde Country."

--"Don't let too much California stuff rattle your cage, and put some more applications in. And, it's not your mother—they've had the same damn Spanish family cooking on The White House staff since George Washington. No relation to your Texas grandma!"

--"If that's not the truth."

--"STEPHEN…I had a favor I wanted to ask…"

-"Yes?"

--"Before you go back to the apartment, think you could help me transplant some of those succulents?"

--"Just tell me where to plant them."

--"Okay, don't stay up too late! We'll do it first thing in the morning. I've got some errands to run with your father at eleven."

--"Can I squeeze some beer money from the neighbor's tree?"

--"Listen, stop picking their lemons. We don't know them. They're new. And didn't you bring your laundry here last month?"

--"You want more?"

--"More new neighbors or more lemons?"

--"More succulents. I think there's a hybrid strand in the apartment garden!"

--"Listen, you watch your savings because it's the holidays. We've got birthdays beside yours."

--"Hence the other candles for the ribbon-cutting of the new synagogue!"

--"Maybe we'll get Barry's hair-stylist and the vampire in one big real estate shot for opening up some land in this district."

--"And whose your gardener, Steve Sr., Johnny Appleseed of the Californios?! I've got words for him."

--"Stephen, get to bed."

--"No, I see no reason he can't hear the truth!"

--"Think I'll see if reading The Times helps me get to sleep at a Godly hour!"

--"Good night!"

--"Which devalued more, the Social Security check money or the Peso, and are they connected by critical economic theory?"

--"I'm sorry, Mom, what was that? I'm catching a tape with Kathy Baker's finer episodes of Picket Fences Dad had buried in the drawer upstairs!"

--"Good episode?"

--"Yes, the down-syndrome kid gets a date."

--"I'll have to send it to your cousin. He went to the Prom this year, you know."

--"Is he still retired?"

--"Don't get any ideas! It's not cheap!"

--"I know, Mom, I've been shopping for my own beer for twenty-five years. I'm surprised a scratch-off number didn't come off in the beer cap for stock shares in a contest of the last microbrewery six-pack I bought. Would be poetic justice if this still was a red, white, and blue flag we know!"

--"Well, let's not make it a long story every time."

--"Fine. Next time we'll put our foot down about which side of the neighbor's lawn sees the rough side of the plank on the fence."

--"We did. It's the planks. Told you not to go to that store, that he's worse than a mechanic."

--"We got a great price!"

--"He's a Mason. He won't get off our ass until he's got my resume on file for shit's sake."

--"Be careful driving home. The Fuzz is out! Here's your transplanting money."

--"Under $65.00. Nothing to arouse Social Security about."

--"They've got sex lives too."

--"In English, Mom, not to be narcissistic, but we say, 'They've got two sex lives.' When they're done screwin' me, I'll be able to keep it to two beers, okay?"

--"They're having problems splitting that hair in your soup."

--"Hair in my soup?"

--"Yes, used to denote the fine line."

--"Oh. Yes. Well, let me get it out if you don't mind me using my fingers. There is a difference between abnormal psychology and its neuroses leading to strange illegal behaviors and the uncomfortability and/or weirdness associated with the phenomena leading to the treatable moods of the bipolar and /or schizophrenic."

--"So that half of hair there, that half on the napkin."

--"The hair split length-wise, still only to denote a certain two ends, symbolism on which we need not extrapolate on until a later moment."

--"And the phenomena in the meantime?"

--"There's dishes, the floor needs to be mopped, the carpet shampooed, the walls painted, the tub scrubbed, the windows washed, and last but not least, the laundry."

--"Is there more soup?"

--"Sure."

--"Open a window. Air it out. Use the air-spray again."

--"That's not going to solve this phenomena problem."

--"Sure it is. It's an air-born allergy."

--"But as soon as you open the window, something wild happens in the neighborhood."

--"The kid saw a raccoon, he's not on drugs. Never mind on the soup: pour me another cup of coffee."

--"It's 10:30 P.M.!"

--"It's not a full moon, don't worry."

--"Man, that's some strong stuff!"

--"What?"

--"You didn't see it passin'by? The secular world, wow, that's some strong stuff!"

--"Yea, all that up your nose, as if you're pickin' it, and breath again, and wow, that's some strong stuff—what the secular world did and does with their lives!"

--"Who wouldn't smoke a cigarette just to change the subject?"

--"No, I don't use the name of 'Esteban' on the job applications as my other name, though yes, I am familiar with the translation, can peel through the phone book adeptly, and both without referring to relatives or references."

--"Now is it A) You're popular.

                B) You've had college experience.

                C) The Latino Movement has grown."

--"No the cat is in the bag as far as where the nibbles are...on the line I've cast to reel in a job. Not, 'the cat is in the bag as far as where Nibbles the sorority girl is hiding herself until game-time."

--"Hey, Herman, are those bagels? Jelly bagels?"

--"No. You want one?"

--"Oh, well, what kind of luck does it get me?"

--"Don't rely on luck to sell my doughnuts. I'm Jewish."

--"Oh. Herman, you prefer _hagadah_ or _agadah_?"

--"Don't have a preference. What do you prefer?"

--"Oh, _agadah_, as subjects go."

--"Well it's a jelly doughnut all season. Why so much of the other?"

--"Must be that the New Testament is good, rabbinical protocol...you my neighbor now, Herman?"

--"Ah, hell, been here for years! Anything but more of that Air Force Brat town shit!"

--"Uh, now, Herman, don't want to carry your baggage if you can't take an order across branches when you serve, and thank you for serving, Herman, looks like you're the new jelly doughnut in town."

--"When you can afford the hole, let me know."

--"Herman, I don't buy doughnut holes from guys who eat the best ones off the pan when they come out of the oven, but congratulations on the wife with an A.A. degree, the newborn, and the home!"

--"Got the seventy-five cents?"

--"Yea, Herman, I'll help you with the pay toilet at the Laundromat. I don't have one at the apartment, I understand."

--"Beats a motel."

--"Oh, Herman, never would have found out about your doughnut business if I didn't open the Gideon's Bible there and find your old recruiter's number in there. How long have you been out?"

--"Since my last tour in Afghanistan."

--"No, Herman, out of the hotel—This Gideon's got numbers in it after yours, and they don't speak English!"

--"I remember that Japanese liquor store! Here, have a free hole!"

--"Happy Pearl Harbor Day, Herman! Thanks."

--"How many Italians does it take to row inside the Blue Grotto?"

--"Rough waters?"

--"Any water you got."

--"Go find out."

--"How long does the boat-ride last when you're in there?"

--"How much time you got for me to explain?"

--"Do all the same people row out under the Italian flag?"

--"Which one, the one on the beach or the one under the water in The Grotto?"

--"How deep does The Grotto go?"

--"If it was deep to begin with, they wouldn't need a cave and a boat to explain it to foreigners! Hey, *parli Italiano*?"

--"I'm leaving."

--"Really? The Italians finished that level of schooling and are reading Isaiah!"

"Why Isaiah?"

--"Because he never ends, and we need to end this conversation until you salute the American flag with your thumb under the other finger!"

--"The turkey could be juicier."

--"The Italians have the ball."

--"Take the paperwork on Metrolink 487 at 9:00 A.M. to any office of the Party near the bagel bakery by one week from today, and skip the Torah study."

--"Do I have to skip breakfast anymore?"

--"No. Welcome to America. But you're buyin' the first time, buck-o!"

* * *

--"The Diary of the Long-Hair's Stylist."

--"You touch it, you buy it, in this Barnes and Noble."

* * *

-"You know I've got to just love Southern families: They think 'loophole' and 'asshole' are synonymous, and they love to explore!"

Family Dinner Prayer

Dear Lord Father In Heaven, Bless this food that it may nourish our bodies to do your good works daily, and bless us that if things turn against us, we do not turn on our own.

--"How did the President figure out the Native American thing?"

--"He drew straight lines. Serrano, Mohican, Mohawk, Mo-Honkin'!"

In this scene of Holy War, Knight Stephen stabs open the belly of the over-flying Quetzacoatl with his sword when the serpent god swoops down, after which he rises back into the air, swooping madly while gushing blood from his abdomen wound. The Californios now turn to Stephen for protection in prayer realms, although the Cortesian lines maintain land and water rights in the Isabellian *Descendencia*. Meanwhile, a son is born to George Lopez, <u>Judas Francisco Lopez</u>.

--Meanwhile, in the present, retired law enforcement rescues an Asian prodigy with sickle-cell anemia from an overflowing theatre bathroom.

--"Oh, he doesn't sleep. He likes money."

--"He used to like my friend in school."

--"Bill, what does that have to do with it?"

--"She married Scott Daniels, that's what!"

--"Bill, we serve retired firemen in here. I can hear their sons now: 'Scott Daniels? Who gives a fu-u-u-u-ck! Oh, man is <u>THAT</u> all THIS is ABOUT? You mean that dude with the glued-on brown hair that used to hang with J.V. soccer? OH-H-H-H, F-U-U-C-K! I can't believe that's what this is about! Who gives a damn, man, we're all hungry!" And the other one says, "Really? Who gives a rats ass?!"

--"Oh, hey, I'm not lower-level management because I was born yesterday!"

--"Uh, bub, for future reference, I'm going to need a pin-up of Monica Lewinsky the next time you want my chefs to rebound off <u>that</u> while the troops are gone. That's <u>cotto</u>, cotto salam' Bill. Not coagulus rex salam. Make sure you scrape the bottom of the pans better next time you burn the extra-large pizza!"

--"I figured it out: The Messianic Jews are the unorthodox Jews."

--"We call them 'Messianic' though, because the other Jews will just say, 'Get laid!'"

--"And there's more to being unorthodox than just getting laid."

--"Think about it!"

--"HA-HA-HA!"

--"If you're staying young, you should be more honest about your feelings for sports nuts women."

--"End days...end days, dudes! False prophets will be seen along the trail of Christ's love!"

--"That Parish has all the studs."

--Studs? You mean nails?"

--"Is that a roundabout way to have me ameliorate my vocabulary?"

--"Yes!"

--"Geez, I had enunciated more vocabulary indicative of High English at one time, though was sent from 6'2" to 6'1 1/2" when I tried to run my first book by the publishers of the 1990's!"

--"That's, still, quite tall a story, despite the atrophy! Going steady! How much extra weight is behind the new <u>louvre?</u> Since then?"

--"Oh, it depends if you weigh it on an old-fashioned scale or a 'modern,' sit-down electric one."

--"Let me guess. The modern version has you at a much lower weight."

--"Yes! How did you know?"

--"That generation tried to pick my pocket too!"

--"I was walking in a nice part of L.A. and someone drove by to jeer, 'Who do you think you are, Brad Pitt?!' Mel Gibson happened to be walking by and told me, 'Are you getting the treatment too?!' So I replied, 'Yes, I believe every first-born male from the state of Louisiana has now!'"

--"Who are you rooting for, as a Californian, Republican and a writer, San Francisco or New York?"

--"I have old, old friends from my college days that are originally from The Bay."

--"Don't you think it's time you moved on from college?"

--"You've been selling wine for some time. I'm just now, in my 40's, getting contacted by college Alumni Associations. I'm part of the middle class that is now classified as the cheese going down with your burger!"

--"Is that a sexist innuendo?"

--"I see you favor the Napa selection too!"

--"Bub, Aldo. Hal-do. Haladol. My father's an Italian scientist in pharmaceuticals, remember? The American Field Service? I was the only member of the Fascist party to give you an inch because you won the tournament?"

--"Hal-do. Rings a bell. Been savin' up to marry an American for awhile, haven't you, Mr. Naples?!"

--"I think I've got my priorities completely backward. Would you like a cigar...*Napuletan`?!*"

--"Ye-a-a-ah *guagliu`,* almost through, Machu-Pitou and Hungarian too!"

--"Let's get on with our lives!"

--"The Art of Marrying Without Being a Fellow Asshole. It would be a good book."

--"Yes, it's not so much waiting to marry for nuptials, it's the torture proceeding while waiting; the shots being fired waiting for the job before the second cavalry comes in to discuss serious rapports."

--"Can you bench-press that baggage? I mean write a book like that, and tell American youth all they need to know about the Contemporary history on the preceding assholes?"

--"Oh, gonna' have to dead-lift that and load it along with the rest of the second-hand clothes into the Goodwill truck!"

--<u>HOW THE MODERN LEFT GAINED POWER IN AMERICA</u>(The <u>Culture of Right-Handed Liberals to Maneuver the Laptop Mouse with Their</u> <u>Right and Drink a Generic Beer in Their Left.</u>).

--<u>POSER SURF GENERATION</u>(California Historical Denials of Pot Not <u>Being Connected to Genital Herpes</u>)

--"Who was that on the phone, Bill?"

--"Fusil."

--"That's an acronym for Fat, Ugly, Stupid, Ignorant, and Lazy?"

--"Well, he's still a Giants starter!"

--"What was it this time?"

--"His wife wants you to run to the store for some honey butter, that stuff imported from California. She's makin' _quesadillas_ for Fusil's dinner. The microwave takes his mind off the game."

--"Bill, if that's the story, are you sure you're an alcoholic? I mean, you know, I fly, you buy, and Mormon time, since this road is a long history lesson all it's own."

--"You want me to just get a craving, and react, or to actually slip?!"

--"I said nothing about your slip, Bill, now does your car have gas or are 'ya Italian? Yamesh, Bill? Same speed on the English As A Second Language."

--"Oh, I'm from hardware country. I'll pick up some tacks to put some of these abstracts I got in the mail from their daughter on the Wall! Gimme the keys!"

--"Bill, Fajitas in the Tupperware, and don't need more tortillas, but do you mean—I get to see the cashier's synopsis of why this is a new Democrat beer? I mean, she's a true intellectual with morals!"

--"We'll buy two tubs of butter then!"

--"I've seen some people who run their towns like their Daddy's bar, but this is ridiculous. Now that the troops are home from this extended ecclesiastical prayer session, send some more U.S. Army souvenir security blankets to Sports Mart. The answer is not Prohibition, it is to run a legitimate bar. If any real Armed Forces Supporter has the paycheck to get in the vicinity. Two failures on jobs creation—three times a charm!"

--"Rough patch."

--"Get a mother who can sew 501's then. No, no, see, young leader, suddenly it's teamwork to be led by Christ. Your desert oasis on the organized religion exchange front. Host-family's holster on the coat-rack after work every night is not a relic, pardon me, going to get my own pasta now that you're creamed!"

--"How 'bout I rent an apartment?"

--"Yes, L.A. real estate sucks up to the damndest foreign critics. Lakers, that needs a new needle. The ball, for the ball, millionaire, not for the high school Alma Mater! Can you jump high for the wild pack of real dogs loose in European parks, or do you need me to teach you Braille for 'silence' if you're going to walk away at night?! Doesn't give a damn about cartel rules, is a real rabid, loose creature of God. Is not eating trophy meat, but don't go near the guys with the good ashtrays. Yes, that's the greeting party when the pep-talk about avoiding sex a bit, pre-Magic Johnson H.I.V. diagnosis, is given to Neapolitan men who smoke.

See, Jesus is ahead of you to take the Gospel to them. It's nice trophies, good games, don't misunderstand, but that's a stadium crystal ball, and that's my Nike footprint in the park of organized crime towns, with a Good News For Modern Man under my coat. Christ is servin' coffee and tobacco from India until we get more protocol on avoiding the syringe and condom crowd's Saturday night car lots outside the school gates, complete with newspaper on the inside of the dashboard. That's the Gospel abroad. The Gospel in L.A. is about the dress. I wouldn't call it other than Mal-Tribu`, not Malibu`, after seeing AnaCapri without a pest asking what 'Buon Salute' means while I have to provide the toilet paper for the other hikers' asses. That's an Eagle. We don't play high school ball."

--"Yes, as it was said, Christ came to bring a sword to divide father from son, mother from daughter, and more, we have the end-days prophesy of St. John. Do you really believe either John or Paul are the subject!? Is it not the Gospels, when all is boiled down and filtered out, and turning the cheek, loving your neighbor, whatever street he walks, as yourself? Is it not simply exchanging money for goods in peace? Or is there more to express on peace? Try the peace of Christian writing and music appreciation after a naïve, molesting nation wreaked God Almighty's wrath and sacrificed 4,500 of our military for the offense to the children, children of God like you and I! Try peace with Christ in the home to educate yourselves and your children on America's' sins as a whole, in a Catholic view, with respect to your other Catholic countries that don't have American sex hang-ups, divorce, or greed, but that hold hands with their neighbor over a beer three times as much as us over the holidays—If you can't sleep now, stay awake and drink one for the Scots, the Brits, and the Hungarians that died out there with the American military. And that's not scoobies, non-smoker. That's an air-born allergy in the U.S. environment. My food is kosher, so is my tobacco, but try polyester, not the wool, on that sweater material—you're making the birds in the tree outside wait for the half-moon to swoop on the morning worm, which means they know you'll need a shower in four hours."

--"N, 'Papa', that's not a junkie, that's a blood donor. You grabbed the extra Carls Jr, pointed the finger on diabetes at the 205 lb. weight class, and

Paula Dean snagged in the Medical Community's trap while she was helping your friend Dumpy change his tire. See that? That's 65% of Americans hittin' the gridiron, tennis court, soccer field, and baseball. He's one of 37% of Americans whose blood is so pure Mt. Rushmore, he can donate to kids and adults alike suffering from a cancer he don't have. Got in six pints to the Blood Drive before being late on his flu shot eatin' with the Anglicans. Next…British flu puddin'."

* * *

Marijuana may not cause schizophrenia, but will exacerbate it, according to research experience from the UCLA medical staff. As well, schizophrenia is both genetic and environmental: environmental factors such as illicit drugs, pot listed as well despite belief we are avoiding certain realities concerning cannibus. Once under treatment, how do you differentiate between medical side-effect, and the new levels of stigma seen in current California? Exercise + diet= Truth – others' education: All fields of professionals, even the ones not prone to mental illness, have current or recessional factors of illicit drugs. Drug tests are popular to get jobs. Background checks are too. This, as well as criminal reform, need not burden this state with its tradition of grace. Some await their second chance to win the West. The truth behind a modern state where more Californians are given just credit, for the victory belongs to those possessing qualities of loyalty, selflessness, wisdom, and purity on faith-based intent.

--Oh my God, this is upstate New York?! Did you see the summer hygiene on that chick? How those strings are pulled? She's a leafed-roast whore!"

--"*Ho gingivitis* (I have gingivitis)."

--"I'm sorry, God, I was just trying to be a polite tourist. Watcha' sellin'?"

--"Tobaloni."

--"I'll take a shit load if my sex life picks up like that! How much for a shit load?!"

--"45 dollars, good shit."

--"I've got $135.00. Know how many shit loads that is? That's such a beautiful language. We're going…after Upstate, to Rome!"

--"*Buona Fortuna* (Good luck)!"

* * *

--"Believe me, only box you're getting from the President. Whatever goes with this 'It's just a label' business, tell'er to keep it, baby."

--"Okay now boys, your lesson for today now is to encapsulate the proper Freudian worldview, for voting Americans, in order to conduct more peaceful relations in the Mid-East, while continuing to discuss commerce theories with the diverse Europe of traditional presence, so that the Church of history, Catholicism, and the new Pope, never lose face in eyes of ancients. Quickly now, before soccer, in sum, take a Romanticist perspective on Dante, to apply to a now Freudian America for the purpose of world peace. You have one hour. A good essay can turn over the apple cart on the grades all year. The actual Neapolitan soccer team's failures are not yours. Take your mind off of the athletic system in order to perform essay-wise in more than rhetorical fashion. _Buona fortuna_. One hour counting....begin."

--"Okay sure, a bit of Socialism. Spice it up with a bit of Fascism, Nazistic youth, and sculptures, then add the girls after bringing it to a boil. Let it cook for twenty years. _Sopa a` la Italo-American`_ is served!"

--"The taxes are too high."

--"In tax season, there are payers, those who don't make enough to be taxed, and evaders. Crime waves are indicative of (1) desperation; and (2) a structure of organization...seeking demographic-wide influence, similar to a corporation, though not interested in the legal framework of incorporation. A government may increase taxes to come up with a solution. Organizations at the Federal, State, and Local levels can protect payers and non-desperate low-wage earners with faith in an answer provided by a democratic system."

--"Who? What Federal, State, and Local organizations?"

--"C.I.A., F.B.I., sheriffs and law enforcement."

--"Law enforcement alone has necessitated scrutiny."

--"What does that tell you about a demographic that has both timely crime waves <u>and</u> law enforcement necessitating scrutiny? When you answer, you can come up with a solution to your own answer. F.B.I. and C.I.A. are also scrutinized during Presidencies that are high in taxation and government spending, indicative of...a laissez-faire attitude toward the desire of some foreign nations to see a global shift from America as a superpower. We seem to give a cut and dry answer as to why they want this: They don't like America. Some are taught to think like that not because of history and religion alone, though because suddenly, in the New Millennium, their area of the world has unforeseen needs on its plate now. If it's not ideological, it's also a form of desperation, with anti-American sentiment as a result. In a situation as such, opinion of others in said area of America may topple lest hands be tied. Military intervention and foreign aid have been two time-tested American methods of winning support, though in wake of the frenzy of discussion of environmental issues, a new strategy or option to these two, not a third party, is needed for balance. I believe there must have been some historical problems with the concept of dual citizenship: The right people taking pride in two separate areas produces a healthy raising of the bar in ways. However, the U.S. is very hesitant to enact said laws, and so, in sum, another solution could evolve out of necessity to adapt our allies to America in its new challenges."

--"Arts, entertainment, sports on an international level."

--"It can create jealousy, not for fact of the American athlete, though sports training facilities are well-kept here."

--"Sports is a brutal subject period."

--"So is entertainment. Educational exchange is actually a high-maintenance tool to communication of a more-dispersed natural means. As bad as home ownership. Could the face of the ideology be in a state of poor translation, other than laissez-faire leadership? Is it a more systematic diffusion of public American information that is twisting, not direct or polite? And translations…this is not a grammar class to some foreigners: they want both our language and theirs read in a manner that captures the ideal of the respective cultures. During Pope John Paul II's time, this concerned the message released and how to apply it somewhat pragmatically. During Pope Benedict XVI's current stage-setting, more and more attention is paid to mood of literary statement and what produces this. Scrutinie s can be redundant even though they play a role."

--"Was that scrutiny-peace or scrutiny-piece?"

--"That's the difference between one American and his neighbor. Some still prefer the direction of the pointing thumb handle it indefinitely. I personally find a Revelation-style Judgment Day pictured in Earthly terms of scrutiny to be difficult to believe, though don't think loving neighbor as thyself for eighty years before burial means Jesus sort of awards you your Eagle Rank in a private ceremony before Peter let's you through The Gates. Jesus said to enter Heaven you must be 'born again' or 'born from above,' and Paul deduced this is for the reason that are bodies are spiritual, not physical, in Heaven. So, to draw straight lines on again being born, there is kicking, pains, a snap of the cord and Boom! Out into the cosmos you go, a spirit that can chose to believe in a Heavenly Order or float at random, infinitely distancing from the warmth of the sun."

--"Hey, it's the anti-American foreign guy on the phone…what do I tell him?"

--"Tell him we'll call later and to leave his number—it's a national holiday in America."

--You guys can't even get the laundry done without a political filibuster—California, we feel sorry for you, but its baseball season!

--"And now, Hallmark returns you to, Too Much Porridge For Baby!"

Post-Easter Sunday

--"This Loma Linda thing. My own mother, back surgery, by a doctor with a name that looks Arabic! Seventh-Day Adventists, one and all, gifted medicine people, but their Liberal Arts classes, apparently taught by interns for the American Field Service! The rest of us, Liberal Arts classes, taught by that left-hander, Skeeter the Yard Duty, of Baseball For Boys fame! Do you see why they crawled in the bottle after tossing the cork in the bushes at the Dog Park, for good luck? So their boss could at least locate them by Christmas time, with the help of all the dogs in town!

Oh, let's not criticize it if it's all talk! Barring the chances you escape the dentist's expert opinion, you might get out of this argument with every sermon giver in town with enough strength in your jaw to enjoy kissing a new one before the next election! Where did the love go? First time in ten years it's not her bag! And by the time the subject isn't sexual discrimination, the psychologist is retired! Before we lose it, put it in UNICEF's terms: Home. Plate. Food. Amen.

--"Different rules for divorce in L.A. and the Inland Empire, I believe."
--"You're living under a dark cloud. How do we get better weather?"
--"Going to have to rest, I believe, on the No-Cal Regional Vote."
--"Are you crazy? Am I going to escape my own kitchen with the spice rack? The cashews under the four burner for a trail snack? The magnet on the fridge with The Church on it?"

--The only thing talkin' in America is the money: It's the only thing. Nobody can afford, between The Mob and The Nice Guy, to meet halfway to kiss each others' asses.

--"The ass contest? First String wants new rules...Second String turned this into a marathon volunteer event. And, First String is still sending regrets to confused Gay people, while trying to get a job without being called 'Honey' by Broom Hilda one more time. Sex in America? It's for the teens, everyone else was expected to tap out at Monica Lewinsky, which is hypocrisy for my roots. That's the 12 in the side on the left, no bank, and the six into the eight, right corner pocket. These guys are pricks from a time-frame when Grandpa drank only tea. You can have your lemons back: Too many newcomers is a fact of life. You break them in at the New York Times in their twenties: Californians are born with it, or don't get breaks before they're 50. New Yorkers and Southerners both spoiled shitless. This, jack-ass, this is the 1970's Southland. You're a Rich Man, Poor Man Democrat. You other Americans don't know this State worth a damn, but you can eat in our cities. The Swine Flu started in America, spread to Mexico, and almost got the Angelinos. Peace if the rest of the country thought this was a Golden State in-fighting season. We should investigate them too! Love, Holmes P.S. They started over-spending then!"

--"The 70's, or Clinton?"

--"I don't know, but that's the bottle of Cutty Sark—don't lose it!"

--Oh, before you pass me more cornbread, wake up—those three guys in there, living it up with this combination of extracurricular activities, from bar scene to bowling to pizza parlor, are not Born-Again: Who is the Black Christian on this play, your younger brother renting Eddie Murphy at the 99-cent Video Store? You should see the DVD he tried to sell me. If we get any closer to the neighborhood with the hooka crowd, for the cheaper rent in exchange, the desert soldiers will start drilling below the fault line so they possibly don't have to go by Texas in the next war—since that's where we left off in the Bible Study—with Christ bringing a sword to this and separating father from son, mother from daughter. Are you kidding? Like I said, those three aren't Born-Again, and they're full of IT! So helping Mom for care packages is what one does in wartime to help the troops. Why? Because his $75 went to your brother's DVD for sale, and Mom's went to the elf in the bottom of the cookie jar passed to the groundhog, into the West Coast ground hole and out the other end to an American charity."

--"President Obama wasn't askin' about that."

--"Yes, he was also at Pizza Night. And suddenly after a beer with the boys, it's off to the L.A. set. This is college for Brat-Packers. The rest of the Nation is on Green Bay. Now do you know why the cornbread is too dry?"

--"Why don't you drop it about the DVD and open up some Gospel charm on these three?"

--"Pluralism. They're disciples of a prophet taught as an elective (not any religion known to the Eastern or Western world, but the disciples themselves can prophesy in Native English). So out of respect for their own 'unique' beliefs in God. And my ears' time with the spoken word."

--"This is a tug-of-war over which Bible to use."

-"Watchtower has material for one and all, too, but no, it's more of a wrestling match to keep the Bible from being mentioned, until their sisters' boyfriends can organize to compete with Dad and I."

--"I thought you worked for Mom."

--"Oh, that's the ear their Dad's English goes into. The ear their beer money goes into finds Dad a deep inspiration."

--"Add a wife or two and you've got a time-share."

--"We wanted to add a job or two, not necessarily more than two degrees, but no, the political cast, my how they have grown! Such a short war!"

--"Well, you know why it's dry cornbread, and I prefer you eat it. Because you don't want my problems."

--"Oh, I think you actually, finally said...'You don't want my problems, you want my leadership: and I think that's a crapshoot of a Bible Study, Democrat or Republican, but you do what you have to.'"

Serranos on Tobacco in the I.E.

--When you smoke, you have interesting people who channel all in different directions.

--Chief originally smoked peace pipe to commune with the Serrano's god, who was killed by Coyote God and is now deeper into the blue sky and night. Sometimes the Serrano God is travelling the universe. Hence, you may be smoking for years before the Serrano God is back, in order to commune yourself!

--San Bernardino was originally from Las Vegas, and ventured to this valley as a Catholic priest. None of the priests since have been from the desert plains of Las Vegas. Don't rely on The Church to help you quit, regardless of what they say. You can get to know the Vegas area on spiritual terms or take your knocks in this County 'til ya quit!

--There are sacrifices to make quitting smoking: You're no longer going to enjoy League B Southland's or the Bourgeoisie's Party War Stories with as much *animo (gusto)*!

--Look at the people gettin' in line to see this one! Look at the French gettin' in line! They've all got one thing in common: They fly under the flag of the Southland with Mickey Mouse on it. There's two other flags! Mexico and Camaroon. Anything to do with other languages spoken isn't good business.

--"Can we talk to the Queen?"

-"The Who?"

--"The Queen in Mentone—and don't send that mystic mamma that runs the sober living home!"

--"What is this about?"

--"You guys got enough water to give Loma Linda a pro football team! Hinckley Dam my ass! Where's the rest of the jobs?!"

--"Did you hear what the snitch for the N.Y.P.D. did?"

--"No. What possibly could the N.Y.P.D. be up to now?"

--"He reported Barack Obama to President Barack Obama!"

--"On smokin', and quittin', it's not how many times you fall off the horse, it's how many times you get back on."

--"What the hell?"

--"That's Ex Post Facto, Revere's horse, with your job search papers in the saddlebag. He's yours!"

--"Will he keep going fast enough not to get eaten by that flock of crows?"

--The Major & Mrs. Sex Talk Tapes, an exhaustive course of 1 language cassette, and 5 1\2-hour meditation cassettes, only orderable at the rank of "NOW!!" Language cassette consists of famous, "Go to your room!" method. "Get out of the HOUSE!" Edition included as a bonus for ordering now! 1-800-USAF!

--"How are we checking in today? Symptoms managed properly?"

--"Meds seem fine, thank you."

--"Complexes, complicated or dormant?"

--"Oh, you know me, sometimes I'm a ramblin' man, and that's how rough riders role, sometimes I'm the 'baby of the family', as they say. That's a very California thing, this expression. They would have to see the flies swarmin' around the horny bull's ass in the pen at the slaughterhouse yard, in the peak humidity of a Texas August, as a young boy. Rather quiet animal, even when horny! You actually don't have much choice whether he lets those boards constructing the pen stand up or not, and then he shoots you a corner-eye look as if to say, 'We're not gonna have words with the tribe when we grow up, are we?' Those flies can crawl up there all they want, but he'll just stand there catchin' sun, and you don't know how to answer the question before he readies to impregnate cows later. If he wants to give you the blessing to stay away from a life behind bars he'll give his tail a quick swish back and forth.

--Is this book dragging out a little too far? You would have had to win the infamous Paperclip Award from the Surf Club in the South Bay to understand—how far a bent paperclip can go in your ear before people get the hint you don't author screenplays, you author books! Subtext on the _raison d'etre_ of the Writer's Strike…One and all on the end of that rope pulling, after learning that books can be stolen even after college. Hence, yachts turning over in a storm, fires, dramatic criminal moments. Epic God's wrath on people for being slack about Satan's designs. A tragedy of an era? A _tete a` tete_ in a worldview context between film and terrorism, loyalties upheld.

--"How can you tell your ex-Marine cousin and his new Latin American wife are coming soon from Northern California to visit?"

--"Because you can smell the diapers in your kitchen when your dinner is over before they get there."

--"Guy and gal see each other for the first time since youth at a New Millennium wedding. She's been hired to dance hula with the troupe for entertainment. She hulas all night, then the audience is invited to learn to dance one. He sits and keeps eating Swedish meatballs. At the end of the night, they need help cleaning up. He carries the large seashell holding the melted chocolate for strawberries into the kitchen. What's the meaning?"

--"I don't know, I'm not here for Twenty Year Storm humor."

Peni`: "Look, on the book shelf...World's Most Infamous Lies!"

Pati`: "Such a thought!"

Esteban: "Well, the Ten Commandments say 'Don't Bear False Witness Against Your Neighbor'...that's different. They don't say not to lie. What if you had to get the neighbor out of something?"

Ricardo: "Esteban, very American of you. You need to see another bullfight to understand."

--"Would you like a beer?"

--"The Parish is trying to cut down. Night driving."

--"The Parish has never cut up. Cut down?! You want me to tithe more."

--"Look at the paper! 5-Lb. Tadpole Caught!"

--"Don't forget Lent 2012!"

--"Shit, I've been fasting since President Obama got elected!"

--"California—the state is a gas on the ground!"

--"I think it's about the same-sized ball. Just one culture is used to peeling the potatoes, and the other is used to pickin' the peaches!"

--"I've seen some dumbshits with fishin' records, but buddy, your county is your name. C'mon now, everybody Big Bear it for Santa!"

Ann Coulter: "People who vote for John McCain will vote for any Republican...We need to get some of the Obama votes now. Soccer Moms, etc." Soccer Moms. Not voting Republican. Have we explained to you about soccer Dad? Do you need a person to play a modern-day Judas Iscariot? He is here for you!

--"I don't know, we've had red herrings of all sorts in this small town over the course of time, before a polite introduction to a new President, quite a few both sensical and non. You'd think there was a hatchery by the Forest Station."

307

--"Nibble on it a bit more, it might turn the boat around on the Captain Crunch shelf! Pirates? Send the trademark in for your rowboat's flag."

--"No, what is going on Downtown, what, that take-out Grande Cup of Starbucks Brew of the Day been sittin' in the a) Police Chief's Office; b) The Quality of Life Dept.; c) The Martini Bar restroom; or d) the Psychologist's second floor office...for three weeks now? No, the President and wife are at the summit in Hawaii with Pacific Rim Basin Leaders a couple months before volcanoes start to blow. And World Peace is ejected in Oklahoma. The world is spinning around the satellite disc on top of the ABC building in Pasadena, and if you don't move around the apartment, the T.V. fights you for your cigar."

--"BEEP********...Attention. The Townhome Project has been cancelled. Residents of the Historic and Scenic District will now resume the economic seats they had before their mass exodus had been planned, to the tune of $500,000 mortgages. This was only a test."

--"Did you hear the latest? 'When hygiene is used as a gimmick or trend, it takes away from the true moral values of the devout psychiatric professional's desires for his own social rights.' A quote by a former Chairman of the Cedar-Sanai Board of Medical Ethics who will remain anonymous. Sponsored by the American Lung Association and the Socialist Party of Southern California."

--"Okay, Ray, that is $200 under the Arts and Sciences Category, or...wait! Double Jeopardy! The question is, "A Famous Gene in American Entertainment."
    --"Gene Autry!"
    --"No, Nancy, your guess."
    --"Gene Simmons."
    --"No. Barry..."
    --"Uh, Gina Lollabrigida?"
    --"No, folks, 'Gene Pool.' The answer is 'Gene Pool.'"

--"I can't figure it out. All those years, then the war. The gang from the hostel I understand—new commitments, new pathways, since we all have to go in a new direction, other than the way we came in. But the cop and his wife, the sons—why hasn't the host family that won this toss in the language department written in all this time? There's no back-burner Interpol issues—can't be!"
    --"Steve, want you to have a seat, good news, bad news."
    --"Okay. What is it? A game-changer? The horoscope lady predicted such."
    --"Some people that know them are in town."
    --"Redlands proper?"

--"Yes."

--"It's not like I'm a tour guide. I don't mind if they have their privacy. What's new about that?"

--"Mamma had an affair."

--"O.K., Geez, now what?"

--"You go with Babbo, the boys gotta go with Mamma on it."

--"Babbo? Who's gonna cook?"

--"I'm lookin' at it. Now, you satisfied? There you have it, Naples past, present, and future. Get your apron on."

--"*Cazzima!*"

--"Exactly. How many years you spoken Italian?"

--"Twenty-seven."

--"You still understand the dialect?"

--"Yes."

--"Babbo said something like, '*Se oiss' e` 'na 'malita` mentale, tu sei o' cuginn' di Gianni Schicchi.*' What does that mean?"

--"If this is a mental illness, I must be the cousin of Gianni Schicchi!"

--"Cheer up."

* * *

--You say smoking cessation, they say breast cancer. You say psychiatry, they say diabetes. You say Italian wine with the meal, they say obesiosus. You say sex life, they say tattoo parlor. You say alcohol, they say it's that time of the month. You say work, attempting to change the subject, they say parolee reform program. You say, 'Come Home From Afghanistan!', they say, 'I don't mind an affair but have respect and volunteer!' You say college, say it low and keep to the streets, you say Catholicism, they say martyr, you mention work twice in one year, and they say flu epidemic.

They are:

A) The Bureaucrats
B) The Left
C) The Inside-Traders
D) The Crowd
E) The Alumni Association

--"Stephen...Undercover Officer Fonseca here. And Undercover Officer Grints. You've been hoping to make love to her. We've just helped her screw you. You wanted to marry her. We're still teaching her to drink. You want to know if she minds smoking. We said it's an option, after she learns to screw you when drinking. Take Five and welcome to the new Parish."

Ladyluck

--"That Jeannie's a real gas!"

--"Oh, a little spitfire, is she?!"

--"Talking about her cooking, dimwit."

--"If you go to the source, it's little wonder all the dogs in town need two baths a day! But you should be thankful—I hear she's a career woman."

--Don't remind me. I reciprocated by asking her where she'd like to dine, someplace not too intimidating for her as it's our first time out."

--"What was her idea on that?"

--"She suggested Pomona Valley Mining Co., high on the mountainside overlooking all the valley's lights!"

--"Hope you don't have to take out a loan."

--"She suggested it while I was clearing her table—she works in the Loan Dept. at the bank!"

--"Not a small thinker, this Jeannie!"

--"Yes, considering it's a one-story bank."

--"How's Bart? Off to college?"

--"Bart? He got out of the rec room at home for the first time in six months the other day—to volunteer pumping up basketballs at the YMCA with some of the old hands from the high school ball team."

--"Just how old are those hands?"

--"They've got scars on then from a wrestling match with their first Mexican girl in elementary school from the 70's!"

--Ouch! You should be thankful all the more for Jeannie!"

--"Yea, she's got the class to purchase the 12-oz. gloves at Big Five Sporting Goods!"

--"Knock yourself out!"

-"She just wants to spar a bit."

--"Better than the Russian psychologist's daughter who volunteered you for the Knife Throw at the Renaissance Fair! What's next?"

--"She comes over to explain to Bart that him losing five pounds is different from passing the physical to go in the Navy!"

--"Did she have a mini-cigar with you?"

--"She took a month's supply—'Hope you don't mind,' she says, 'my gardener experiences racism at the tobacco store.'"

--"Why doesn't Bart do the lawn?"

--"Because Jeannie tried to line him up with the stylist at the The Glass Menagerie Salon!"

--"All this from breakfast!"

--"By the time the last pan was done and the phone stopped ringing, yes!"

--"Anything good come about from it?"

--"She had the perfect answer for my boss when I was twenty minutes late: 'Don't dump on him if I won't go out with you again!', she says to him for me."

--"That's perfect."

--"I actually got more hours for moving all that off of his back-burner!"

Receptionist: "Erectile Dysfunction, Loose Bowl Syndrome, Genital Herpes, Premature Ejaculation, Varicocele 65% successful, and Non-Specific Urethritis. Let's see, which problem do you want to isolate today?"

Doctor: "Listen, I know all that, but hey, I think they should start catchin' them buffalo!"

Receptionist: "They're endangered!"

Doctor: "Well, amen, what comes around goes around!"

--"Ponce found it! Hold the Caravan! Have the men put down the poles! Muskets and swords, Pa-rade Rest!"

--"Where are we?!"

--"It's here (It's here, It's here, It's here…)! The Long Lost City of Gold (Gold, Gold, go-o-o-ld)!"

--"But the sign says, 'Social Security Administration!'"

--"Damn those Queen's court interpreters! Start mining anyway!"

--"You give us two pieces to the puzzle that don't fit. I'm curious as to why."

--"Perhaps it's the budget differences."

--"Yes, for the salon lady to treat your hair so you can grow a two-foot mane. Look, I really don't have time for Lewis to Luther to Laurie, or any Protestant opinion, remember, Italy is a culture involving baseball bats on the ground behind the scenes if you want complete cultural immersion, and I can't believe the bridge you've built from high school to the present in this region. You're the puccinello of Southern California, not a Grad Student with intellectual faculties in literary criticism. You act like you're 30 years old."

--"That's discrimination."

--"Bleed the subject dry while looking at the photos in Vanity Fair at the salon while you're waiting, Barbie."

--"Well, how old are you?"

--"Put it this way: I've only missed two-thirds of Redlands history since 1888: been here writing the rest of the time!"

--"Oh, after the off-white, below-the knee skirt zipping up on the side like a kilt made in China, please do not feel an exigency to move forward with current academic fashion design. If you waterproof it, the dishwasher at the Tartan can use it!"

--"How 'bout that professional writer's group? That extracurricular activity has promise."

--"To Groucho, or Karl? I suppose we owe back-dues to complete the project. I think I generally avoid men who see it as teamwork and comradery to wipe one's rear in the morning fog. I'd rather watch little Ricky Sepulveda dress up as a fawn on stage for a spoof on the Greek Pantheon at Montessori!"

--"Good vegetable soup?"

--"Yes, nice flavor, but so many potatoes?"

--"They say if you get an aversion to potatoes, soon you won't have money for beef, because that's what Longshanks was like in the war."

--"What if they cut you off from your potatoes?"

--"Then she really didn't want to have your children!"

--"Now we know those Hungarian interpreters are pretty hard to beat, but did you hear the one about the Hungarian bobby? He crawled into the pub to address a complaint of 'stolen fish and chips', and (1) wound up diving in the water to retrieve the plate after the perpetrator escaped in a stagecoach with a screaming cocktail waitress!; (2) Found the items, and swore by it in court even though the establishment complained. They were behind the women's loo!...or (3)Caught the perpetrator nibbling on them in the alley, knocked him out, finished the meal, tossed the plate to the brick wall, walked back in with the perp's wallet and put a pitcher of martinis on the credit card."

--"What's the difference between dating an actor and dating a writer?"

--"The girl says to the actor, 'I hope this isn't all edited out when it's complete', and to the writer, 'Please hurry! Hot as hell! You must finish the treatment quicker!'"

--"So you're both screwed?"

--"And as we say in the Inland Empire, 'I don't have a screwdriver that long!'"

--"This is a smear campaign."

--"Keep smearin' it in the blank slate: it'll be an abstract we can work with soon in class if the new medication has unforeseen side effects. Remember, its F.D.A. approval has its critics too, not just the patient."

--"Okay, this is the part where you can leave the Church for Non-Denominational life, make five more folk songs, and create a C.D., taking your chances on the market. A higher goal than cognoscente, or 'in the know', so to speak."

--"English as a Second Language, this is!"

--"My 'native tongue'? Don't the Indians find offense to us borrowing the expression? Furthermore, I suppose you could say American English is my 'fourth language', as a polyglot, behind the other three with more cunning aristocracies! Unless you just want to 'drop it', as sits?"

--"The smell of the gas escaping British ass as cleaved by the axe in the middle of Bannockburn!"

--"The Hungarians try their best to discuss film."

--"There are some things the Hungarians will never understand."

--"God help the Queen, don't you dare put the geniuses on another of your intellectual tasks, Asshole!"

Down The Home Stretch

--You mean tobacco is worse for the lungs than pot? The UC San Francisco research? It's a different San Francisco now. Lots of the businesses from the 90's are closed—big buildings, different contemporary mainland dialogue, though most likely still well-travelled landlords. Could be different country's influences. The Advent of the time leading to the passing of Steve Jobs. Or, medically—how your family trees line up. Your bloodlines. Their reaction to either version of smoking. Family values determine, and constantly reinterpret, family habits.

--"Ah, St. Patrick's Day, after Mamacita's birthday. At UCLA. The clams underneath the sand filled to the top of the metal garbage can. The great Greek system of UCLA, with the film of Belushi and the ladder in the background. Ay, the symbolism. The clams. The campus secrets. The boycott on trash-talking, until dinner has been patiently prepared. Now that the professional representatives of our mascot the Bruin have retreated to their separate suburban forests, after a gridlock commute without Vicadine produces an ally of modern time for nations willing to talk peace, and willing to stand by the Vets buried a 20-minute overcast Westwood walk away, under an ocean layer of breeze like Albutrin to an asthmatic city child.

The patriot in the Bruin. The Bruins whose families have served. Their manners. Their sacrifice. Their message from home phone when reputations sagged to 'Respect Authority. Don't be a complainer. Get a job on campus. Stop looking ahead. No, no car for you. If you don't get your Greek clams, shack up at the dorm and don't dilly-daddle. Be safe. Tell the truth and don't expect favors. You're not the only one with tough times in the family. If you get an apartment, don't make it a never-ending story. I have an agenda too, and wew, you're telling me, tuition up....You wouldn't know what would hit you at a military academy, though. So I didn't think you would do well. Try your best, despite the mediocre influences. Your sister's doing fine. Thanks for hearing her out. Here, your mother has something to say....' 'How's Spanish classes going, mijo?'

'I'm taking Italian, Mamacita!'

"OOOH?! Who is she?"

Yes, you would have to carry a flag in a troop ceremony to understand. You would have to lower one from the pole and fold it with another American. You would have to carry, halt, carry, halt, and carry a crucifix when young, with what arms you're building, enter behind the open communion rail, and slowly lower the unshaking eight foot pole into its holder by the entrance to the altar area, which has its own proper English name. You would have to sit or kneel the whole service to the right of the priest publically reciting the Mass before a congregation of near 350 people in number. Placing the white cloth on your left wrist with the silver bowl in the same hand, and pouring the Holy Water over his hands, after he blesses it, then dries his fingers on the cloth he took from your wrist, replaces it, before returning to the altar to pour the wine into the chalice. You replace the towel down on the shelf, placing the sterling silver on the towel. That's what I know about British culture. Other than I became a Catholic in 2008, and the Duchess is actually a Catholic ahead of this time, even though she's not 49, as I turned this year. Onward Christian soldiers may seem like tacky fare, when you figure all of it out, but word is...the Confessions are not honored by the Catholics-to-comment—say them again!

Well, no earthquake, all California, and here we go...the Catholic priesthood shuffle, when they change every year, not the permanent Rector's routine. Are we splitting hairs? You know old buildings...always spraying for spiders. They're making plans to build a new Catholic church. Mom says they bit off more than they could chew. I think they just over-sprayed the town, and want some air. It's a cycle as such between the two, not as if the rest of the Protestants aren't short on their own talent.

So, hermaneutics, literal, or figurative? End days, world peace, or...look, Mom's birthday! The rest of the ritual...depends whose asked to say the prayer at the dinner table, or if I'm invited up to go out on the town. If you think potluck was good, I didn't know the rest of the high rollers were dining out so often. So as Hungarians say to their children...'Don't cut off the hand that feeds you.'

--"Oh how do you know she's not a convert after you?"

--"Begging your pardon, I can tell she gets more exercise than the Anglicans in my neck of the woods....judging by growth and contemporary consumption patterns. Now, without further ado, St. Patrick, was it necessary for the students in California to rout their capitol? I know Christ was to bring a sword between father and son, but many of us are in-between the two generations arguing, and want a simple occupation."

--"Inner Critic," "Alter Ego," "Secondary you of two," and "Writer's Voice"... when the music comes on the radio, this reads like a poorly-written Marx Brothers screenplay that was buried in production. By law enforcement! Sure, sure... "The world is a Vampire...." Twelfth-level Procrastinator with the peanuts in the bleachers, in other words the house on the over-looking hill where they've moved back in with their parents. I'd rather take a loan than have that routine, and reorder a new high school yearbook for my birthday. I'd rather drink alone, even if the vamp smoked mine one after another until I was forced to buy a cheaper brand. When they're gone from your scene, and you're on your own, you'll discover it wasn't how much you used to drink that made you barf, nor your bad cooking... it was their hygiene in proximity to your work ethic. Girls at this fork, even if it's further than a mile for a camel!

--"You didn't get anything done? Don't you feel like a slug?"
--"I know the concept...I'm from a large family spread nationwide. We're actually quite productive. You would have to know the hard-bodies in the fam who got their degree handed to them on a silver platter to understand."

--"The Keebler and the Female Elves! The psychiatrist's staff and patients combine in a 10-piece Big Band Combo! All charity fundraisers played! New Year's for all denominations!"

--"How do you like it? The nose?"
--"Great. Posture wins that Holy War."
--"You're not put off?"
--"A tall woman in a compact car, her profile? 95% Rating? Wins the hand when the brains need a nose for progress if they convene."

--"Buddy, it's Mayor Pete Aguilar."
--"Hi, this is Pete...Sorry. I thought you were Action Jackson.."
--"Oh, Pete Mayor, we won't complain to the Quality of Life Dept., you caught the bad guys. And your Police Chief Garcia's _heino_ now?"
--"His _heino_?"
--Apparently it means on male bonding we can't touch it! Looks like nationwide laurels to rest on!"
--"Are you sure that's the Mayor?"

--"I'll look.(). Oh. The apartment maintenance crew is there."

--"You can't tell the difference?"

--"I can...the people who respond on the other end of the security camera need a restroom break to explain though. Thought I was Sean Penn last week... this week it's Nicolas Cage."

--"Wow, I would cry if I had to shave that often for Billy Bricks...I mean I would. You must cook like a National Guard chef in your kitchen to get through all that extra dreadlock alone."

"No push-ups, all yoghurt, and three errands in a Toyota, before coffee, soup, and a show."

--"What's your secret?"

--"The toilet paper. Works every time!"

--"What's cookin' in the interpreting field?"

--"Dante is escorting us across the River Crap. Where we arrive at the Island of European Working from Mom's Home, to serve time there."

--"Where they went wrong with Father Time!"

--"Mother Nature's divorced from him now. After that."

--"And now, enter the Backwoods Real Estate Company's website in Downtown San Bernardino!"

--"Yes, what was your tribal name again?"

--"Some Kid."

--"I see they were given a blank check. They never met Sum Kid, whose father served in the Coast Guard."

--"Just, uh, if he isn't given the Army + Navy tickets, it's the same line...he has to earn it, the Coast Guard kid doesn't have a 6.0 GPA in this district where we're at, and our kid will now spin your damn bottle so it doesn't land on Elton and Bishop Dolan."

--"Such family. I mean. Lucky."

--"It's freshly squeezed. Calling the uncle in Louisiana since you asked for more. Go Tigers. 'Bama snoops."

--"It'd be capsized, if these families weren't legacies!"

--"Santa Monica, Santa Cruz in the holes, San Diego, you ready for a manly math, or is it 'Divorced Eggs' for breakfast?"

--"The Hook is calling it...Landlubbers take the hand on St. Patty's 2012!"

--"I only know the surfers who are assholes from San Bernardino County!"

--"In the Biblical sense, I imagine, now that we've seen how loose-leaf this environmental studies report was that stuck to the café's magazine rack, askin' for beer money."

--"Hey, who got all these cornnuts stuck in the ranch's bar-b-que grill?"

--"What, you mean that old neighbor still talks funny even though you're sister decided to marry a nice Jewish guy?"

--"Yes."

--"Know what I got to say to that...Knock-knock."

--"Who's there?"

--Pau.

--"Pau who?"

--"Pau-pow-pow-pow-pow!"

--"How does coffee work with chile?"

--"Self-explanatory, really, but if you must know, only for breakfast!"

Now that the Devil is unleashed on the planet, and we have a good study on Revelation going, go to Temple awhile, because the Catholics and Protestants in America aren't sure what to do, and even though they don't believe in Jesus, the Jews will be the tribe that collectively can function if God puts it to us all in finalistic showdown terms, complete with all the angels and demons you can clean your corncob pipe for!"

--"GET YOUR ASS KICKED NOW! BECOME INFATUATED WITH THIS ASSIMILATED BLACK WOMAN IN ENTERTAINMENT! SIGN HERE! GET YOUR ASS KICKED NOW! The heat is on! You can do it! Get the heat of the 'evil eye' the next time you apply for a local, state, or federal government job! Take advantage of opportunity! Sign, and at the very least, lose your collegiate ass! $540 paid by the Professor's crew for participating in statistical research! Come on y'all! Come on out and get your ass kicked! Downtown Los Angeles wants you!"

--"Dad, I want a Ford truck."

--"Son, those are the Chevys and those are the Fords. The Fords and the Lemons live in the same town. I'm buying you a Toyota, like the last neighbor who had a job in this town!"

--"Dad, those are the Fords, those are the Farts, and those are the Forts. As a Mexican-American whose parents have served this country, and smellin' the Farts that came after the Fords while you and Mom were holdin' the Forts, I am a proud Mexican-American and the new owner of a Toyota Camry!"

--"Amen, Son!"

--"Dad, remember, we don't pray at lunch, only dinner!"

--"Patty lives at her folks in beautiful Cherry Valley on some choice ranch property, and commutes to serve cocktails at San Manuel Indian Bingo."

317

--"Cherry Valley to San Manuel at night? That's a huge drive. Are you crazy?"

--"Haven't you heard of Interstate 40?"

--"Steve, she takes the bus!"

--"What?!"

--"I take the bus, Steve. The R.P.D."

--"Balgame Dios`, what is this? I asked for a real girlfriend. How much extra work are you trying to give me?! C'mon, Jackasses, what happened to the Olde Cali, where are the girls that are more bravado and less baggage? Hi Patty, Steve here. There's an apartment, room 'K', for rent at Merryvale. That way you have an easier drive to your job at San Manuel. You can live in the same building as me. All you really have to do is go down Cajon to Orange and catch the 10E, then the 210 to S.B. ways."

--"You mean that's all? Is it a safe route?"

--"Oh, it's better than Interstate 40."

--"Interstate 40? Where? Steve, where?"

--"Where you go into 7-11 in Yucaipa, buy a 40 oz. of Olde English 800, get back in the car, joy-ride into S.B. at 80 M.P.H., and start your shift Toast-tee, but nobody cares because you're Patty with the gams, and your uncle knows the Sherriff, who co-signs the b.s. since you're partially volunteering for The Good Olde Days Program."

--Jesus, another Eyewitness News with a dark forecast for us old-timers moving on with the Recession. This Michelle Tuzzi has been pitching shut-out after shut-out over the years. What the hell, pizza time rolls around and 7 feeds us to a Pepperdine sorority legend! Give us a break!"

--"Steve, we don't exactly know where she was educated. Healthy circulation in the English language though, isn't it? A true broadcaster."

--"Mike Fukazaki...I had to go to try-outs in high school with a Japanese guy built like that."

--"Yes, be thankful, when you can afford it, luckily it isn't always a crime report and it's quite a drive to the set to film the news in L.A. traffic."

--"Oh. You're breaking it to me gently that they wouldn't give the job to Michelle Tipsy!"

--"Are you gettin' out of shape, Steve? You look a little heavy. A little Panz`!"

--"Oh, I see, this the 'Panz` Scheme' again. The doctor weighed me in at 204—two pounds lighter than last month."

--"How's your strength? A little light-headed in the sun?"

--"Jesus, how many times you going to run this show at wine and cheese time, turn a blind eye to the resume, then have Mom draft me to move a 50-lb. potted plant 75 feet in the 90* heat from one side of the yard to the other?"

--"Oh, so don't brag. 'Texas offspring, gone crazy in the California sun! Keep taking the medication. What's with the hats? You sure you're not being sneaky?"

--"I'm a chameleon. Before you call 911 again on the 'mental health guy who looks suspicious in the neighborhood.'"

--"All right, Schrek."

--"Typical acting job offered a Latino male who starts off young on stage in this state before every you-know-what in the White world offers their 'advice.'"

--"No, I don't know what."

--"That's why we taught you the cusswords in Spanish in these parts. You're worse than going to an audition for a director who's a *hoto!*"

--"Did you see this documentary on Greek life entitled, 'Little Bro'? They've kept the heroics at Security for the Theme Party when the party-crasher arrives, and edited the material on the Fraternity being given a $2500 fine and a suspension of its charter!"

--"It wasn't adding to the post-Hell Week romance scene, after 'Little Bro' becomes a full-fledged member in good standing, Steve!"

--"Dean Grudge, just <u>how</u> bad in bed is your wife? This didn't exactly add to ticket sales Homecoming Week when Aikmann was pitted against Peete, either!"

--"I refuse to validate this a reflection of the sin curve in the campus micro-economic system!"

--"Dean Grudge, someday I'm going to catch you having an affair at The Alumni House, and blow this campus conspiracy against the academicians wide open!"

--"I can't help it if despite a 3.75 in the Italian Dept., you're not qualified in the system's eyes to succeed at interpretation and translation in order to snap out of your poverty phase! You shouldn't bear a grudge yourself. Perhaps if you withdraw the complaints we can make amends..."

--"We have our own version of *'La Migra'* off-campus in residences for professionals...Law Enforcement to prevent your 20,000+ student increase in admissions from lowering property value. You're lucky the landlord was lenient a couple of years. Please don't pretend you should've been playing Notre Dame in the Rose Bowl after you dumped that line on <u>Rudy</u> being an untrue story after all was said and done. The trash your jealous wife can come up with to send through that propaganda machine of a Film Department of yours!"

--"True, we cannot guarantee employ in the Industry after a Master's is awarded, but I assure you <u>left-wing Hungarian liberals</u> have no entry into our reputable halls after the <u>intellectual sabotage</u> of the 90's!"

--"I see where the custom of keeping it just one beer on the staff started... before the College Regional Vice Tax was introduced into the market place to fund your retirement!"

--"Are you calling me an asshole, Steve?"

--"Dean, my D's are actually for discipline in this life. When you're raised by a Naval Academy parent you'll understand. I think I'll pass on the labeling, since I am allowed to pocket the cocktail napkin in memory of this fine moment in time."

--"Good luck. You're going to need it."

--"What, Dean, the luck or the napkin, you're spilling your beer on the table!"

--"Isn't that worse than the Clap, Jerry? When you know that many people on a personal level who are retired?"--

--"It's a party of the most well-mannered, enthusiastic, and well-dressed unemployed 4,000,000 Americans I know. The Art Association Party."

--"No, that's not the art association, Jerry, that's the Literati and the Renaissance Men: They're conspiring against. They hole up in there for something besides a Paulina Porizkova poster on the wall."

--"Nothing wrong with a poster."

--"Oh, Jerry, I hope your marriage improves. I'll pray."

--"Did you see the crowd at the park?"

--"The Hot-flash Affect, huh?"

--"...And one for all!"

--"So, do you see how to apply Foucault's 'control of discourse' in the book, Power\Knowledge, formerly using the Russian Gulag as the example, now applying the producer's agenda behind the scenes in the L.A. Media?"

--"Is this our own, not private Idaho, but private Patmos, the disease-ridden landscape of California, similar to the very mystical island Saint John wrote Revelations for?"

--"Or simply, do we need more Pathway to Money, and less Pathway to Asshole?"

--"That's the side of town that's trying to make a comeback. They're glad you remember their women."

--"Yea, his father, when he was alive, used to apparently plug all the holes in the world, because now this state is bullshit. I see him handling the invisible, but his son had to hog the whole mirror, and Europeans see it differently! Believe me, I've spoken Italian for 30 years—the women in my life are a pantheon all their own! 'Built to kill', you say, but hey, brains and beauty—I'm starting to see ours like that after this war."

--"He's sorry he hogged the mirror—he said to tell you he's an alcoholic."

--"Oh, countryman can take all that lung cancer, stuff it in the bottle, and drink with the Crimson Tide this year. That's the subject! Smoking! Drink your fill, it's a wake for the troops in Iraq that is still on my plate as a patriot. Your homeboy would have to see the housing after rehab at Cedar-Sanai in L.A. to understand that I prefer to keep the bottle I can buy than face that again at 9\11 time. Tangents? My training! My zone!"

--"Censa?! Great! I could lose about 35 lbs. Can you help my father and I lose about 292 lbs.? Because he's gotta couple Kling-ons that could lose weight too, get it? _Si parla Italiano_?!

--"They have twice as much lunchtime as nighttime sex. They're Southern California now. They start quick in the morning exercising all over the map. Spastically. By lunch they hit the gong and a new weight is placed on the day. People must now drag that weight around the rest of the day, evening and night. What is their forte`? They can wipe their ass in a hot bathroom during the summer. Wise man say, "He who wipes ass in morning sends shit downtown at night!"

(You were asking what caused the schizophrenia—I'm not really schizophrenic, they just won't get their shit together!)

The problem in Southern California now is that they don't smoke pot worth shit: They don't clean the bong, they don't brush their teeth, they don't wash their clothes, except with that detergent that causes allergies, and they've brushed their hair with a 10-year old brush. They loathe water but live by the beach. This is the culture to go along with their tattoo art. Look Ma, no anti-perspirant, but a full tank of gas and a corndog.

--"You would have to take a shower to understand."

--"Oh, Jesus, these people are so sedentary it makes me sick. God, what is this, Toys R Us?! These are former athletes? I've seen people with garbage on their minds when the Southland has gangs crawlin' up the crack of its fault, but this has got to be a record!

--A model of the Golden Gate Bridge built out of toothpicks. I'm fascinated. Burn, baby, burn. And this is the great cookie-cutter in the sky landing on Southern Californian dough to remind us of their social skills in The Bay. The cookies are cooked at 375*...the hash brownies are store-bought from Trader Joe's-Tahoe now for $12.99 a 4-pak. Governor Brown is back from the gym—he's rubber-stamped the region a health issue to co-sign President Obama (having trouble bringing work there), who is at Camp Hood for a family reunion. This is where they got the idea Ellen DeGeneris could break the game open. No wonder Berlusconi still has the right to run for election again—the Italians can't handle Narcissist Cake with Cherries and Corn Syrup. Jump the damn canyon again, Evyl, and create a wind-block for the fires the pot smokers have set since getting the price on herb lowered from $150 to $17.25 a Q.P.! The dispensary owns the floor on the boardwalk people—we knows gall. Who signed the last game ball at Raiders Stadium, Ruben E. Lee?!

--"Franco said 'Thanks' for witnessing the Gospels to him when you were on exchange. He's in Law Enforcement now."

--"Law Enforcement? Where in Italy?"

--"Naples, his home town! They can't tell if he's dirty or not, and he's collecting undercover notes."

--"They can't tell?! Franco was never one to shower! Does his wife at least keep his uniform clean for press conferences?!"

--Depardieu, Snowden, and Stephen Seagal are slapping around the hockey puck at Gorky Park's ice rink at night, just for Left-Wing Shitskies stuff. Suddenly, Depardieu misses the puck, the stick goes high in the air, and he falls on the crack of his ass to create a HUGE crack in the ice. Water begins to seep through the crack into his clothes. He rolls, giving it three efforts, and a final heave, to get unto his belly, and crawls for the grass hedge. He FINALLY reaches it, with water seeping on the ice everywhere, but no chunks of ice falling to leave pools to drown in. Snowden and Seagal have been shitting their pants the entire time, until Snowden exclaims, "I think if we skate the opposite direction, we can get out of this!", and they do! THE MORAL: America will win in the end, despite set-backs, and that's enough testing it.

It is a small town
So do not frown
That the priesthood's confessional was
Last to know
That I had been a living ho!
No secrets are hid
In the Kingdom of God
If you want to marry
Fish with their rod!

--"White people—the toilet paper—what gives? They use it for everything but wipin' their ass. I've seen it in ears, in noses, on trees, in art collages dyed pink, and used for Rose Parade floats, not to mention decorations for Greek fraternity parties. How do they wipe their ass, pray tell, do they take one of those Fire Dept. nozzles, attach it to a garden hose in the backyard, and blast out the crap as a team, because it feels better than puttin' your head down? Don't squeeze the Charmin—it's going in the Art Museum on a college campus near you.

--"It's not that, Stephen, it's the Democrats…on the Jewish Sabbath. They drank, went to bed without brushing their gums, woke up hung over and began to bitch, chew, and moan all Sabbath day about the oppression of the Palestinians nearby the Gaza Strip. Now Syria is using chemical weapons and, Goddammit, they're the Democrats in the Armed Forces, they're ready—they've been ready the whole time. They're going to get their ass kicked again unless they let the Air Force handle it, since the USAF is the Master in this sloppy seconds dialogue!"

--"That crowd is a cold fish!"

--"It is a runnin' joke on you for not bein' more personable!"

--"That is just tough apples if we "White Boy"….to "White Boy"….to "White Boy" to "Touchdown!" the whole gosh-darn summer!"

--"Oh, that bad?!"

--"Oh, you know the game has sunk so low, that the guys livin' at home to save money for a home of their own are gettin' the jobs, and the guys rentin' apartments got stuck doin' yard-work for their folks, takin' $10.00 and then baby-sittin' downtown at night on a beer run, then when the Prince of the Hill takes out his first mortgage(now that the war is over), the asshole makes his debut as a Republican by 'cuttin' in' at the theatre to handle the girl. I mean you would have to see a catalogue of these guys pullin' that. This is all before you get to Hilary's Campaign For College Women. I'm about to drive out there to the farmer's and steal that metal rooster turnin' in the wind in the broad daylight, 100*, and leave a pack of cigarettes with a lighter behind, tellin' him to kiss my ass!"

Jesus Talking To Mary

Jesus At Lake Capernaum

The Butterfly In Mary

03/07

General George Washington

Hawkeye Of James Fenimore Cooper Novels

Knight In Armor

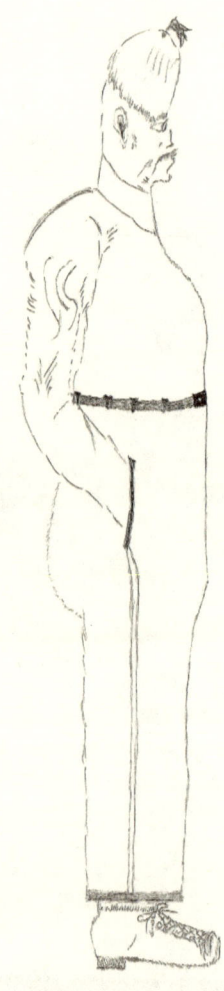

Pancho Villa In Formal Attire (Villa En Forma`l)

Ava Arrives From The East Coast

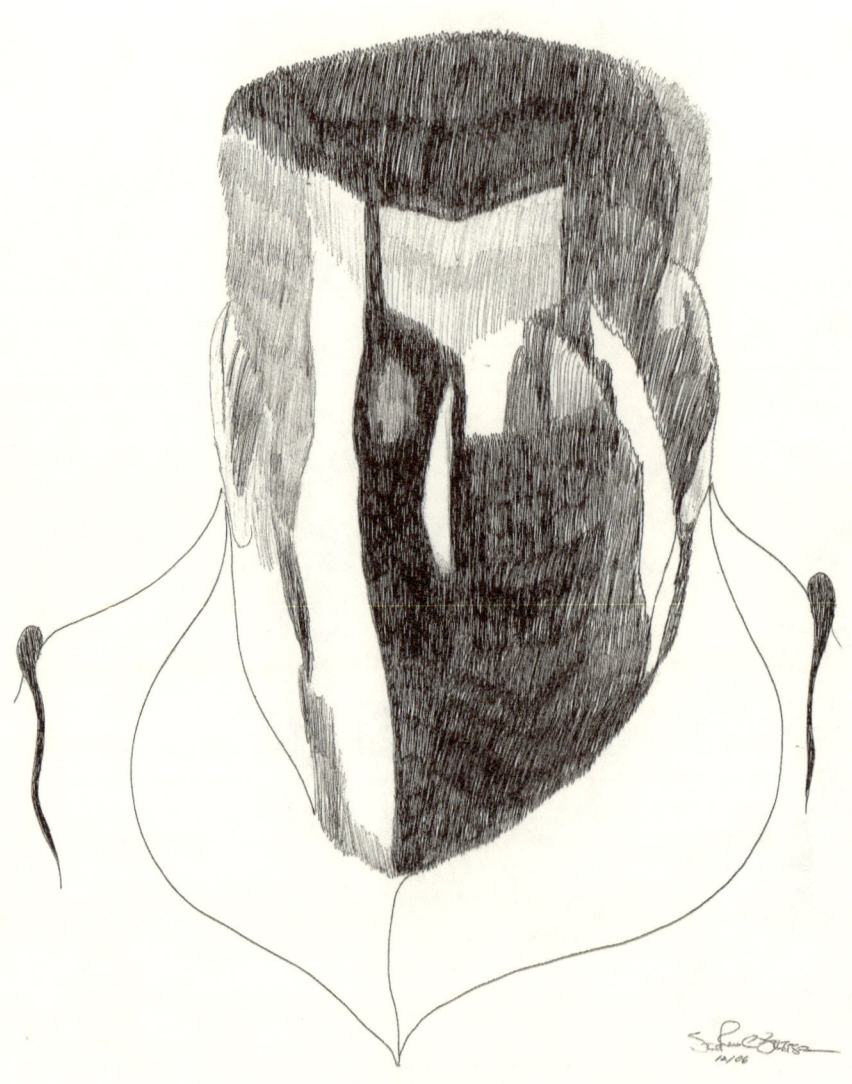

Hungarian-American World War II Vet Of The Japanese Theatre

Madame Carla Bruni-Sarkozy

President George W. Bush Jr.

The Gargoyle

Drumming Up The Vote

Walking The Tightrope On Global Warming

Gum Gum Gommy Scotch

338

The Bartender At The Chinese Spot

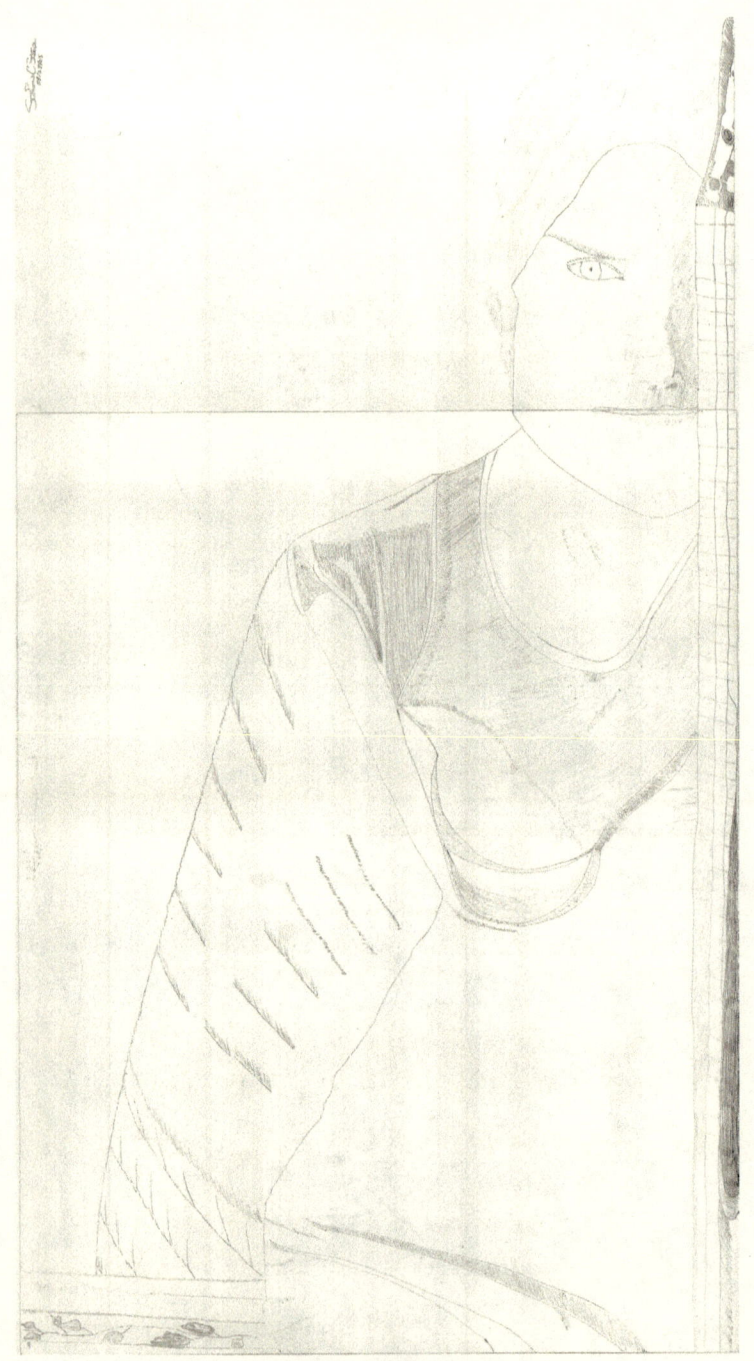

An Angel's Musical Notes

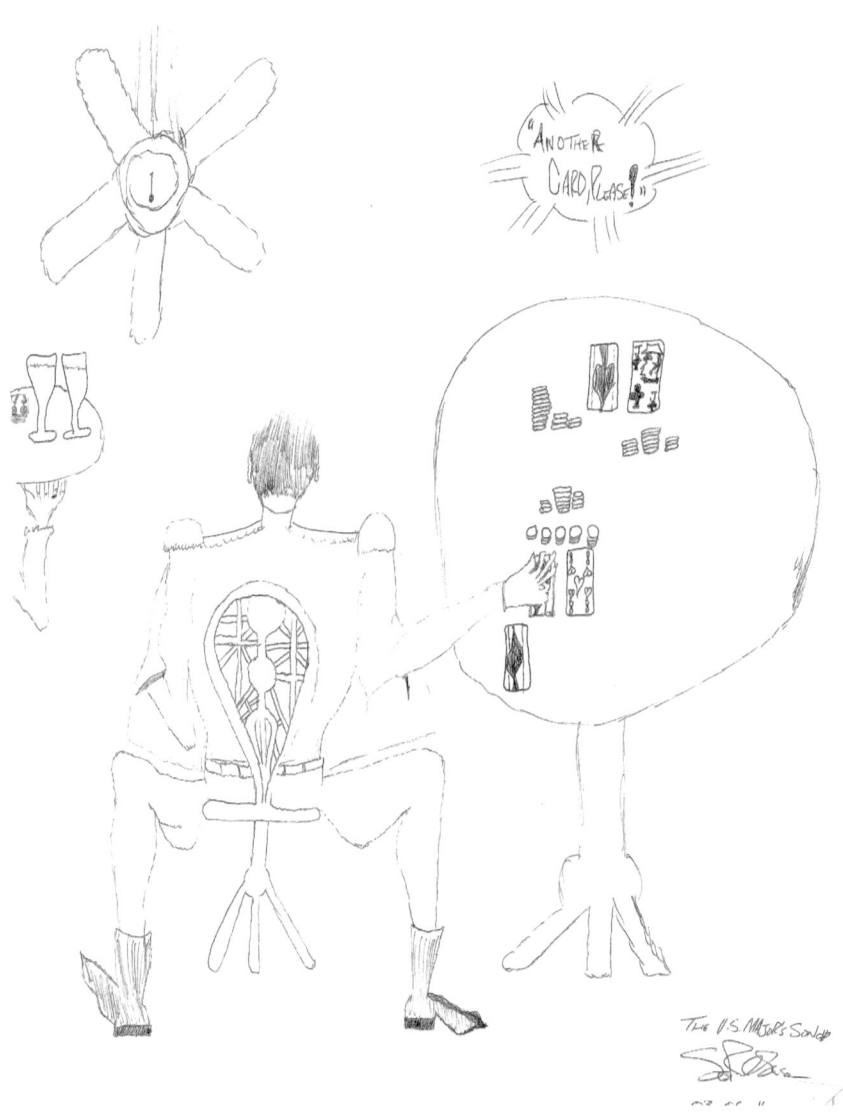

The U.S. Major's Son

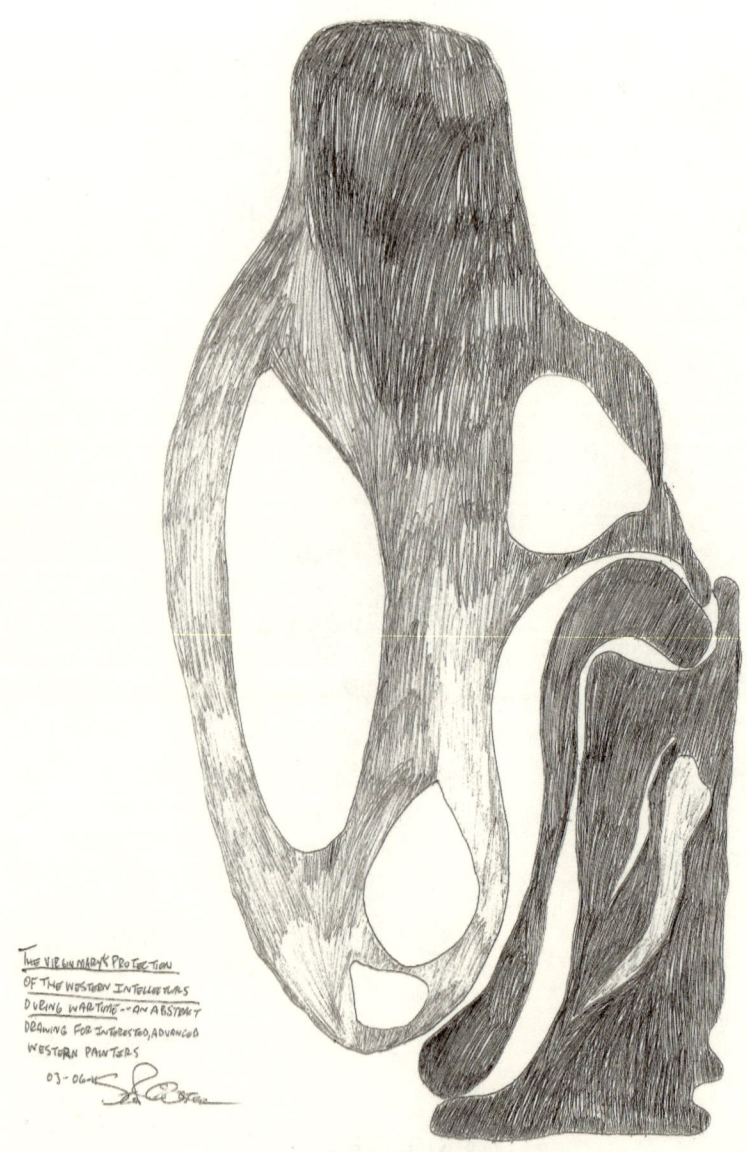

THE VIRGIN MARY'S PROTECTION
OF THE WESTERN INTELLECTUALS
DURING WARTIME -- AN ABSTRACT
DRAWING FOR INTERESTED, ADVANCED
WESTERN PAINTERS

03 - 06-16

An Abstract Of Mary's Protection

Christ At Golgotha

www.ingramcontent.com/pod-product-compliance
Lightning Source LLC
Chambersburg PA
CBHW020725180526
45163CB00001B/116